Beyond Energy Efficiency in Architecture. New Challenges and Research Trajectories for Buildings and the Built Environment

Beyond Energy Efficiency in Architecture. New Challenges and Research Trajectories for Buildings and the Built Environment

Editors

Tiziana Poli
Andrea Giovanni Mainini
Gabriele Lobaccaro
Mitja Košir
Juan Diego Blanco Cadena

MDPI • Basel • Beijing • Wuhan • Barcelona • Belgrade • Manchester • Tokyo • Cluj • Tianjin

Editors

Tiziana Poli
Department of Architecture, Built Environment and Construction Engineering, Politecnico di Milano
Milan, Italy

Andrea Giovanni Mainini
Department of Architecture, Built Environment and Construction Engineering, Politecnico di Milano
Milan, Italy

Gabriele Lobaccaro
Department of Civil and Environmental Engineering
NTNU Norwegian University of Science and Technology
Trondheim, Norway

Mitja Košir
Faculty of Civil and Geodetic Engineering, University of Ljubljan
Ljubljana, Slovenia

Juan Diego Blanco Cadena
Department of Architecture, Built Environment and Construction Engineering, Politecnico di Milano
Milan, Italy

Editorial Office
MDPI
St. Alban-Anlage 66
4052 Basel, Switzerland

This is a reprint of articles from the Special Issue published online in the open access journal *Applied Sciences* (ISSN 2076-3417) (available at: https://www.mdpi.com/journal/applsci/special_issues/Energy_Efficiency_Architecture).

For citation purposes, cite each article independently as indicated on the article page online and as indicated below:

LastName, A.A.; LastName, B.B.; LastName, C.C. Article Title. *Journal Name* **Year**, *Volume Number*, Page Range.

ISBN 978-3-0365-7136-2 (Hbk)
ISBN 978-3-0365-7137-9 (PDF)

Contents

About the Editors

Tiziana Poli

PhD and MSc Architect from Politecnico di Milano; Full Professor at Politecnico di Milano – ABC Department; Coordinator of Scientific Board DABC – Politecnico di Milano; Coordinator of research Group SEEDLab@Polimi; PI or Associate PI in national and international research projects, coordinator of several contracts for research and industrial development and several scientific collaborations in research projects, co-owner of two patents. Research interests concentrated on daylighting design and adaptive comfort for smart and cognitive buildings; building envelope design and performance evaluation (including variations over time); development, modelling and testing of new building envelope materials and components; climate-based design and impact of building envelopes on urban microclimate mitigation.

Andrea Giovanni Mainini

PhD, MSc and BSc Engineer from Politecnico di Milano; Post-doc Researcher and an Assistant Professor at Politecnico di Milano – ABC Department; Collaborator in nationally and internationally funded scientific and industry projects, and co-owner of three patents. Research interests are focused on building envelope technology design, innovation, technology transfer and integration of IoT devices for building system performance. Such research interests are also directed towards building energy performance analysis targeting high-performance envelope design, and intensively on thermal and visual comfort in indoor and outdoor environments.

Gabriele Lobaccaro

PhD, MSc Building Engineering and Architecture from Politecnico di Milano; Coordinator of the Research Group Building Technology – Dept. Civil and Environmental Engineering, Faculty of Engineering, Norwegian University of Science and Technology; PI of NFR FRIPRO FRINATEK project HELIOS - enHancing optimal ExpLoItation of solar energy in NOrdic citieS through the digitalization of the built environment. Research interests and technical expertise include: integrated solar energy systems at district and building levels; energy, urban planning and climate-based solutions, mitigation, and adaptation strategies; environmental, microclimate and energy analyses; monitoring data; building simulation tools; building technology; daylighting; sustainable architecture; energy-efficient buildings; resilient cities; parametric modelling.

Mitja Košir

PhD Civil Engineering, MSc Architecture from University of Ljubljana; Associate Professor at the University of Ljubljana, Faculty of Civil and Geodetic Engineering, Slovenia, head of the Chair of Buildings and Constructional Complexes. PI of YCPdeep - Calculation of Yearly Circadian Potential in Buildings Using Deep Learning Techniques project funded by the Slovenian Research Agency and PI of the Slovenian part of the ECOnanoVIP— Study of thermal properties and reduced life-cycle impact of alternative hybrid eco-nanomaterials under low pressure–a project funded by the Slovenian and Czech Research Agencies. Research interests include building physics, focusing on building energy performance and daylighting in connection with the design of high-performance buildings, study of climate change impacts on buildings' past, present and future energy performance, utilization of bioclimatic adaptations to counter climate change impacts on building performance and life-cycle assessment of buildings.

Juan Diego Blanco Cadena

PhD and MSc Engineer from Politecnico di Milano; BSc Civil Engineer from Pontificia Universidad Javeriana; Post-doc Research Associate at Politecnico di Milano – ABC Department Member of Research Group SEEDLab@Polimi. Research interests include building performance analysis, operational boosting with digital technologies, sustainable design and operation, personalized visual comfort and thermal comfort.

Preface to "Beyond Energy Efficiency in Architecture. New Challenges and Research Trajectories for Buildings and the Built Environment"

The building sector has undergone a significant transformation over the past few decades, with a focus on energy efficiency and environmental sustainability. This has been driven by the need to mitigate climate change and the realization that buildings contribute significantly to global energy consumption and greenhouse gas emissions. Building designers and engineers have responded to this challenge with the development of innovative architectural concepts, building technologies, and management systems, resulting in cutting-edge and high-performance buildings.

The aim of this work is to provide insights into the latest developments and innovative strategies in sustainable building design and operation, with a focus on practical applications and solutions. It is intended for professionals in the fields of architecture and engineering, as well as students and researchers who are interested in the topic of sustainable building design and operation.

The motivation for writing this scientific work arose from the growing awareness of the need for sustainable building design and operation in response to the challenges of climate change and resource depletion. It is our hope that this work will contribute to the ongoing efforts to create more sustainable and resilient built environments, and to inspire new ideas and approaches in this important area of research and practice.

This book is the result of bringing together a collection of 13 published works by 58 authors, affiliated to 25 different institutions. All of them have contributed to the knowledge update and extension on the sustainable, digital, and efficient design of built environments. The guest editors of this Special Issue thank all involved parties, without whom it would have been impossible to produce this end product.

Tiziana Poli , Andrea Giovanni Mainini, Gabriele Lobaccaro, Mitja Košir, and Juan Diego Blanco Cadena
Editors

applied sciences

MDPI

Article

Current Trajectories and New Challenges for Visual Comfort Assessment in Building Design and Operation: A Critical Review

Juan Diego Blanco Cadena [1,*], Tiziana Poli [1], Mitja Košir [2], Gabriele Lobaccaro [3], Andrea Giovanni Mainini [1] and Alberto Speroni [1]

[1] ABC Department, Politecnico di Milano, 20133 Milan, Italy; tiziana.poli@polimi.it (T.P.); andreagiovanni.mainini@polimi.it (A.G.M.); alberto.speroni@polimi.it (A.S.)
[2] Faculty of Civil and Geodetic Engineering, University of Ljubljana, 1000 Ljubljana, Slovenia; mitja.kosir@fgg.uni-lj.si
[3] Department of Civil and Environmental Engineering, Faculty of Engineering, Norwegian University of Science and Technology, Høgskoleringen 7a, 7491 Trondheim, Norway; gabriele.lobaccaro@ntnu.no
* Correspondence: juandiego.blanco@polimi.it

Abstract: Visual comfort can affect building occupants' behaviour, productivity and health. It is highly dependent on the occupant and how they perform a task indoors. In that regard, an occupant centred approach is more suitable for evaluating the lighting perception of the indoor environment. Nevertheless, the process of rating and estimating the visual comfort makes a limited distinction between physiological differences (e.g., ageing eye, light sensitivity), field of view, and personal preferences, which have been proven to influence the occupants' lighting needs to complete their tasks. Such features were not considered while establishing the visually comfortable conditions; perhaps due to the challenge of coupling the assumptions made during building design to the performance indicators monitored during building operation. This work focuses on reviewing literature findings on how the common design approach deviates from real building performance, particularly failing to prevent visual disturbances that can trigger the inefficient operation of building systems. Additionally, it is highlighted that redesigned visual comfort assessment methods and metrics are required to bridge the gap between the lighting environment ratings computed and surveyed. One possibility is to consider such physiological features that induce lighting experiences . Finally, it was deducted that it is important to target the occupants' eye response to calibrate limit thresholds, propose occupant profiling, and that it is convenient to continuously monitor the occupants' perception of indoor lighting conditions.

Keywords: visual comfort; lighting; daylight; user centred design; modeling; monitoring

Citation: Blanco Cadena, J.D.; Poli, T.; Košir, M.; Lobaccaro, G.; Mainini, A.G.; Speroni, A. Current Trajectories and New Challenges for Visual Comfort Assessment in Building Design and Operation: A Critical Review. *Appl. Sci.* **2022**, *12*, 3018. https://doi.org/10.3390/app12063018

Academic Editor: José A. Orosa

Received: 1 March 2022
Accepted: 12 March 2022
Published: 16 March 2022

Publisher's Note: MDPI stays neutral with regard to jurisdictional claims in published maps and institutional affiliations.

1. Introduction

The concept of a building was initially used to identify a human shelter (i.e., enclosure) that protects its occupants from weather, and is built for security as a livable and private space. The understanding of the concept was later expanded to also include the notion of a comfortable living and working space. Nevertheless, the notion of comfort has been interpreted in numerous ways. In the beginning, the building was understood as somewhere that had sufficient weather protection and security from any outdoor threats, and later evolved into a healthy space in an unperturbed state; thus, making the indoor environment the space where humanity spends most of its time (>80%) [1,2]. The indoor environment can be categorized as an anthropogenic artificial habitat, in which, at least to some degree, its conditions can be continuously monitored by sensors and controlled by integrated building systems. These conditions were envisioned to promote comfort, health, and well-being for most building occupants. Consequently, ranges of physical parameters that describe the indoor environment (i.e., air temperature, air humidity, illuminance) were established in such a way that, if respected, would guarantee satisfactory indoor conditions, promote the reduction of undesired human body responses (e.g., onset of sick building syndrome) and enhance occupants' quality of life [3,4].

To yield such a controlled indoor environment, active building systems are unavoidable (e.g., air handling units, lighting appliances), which constitute a significant energy utilization rendering the building sector responsible for a significant share of the worldwide energy consumption and CO_2 emissions to the atmosphere [5]. Therefore, to contrast with current trends, it is paramount to understand which are the drives for such energy consumption and CO_2 emission growth, and their links with the building users' comfort, health, and well-being.

Currently enforced building design policies [6,7] tackle climate change by encouraging the realization of highly resilient, sustainable and energy efficient buildings with low carbon footprint and reduced energy intensity while, sustaining occupant satisfaction with the indoor environment. In this regard, new and restored building quality has been enhanced. This is reflected in a substantial increase of certified buildings, either by national standards, organizations (e.g., Passive House, US Green Building Council) or public bodies/institutions (e.g., BREEAM buildings). Such certification systems give great importance to green strategies related to embodied energy and CO_2 emissions of materials, installation of efficient appliances, and promotion of green behaviour of buildings' users. Nevertheless, they downplay the relevance of considerations taken regarding occupants' comfort, health and well-being [8]; even when a building's overall climate burden is more intense during its operation phase (i.e., delivering set indoor conditions). Besides, the actual building energy performance has deviated from the desired designed performance (see the case of the European Union (EU) deconstructed in Figure 1 and further explained in Section 3.1), which has been partly attributed to the users' interactions about the planned building functioning [9].

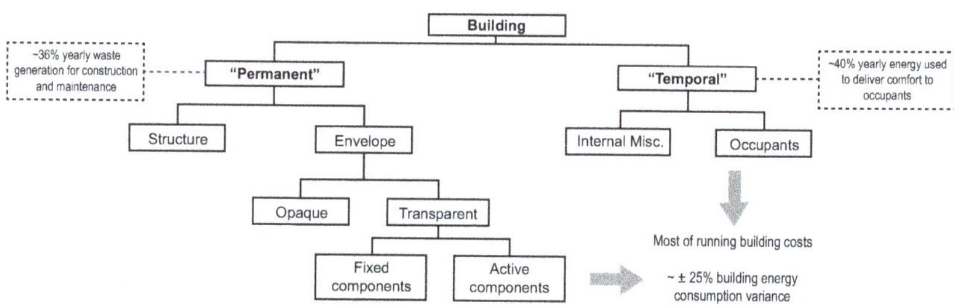

Figure 1. Building components deconstruction outline, and building-occupants interaction potential outcome on climate burden. Based on data collected for the European context.

Hence, if the efficiency of energy use is interpreted as managing it appropriately to serve the occupant, it is necessary to better understand the preferences and needs of the individual as well as of the group of occupants. This is still an unresolved matter, even when comfort ranges were established with a high level of acceptance (see Section 4.1). However, having a better suited definition of true comfort conditions based on a more representative occupancy type would allow superior accuracy in monitoring building operation and design [9–11].

In this framework, the presented work has concentrated on, firstly, collecting literature insights to understand why building designers and facility managers are not succeeding in narrowing the gap between modelled and actual building performance (i.e., performance gap) and secondly, highlighting the importance of acknowledging the building systems capabilities, the occupants' real comfort needs and environmental conditions. While finally, proposing potential solutions to improve the accuracy of comfort assessment which can boost building operation performance by reducing unpredictable occupant–building interactions.

2. Literature Collection Criteria

To capture the aforementioned needed understanding, literature was collected in different bundles. Firstly, literature was gathered referring to actual building operation performance, and how closely is it aligned with the expected performance estimated during the building design phase. Consequently, information was screened to identify the main reasons why such performance gap exists, concentrating in the factors that can be further assessed by scientist and designers. The magnitude of the influence of the effect of one of the identified factors was grasped (i.e., visual comfort perception variance) to support and recall the importance of its evaluation (Section 3).

Secondly, literature has been collected and analyzed on different methods applied for optimizing lighting-related building design and operation. In this context, literature was collected with three purposes: (i) gather, analyse and identify the most frequently used metrics, indexes or KPIs, to evaluate and/or describe the potential perception that a building occupant could have of the provided indoor luminous environment (Section 4); (ii) compare the metrics based on the guidelines which have embraced them, their strengths, weaknesses and challenges (Section 4.1); and (iii) cluster the different strategies employed using such metrics for optimizing building shape, building systems and building operation (Section 4.2).

Finally, based on the challenges identified on the most commonly utilized KPIs and followed lighting design approaches, literature on distinctive visual comfort assessment methods are summarized, presented and described as potential solutions to the identified gaps (Sections 5 and 6). Solutions are presented for different phases of the study of building performance (i.e., design and operation), highlighting their contribution and the current barriers that could be addressed in the future.

3. Motivation for Occupant-Centered Assessment

Sufficient evidence has been documented to assume that the current visual comfort assessment approach considers only marginal, or tends to underestimate, the effect of influencing factors resulting in unreliable comfort provision design/operation and thus, higher variance between the expected and the actual user experience of the indoor luminous environment [12–19].

3.1. *Effect of Occupant Interactions on Building Performance*

Generally, in the building design phase for visual comfort assessment, occupancy is considered on the level of a standard occupant (following an occupancy schedule and space type), with an idealization of their average conditions (i.e., healthy, average age and height). Sometimes also additional parameters are set, like, a trigger for shading devices (i.e., shading area ratio) depending usually on the designers' defined optimum indoor temperature and more really on the set lighting conditions (i.e., monitoring illuminance level and radiation intensity) [20].

However, building occupants should be contemplated more holistically and as active participants. If they are dissatisfied with the delivered indoor environment, they will interact with the building and modify its operation, resulting in altered performance (Table 1). In addition, occupants can tweak the building performance for the non-performed actions within buildings lacking automation (e.g., forgetting to turn off lighting appliances or deactivating shutters) when the surrounding conditions would have provided a sufficiently comfortable environment without an intervention of a building system (e.g., shading device, lighting appliances) or when occupants leave the previously occupied space [21]. In this context, Kamaruzzaman et al. [15] surveyed occupants in different refurbished buildings in Malaysia, finding that daylighting, electric lighting, and glare were among the most dissatisfying aspects of the building's operation and most of the time the blinds were down to avoid indirect disturbing glare from surrounding surfaces (i.e., computer screens) incurring in higher lighting energy use. Likewise, Masoso and Gobler [9] presented an example in which this misinteraction between occupants and building systems, led to an extreme of more than 50% of a building's energy use.This excess mainly happened during non-occupied hours, identifying that lighting appliances are ranked second among the electricity waste in the case of a university building. On the contrary, Reinhart et al. [22] screened day-lit offices in Germany and concluded that building occupants would have a more positive disposition for turning off lighting appliances and blind modification towards a brighter environment with no glare risk, rather than a darker indoor space, which could result in higher cooling loads.

Table 1. Summary of the effects (positive [+] or negative [−]) of occupant–building interaction on building energy performance.

Interaction Type	Effect	Description	References
Blind operation	+	Occupants tend to open blinds to favor brighter indoor environments without glare risk. Which often leads to lower energy use, when solar radiation is not too intense. A manual control over lights and blinds could produce annual reductions in lighting loads up to 80% compared to a constant lighting activation.	[22,23]
	+	Occupants tend to close blinds in instances of incident outdoor illuminance above 50 klux or to avoid direct sunlight above 50 W/m^2. Protecting from both excessive light and radiation influx.	[22]
	+/−	Occupants tend to open blinds to favor brighter indoor environment without glare risk. Having a larger glare tolerance, or avoiding glare by different means, could lead to higher cooling (or lower heating) loads compared to estimated performance.	[22]
	−	Occupants tend not to open the blinds to increase daylight influx when sufficient outdoor illuminance is present. Which often leads to larger energy use to compensate for the lack of adequate illuminance levels.	[15,21,24]
	−	Occupants tend to close the blinds upon arrival regardless of the outdoor illuminance intensity, even when proper or insufficient for indoor lighting. This often leads to larger energy use of lighting appliances.	[21]
	−	Glare control measures can negatively affect the illuminance level and illuminance uniformity in an office.	[25]
Light appliances use	+	Automatic daylight linked dimming of lights is acceptable to occupants. Although, manual dimming results in higher occupant satisfaction levels. Moreover, these systems motivate occupants to use more daylight.	[24,26]
	+	Active automated lighting control (ON/OFF light switching with ideal photocell-based dimming and occupancy-sensing OFF switching) could lead to an annual reduction in lighting loads up to 90% if compared to a constant lighting activation scenario.	[23]
	+	For most users the decision to turn lights on when arriving depends on the daylight level in the room at that moment since, on average, occupants switch electric lights on more frequently in the case of low daylight induced illuminances.	[27,28]
	+	Manual control over lights and blinds could lead to an annual reduction in lighting loads up to 80% if compared to a constant lighting activation scenario.	[23]
	+	Passive reminders (i.e., stickers installed by the light switches) substantially increase the occupants' turning off activity. Avoiding energy waste	[29]
	−	Forgetting lighting appliances on, when not needed can lead to an excess up to 50% of energy use. Indeed, up to 90% of manual controls occur just after the occupant enters the room or just before they left.	[9,27]

3.2. Occupant-Building Interaction Asymmetry

As indicated in the previous section, and depending on the occupancy type held in the building (see Section 5), the occupant–building interaction can be towards opposite preferences, and in some cases, decisive on building operation (Table 1). In fact, undesired human–building interaction, or idleness, can lead to 25% variation in yearly energy consumption compared to traditional building energy modelling (BEM) (Figure 1) [30].

Such variance can be attributed to the physical conditions which affect occupant perception, hence deviating from the standard visually comfortable ranges. People could suffer from eye sight issues, or can be significantly affected by the ageing of the eye, demanding more specific lighting provision requirements; few may be more sensitive to high illuminance levels due to their eye colour or to particular light colour given their eye physiology. Whereas, some others with the use of contact lenses or glasses, might be more or less tolerant to low or high illuminance or luminance levels depending on the case. Finally, other issues can significantly influence the comfort ranges or limits, on which building occupants would provoke (or would have liked) interference with the programmed building operation [11,31–36].

Unfortunately, it is not only about the physical conditions, but also mentality, awareness, education, or habits that can modify the indoor environment perception and rating [11,14,36–38]. Currently, energy modelling and regulations do not include some of these factors for relevant updates that would guide designers and facility managers towards better buildings for clients and users.

Although researchers have worked on predicting the occupancy effect (e.g., the number of occupants present, interaction trends with building systems) [30,39–49], the accuracy is still unsatisfactory. Some relevant factors are still undermined, and part of them are closely related to the type of occupants in the building, affecting comfort perception and more frequently triggering their interaction with the building. Detailed analysis of the occupancy type in buildings is rarely conducted, even when some building design guidelines include an optional methodology [50]. This happens mainly due to the complexity and the actual research extent of such a process.

Specifically for indoor visual comfort, Pierson et al. [36] presented a collection of environmental and occupancy related factors that are currently under scrutiny to improve visual comfort assessment and the extent to which they have been studied and proved to influence the perception of the indoor environment. Those aspects which are closely linked to different occupant preferences and physiological needs are described further in Section 5.

Such contrasting results, together with the potential positive increment in occupants' mood, satisfaction and productivity alongside lower building energy consumption [51], motivated this work to concentrate on how the scientific community is dealing with this discrepancy, generated by the significant deviation of occupants' preferences in the indoor lighting environment. Specifically, the focus is on how to better understand the drivers of users' interactions with the building aiming to reduce performance deviations.

3.3. *Magnitude of Occupant Variability*

The type of occupancy to be considered in building design and their comfort settings for building operation are set for healthy occupants of a specific age range and race or origin. These might not be in accordance with the actual or designated occupancy of the studied building. For instance, ISO2004 [52] has been defined as standard occupants: 35 year old, male and female, which vary slightly in height (1.70 and 1.60 m), weight (70 and 60 Kg), body surface (1.80 and 1.60 m^2) and basal metabolic rate (44 and 41 W/m^2). Such assumptions might be misleading, as the age and height (which have been reported sensitive for visual perception variance [32,36]) could vary significantly depending on the building function type and the context within which the project is placed.

In addition, while in the instance of thermal comfort, a body-temperature regulation model has been included to predict thermal comfort perception (e.g., Predicted Mean Vote), on the contrary, an analogous model for visual comfort is still missing. A procedure is needed to include the physiological features of occupants to assess visual comfort, to predict visual perception or luminous environment rating.

Researchers have exposed noticeable differences in occupants' eye adaptation capacity, physiological diversity, personalities, education, income and even attitude. Sustainable awareness among the world population is increasing, average population age is growing, as well as the number of people with eye-sight impairment (World Health Organization, https://www.who.int/news-room/fact-sheets/detail/blindness-and-visual-impairment (accessed on 5 November 2021)). Simultaneously, the income per capita and education level are diverse among world regions (Figure 2). The exposed demographic characteristics will affect occupants perception and rate of the indoor luminous environment, claiming an update to the current established visual comfort assessment methodologies for buildings.

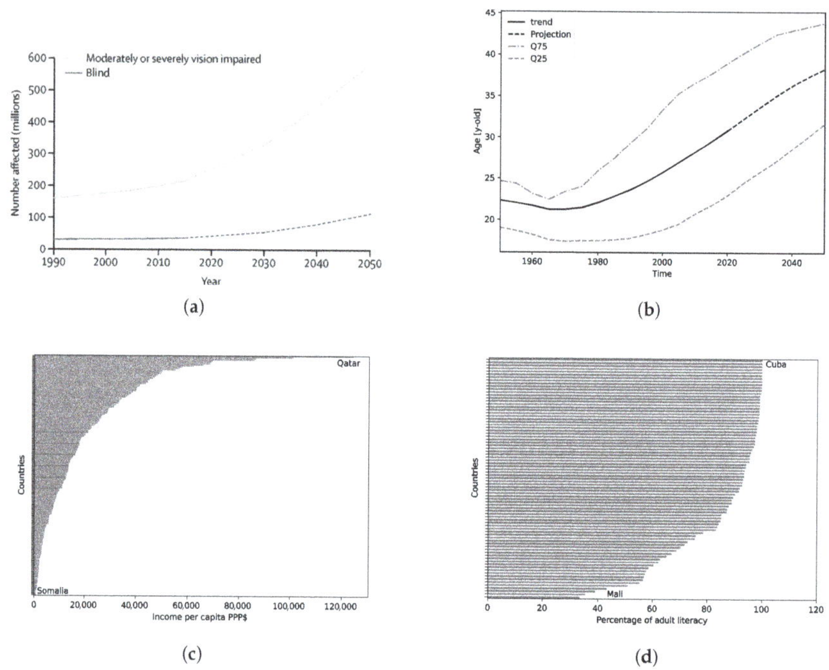

Figure 2. World's population differences probably affecting indoor comfort perception. In particular, that could skew the results obtained through qualitative questionnaires. (**a**) Eye-sight affections (extracted from [53]). (**b**) Population average world's age trend; (**c**) Projected 2020's per capita Purchasing Power Parity (PPP) differences; and, (**d**) 2011's Percentage of adult literacy breach based on free material from GAPMINDER.ORG (accessed on 23 December 2019) , CC-BY LICENSE.

4. Evaluating the Indoor Luminous Environment and Visual Comfort Perception

The indoor luminous environment is the result of the interaction between the building surrounding context and the integrated strategies planned during the design phase. Building passive and active performance strategies can be summarized in optimizing building shape and orientation, glazing and shading properties, integrating and enhancing dynamic (i.e., climate responsive) shading devices and artificial lighting appliances. Upgrading the performance and operation of passive and active strategies, as well as boosting their interoperability, would significantly render a building more efficient in energy use and comfort provision. Accordingly, it could be derived from the identified occupants' preferred indoor conditions to formulate an efficient control system algorithm. However, the proper performance criteria needs to be selected; that is, a comfort-based performance indicator. And, given the variability presented in Section 3.2, visual comfort occupant-centred approaches would be preferable.

Indicators can be either a unique indicator that communicates the state of a particular parameter, one that incorporates multiple parameters to rate an overall condition, or one that comprises multiple contemporary effects on the occupant. However, when considering only the environmental parameters, and based on the robust review made by Pierson et al. [36], the selected performance criteria for an accurate occupant-centred visual comfort assessment should undoubtedly account for:

- luminance of the glare source;
- illuminance on the eye-plane;
- adaptation level;
- contrast effect;
- size of the glare source;
- position of the glare source.

While, it should likely consider:

- saturation effect;
- light resulting spectrum;
- light colour temperature.

The impact of the quality of the view through the window has not yet been deciphered by a comprehensive research approach. Nevertheless, according to Ko et al. [54], the view quality can be defined as a combination of view content (i.e., what can be seen), view access (i.e., the amount of view seen by the occupant) and view clarity (i.e., how clearly the content can be seen due to the properties of the window).

4.1. Visual Comfort Evaluation Metrics

To estimate the comfort degree of occupants, or to understand if they are comfortable with its surrounding lighting environment, many indicators have been proposed and numerous have been established within the design regulations and guidelines.

Visual comfort is mainly evaluated based on the intensity of light falling on a surface (i.e., illuminance, E), either vertical (E_v) or horizontal (E_h), measured in lux. Thus, the visual comfort metrics are generally based on the light intensity, distribution, source and directionality. The quantity and quality of light should be enough to allow building users to perform the assigned area's correspondent activities safely and without induced visual fatigue. Artificial lighting should only provide the missing light to reach the minimum requirement (specific for each task) only when natural light is not intense enough, aiming to avoid the development of eye sight diseases.

Depending on the regulations governing the design and the building's intended use, the lighting requirements can be evaluated by different indicators and acceptable thresholds or ranges of such. Nevertheless, regardless their time domain, they can be summarized in two main groups: (i) quantity and uniformity of light (Section 4.1.1) and (ii) direct light discomfort (Section 4.1.2).

4.1.1. Quantity and Uniformity of Light

Considering mainly the luminance (L) of the light sources and the ambient distribution that condition the actual quantity and uniformity of light falling on a surface, the following metrics have been widely used to express the lighting quality of an indoor space:

- **Daylight Factor (*DF*):** compares the amount of light inside with the exterior during an overcast sky condition [55], the average DF across the analyzed surfaces of the building is typically requested to be greater or equal to 2% [56]. Alternatively, the appropriate DF value can be defined concerning the specified target illuminance and geographic location, following the method proposed in EN 17037 [57].
- **Daylight Autonomy (*DA*):** represents the percentage of occupied time in which a point within the analysis grid is over a defined minimum illuminance (discrete values only, e.g., complies or not). Additionally, the percentage can also include partial values when the conditions are below the established minimum threshold (ratio between obtained and minimum threshold), and are expressed as **continuous** DA (cDA) [58].

- **Spatial *DA* (*sDA*):** gives a notion on how much area is over the minimum illuminance for more than a certain portion of the occupied time, typically 300 lux is used for basic reading and writing activities, and these values shall be met for at least 50% of the analysis grid area [59].
- **Useful Daylight Illuminance (*UDI*):** calculates the percentage of the analysed time in which the illuminance in a point falls in a certain established range (normally between 300 and 3000 lux) [60]. Subcategories are also used to communicate the time in which the illuminance values falls below (underlit) or above the suggested ranges (over lit). Specifying the over lit areas can also suggest direct light discomfort risk.
- **Frequency of Visual Comfort (*FVC*):** percentage of time within the analysis period during which appropriate values of average illuminance are accomplished [61]. Similar to UDI, proposes 3 ranges: Comfort, Under and Over.

$$FVC = \frac{\sum_i wf_i t_i}{\sum_i t_i} \tag{1}$$

$$wf_i = \begin{cases} 1 & E_{Under} \leq E_{Daylight} \leq E_{Over} \\ 0 & E_{Daylight} < E_{Under} or E_{Daylight} > E_{Over}, \end{cases}$$

where:

$E_{Under/Over}$: minimum/maximum illuminance threshold
$E_{Daylight}$: computed/measured illuminance value
t: analysis period

- **Intensity of Visual Discomfort (*IVD*):** time integral of the difference between the spatial average of the current daylight illuminance and the upper limit of visual comfort or the lower limit of visual comfort [61].

$$IVD = \int_p \Delta E(t)dt \tag{2}$$

IVD_{Over}

$$\Delta E(t) = \begin{cases} E(t) - E_{Over} & E(t) \geq E_{Over} \\ 0 & E(t) < E_{Over} \end{cases}$$

IVD_{Under}

$$\Delta E(t) = \begin{cases} 0 & E(t) > E_{Under} \\ E_{Under} - E(t) & E(t) \leq E_{Under}, \end{cases}$$

where:

$E_{Under/Over}$: minimum/maximum illuminance threshold
$E_{Daylight}$: obtained illuminance value
t: analysis period

- **Daylight Uniformity (U_o):** ratio in a given moment, or time frame, between the minimum value of illuminance on the analysis area (E_{min}) and a reference value (e.g., maximum or average) [62].

4.1.2. Direct Light Discomfort

Direct light discomfort is commonly referred to as glare. It is meant to quantify the disturbance that a building occupant might perceive based on how the occupants' location and view/sight are exposed to excessive light. It mainly considers the *L* intensity, concentration, location and the effect of its generated contrast. It can be assessed either directly or indirectly:

- Directly with **Daylight Glare Probability (*DGP*):** which represents the percentage of people that would be disturbed by the level of vertical illuminance (E_v) at eye level and the contrast of luminance sources within the occupants' field of view (defined

by [63]). However, it is not valid for DGP values below 0.2 or $E_v \leq 380$ lux; thus, to extend its usability range, an s-curve corrective factor was introduced to compute the glare probability when E_v varies between 0 and 300 lux (DGP_{low}) [63].
DGP can also be computed using validated simplified procedures:

- Using direct correlation with E_v on a vertical plane corresponding to the occupant's eyes height, lo location and orientation, neglecting the contribution of local quantities as presented by [64,65]. In Equation (3), the probability was re-calibrated with the vertical illuminance values only estimated from rendered images [64], while Equation (4) validated in a virtual environment a simplified DGP definition computing illuminance values only using ray-tracing methods (contribution of contrast is neglected).

$$DGP_{Wienold} = 6.22 \times 10^{-5} \times E_v + 0.184 \tag{3}$$

$$DGP_{Hviid} = 5.87 \times 10^{-5} \times E_v + 0.16. \tag{4}$$

- Using correlation with the computed L from a radiance [66] simulated fish-eye image captured for the desired location, eye-height and gaze orientation using tools such as *evalglare* to compute it [63].

- Glare can also be directly computed using the **CIE Glare Index (*CGI*):** born as a correction of the British Glare Index (BGI), presented by Einhorn [67]; and then upgraded into the **Unified Glare Rating (*UGR*)**. It is a short-term, local and one-tailed glare index based on the split contribution of the direct and diffuse E_h.
- Finally, glare can also be addressed indirectly by identifying the areas of a building which are subjected to E_h levels over certain threshold, that is either:
  - Over an upper E comfort threshold (>2000 ÷ 3000 lux is typically used), which can be described using the overlit portion of UDI;
  - Over a maximum E comfort threshold, but sustained for a determined number of hours, as the defined in the case of **Annual Sun Exposure (*ASE*)**, indicates the possibility of glare occurrence (>1000 lux over 250 h per year, is often used) [59].

Tables 2 and 3 represent a comprehensive summary of the guidelines that have embraced the aforementioned metrics, together with their advantages and disadvantages. These metrics are generally utilized and monitored (if possible) on visual comfort assessment methods under a high degree of building design development, or level of detail (LOD).

Nevertheless, all the analysed metrics and suggested values, already established and included in different regulations and design guidelines, do not consider the adaptation level (or capacity) of the eye, nor the variability of occupant's sensibility to light. Thus, they are inaccurate and occupant-centred. Moreover, there are no clear suggestions or procedures developed on how to include them in the building operation phase when monitoring visual comfort (which in some cases is unpractical). Consequentially, a need to further include the occupants' true perception and response to visual stimuli within the loop of building performance design and operation strategies is clearly identified.

Table 2. Critical analysis summary of collected visual comfort condition metrics based on daylight intensity and distribution, highlighting their benefits and drawbacks.

Metric	Guidelines	Advantages	Disadvantages
E_h	LEED, BREEAM, WELL, EN12464-1, EN17037	Easy to monitor, measure and model; spatial lighting conditions provided.	Point in time dependent, analysis grid selection ambiguous, no human adaptation, light intensity at eye uncertain.
E_v	WELL	Easy to measure and model, light intensity at eye known.	Point in time dependent, analysis grid location ambiguous, no human adaptation, complex to measure with occupancy, no contrast effect considered.
DF	BREEAM, BS8206-2, DGNB, DIN5034-1, EN17037	Fast and easy assessment; spatial conditions provided.	Unique low light intensity condition evaluation, no directionality assessed.
DA	n/a	Annual and spatial analysis, easy to model.	Dichotomous variable (1 or 0), No high light intensity risk considered, light intensity at eye uncertain, biased by the limits set.
cDA	n/a	Annual and spatial analysis, easy to model. Ordinal variable (from 0 to 1).	No high light intensity risk considered, light intensity at eye uncertain, biased by the limits set.
sDA	LEED	Annual and spatial unitary index, easy to model.	No high light intensity risk considered, light at eye uncertain, no information on problematic areas, biased by the limits set.
UDI	n/a	Annual, grid-based and spatial index, easy to model, hint on under-lit, day-lit and over-lit areas.	Light at eye uncertain, no directionality assessed, biased by the limits set.
FVC	n/a	Annual and grid-based unitary index based on E_h, easy to model, information on the amount of time at discomfort and comfort.	Light at eye unknown, spatial distribution not known, human adaptation or variability not assessed, biased by the limits selected.
IVD	n/a	Annual and grid-based unitary index based on E_h, easy to model, information on the intensity of discomfort.	Light at eye unknown, spatial distribution not known, human adaptation or variability not assessed, biased by the limits selected.
U_o	AS1680, DIN5035, NSVV, CIBSE, EN12464-1, CIE29.2, BS8206-2, WELL	Information on lighting distribution contrast.	Biased by extreme values, no other information on light, has to be coupled with another metric.

Table 3. Critical analysis summary of collected visual discomfort condition metrics based on daylight intensity and directionality, highlighting their benefits and drawbacks.

Metric	Guidelines	Advantages	Disadvantages
DGP	EN17037	Robust and reliable, considers both light intensity and contrast, accounts for directionality, gives rating and sensation information, light at eye known.	Complex and lengthy to model, not reliable under low illuminance conditions and less reliable with large contrast, unique moment in time and position, no human adaptation considered.
DGP_s	EN17037	Still robust and reliable, accounts for directionality, gives rating and sensation information, easy to model and compute, can be computed annually.	No contrast considered, less reliable under low illuminance conditions, analysis grid selection ambiguous, no human adaptation considered.
UGR	CIE177, EN12464-1	Accounts for light intensity, contrast and directionality, gives rating and sensation information, direct and diffuse light at eye known.	Unique moment in time and position, no human adaptation considered, not reliable when the sun is within the field of view.
ASE	LEED	Annual and spatial unitary index, easy to model, hint on high intensity problematic areas.	Light at eye uncertain, no directionality assessed, biased by the limits set.

4.2. Current Lighting Design Assessment Strategies

In fact, most simplified design procedures are applied to estimate visual comfort by monitoring only E_h or DF. More granular analyses have been proposed showing better results in pursuit of enhanced building occupant conditions and efficient energy utilization. These depend on the building design LOD and the calculation tools' capabilities [68], or the buildings' smart readiness [69], especially as the higher the LOD is, the easier it is to individuate where occupants would be spending most of their time.

4.2.1. Assessment Methods under High Building Design LOD

Assuming that the building systems are not yet defined, but the LOD of the building design phase is already high (LOD approximately between 300–400), the indoor visual comfort assessment is more likely to result in more accurate results (i.e., realistic) while using the metrics previously presented. Therefore, under the known layout of the points of interest (e.g., workstations in office buildings, study desks in school buildings), or entire areas (e.g., living area in residential buildings):

- E_h is monitored at the task area of each occupant or the most probable occupied area.
- U_o the variability of the light intensity at the task area is compared for each occupant or the most probable occupied area.
- E_v is monitored at the eye level of each occupant position, resembling illuminance at the pupil (E_p), for one prevalent line of view or more than one direction.
- DGP is monitored at each occupant position or glare-risk locations, for the most critical moment of the year (based on designer's criteria) or on the location where and point-in-time when higher illuminance intensity was identified (either through E_h or E_v).
- DGP_S monitored at each occupant position, computed from estimated E_v values.

4.2.2. Assessment Methods under Low Building Design LOD

When building systems and occupant space distribution is unknown or unreliable, detailed analyses are hindered and more robust approaches are generally applied. Some metrics become less effective (e.g., E_h and U_o) as some regions of the potentially occupied space could be dedicated to transitional areas (e.g., corridors) and not as task areas. Among literature, the following were identified to have significant potential to better address visual comfort from an occupant-centred perspective when low LOD models are available:

- A modification of the process to rapidly estimate yearly point-in-time glare with *DGP* from rendered HDR images was presented by Liu et al. [70]. It entails yearly predicted HDR images using deep neural networks from rendering only 5% of the analysis period for every line of view. This methodology showed decent accuracy compared to the results produced by Radiance - *rpict* function, enabling faster yearly calculations of image-based metrics (e.g., *DGP*).
- To spatially cover the visual experience and map the daylight glare class in the room on an annual basis, Giovannini et al. [71] proposed setting lower, intermediate and higher threshold limits of E_v that *DGP* would rate glare as imperceptible, perceptible and disturbing. These threshold limits are meant to be specific, as they are extracted after identifying the worst condition in the analysed room. The latter is mainly imposed as an occupant located close to the window and with a line of view perpendicular to the window surface plane (few initial point-in-time and image-based simulations are expected).
- To reduce computer simulation time while performing point-in-time and spatial analysis, optimized simulation workflows have been proposed and validated. These optimizations comprise the use of cloud computing services and the integration of graphics processing units (GPUs) [72] . For instance, Ladybug tools [73] has created a new simulation platform service based on cloud computing (i.e., Pollination [74]) able to speed up the simulation process. In particular, when a large number of design options should be tested by running them all in parallel. Jones and Reinhart [75] have developed a suite of GPU-enabled tools that implement ray-tracing functionality that speed-up both Radiance and DAYSIM process by one order of magnitude.
- To avoid extensive computer simulations for monitoring glare and lighting sufficiency, more efficient simulation workflows have been proposed. These have managed to supersede the use of rendered images. For instance, Jones [72] presented a method to fast-compute glare (133,000 times faster) based on the calculation of view factors to a discretized sky dome for estimating direct lighting. Alternatively, cubic illuminance can be computed and re-elaborated [76]. Using E_h and E_v to determine lighting sufficiency and computing DGP_S, does not require image-based-point-in-time simulations. A cube-grid-based illuminance simulations with the centre on the hypothetical location of the occupant. It will be directional if ray-tracing engines are used (e.g.,

Radiance). It can be run on an annual basis and, from this, by specifying a direction of view, eye illuminance (i.e., E_v or E_p) and work-plane illuminance can be directly computed (E_h). Thus, a room can be spatially populated with cubes for mapping all conditions within a room, and posterior data post-processing by exploring multiple view directions can be carried out. This procedure can be carried out more quickly by using a practical approach presented by Cuttle [77]. Otherwise, a much simpler procedure was proposed by Raynham [78] by employing the mean indirect cubic illuminance (MICI).

However, none of the exposed modelling methodologies found were related to the variance of occupants' visual experience, resulting in different comfort perceptions, sensations and ratings. None of them has included the physiological or anatomical variations on subjects that would largely alter the subjects' ratings of a given luminous environment. Moreover, these methodologies become unpractical or infeasible to be surveyed within an existing building. This is because no actual occupants' field of view can be obtained without intrusive methods, and no unbiased visual comfort rating can be obtained without considering further occupants' inherent features.

5. Challenging Methods to Include Occupants Visual Comfort Perception Variance

Various comfort indicators have been proposed and established. Nevertheless, the main problem remains how to address the continuous and large variance of occupants' comfort preferences, without largely deviating by applying robust approximations.

Yamin Garreton et al. [79] highlights that "*DGP* models have some limitations for predicting glare in sunny climates with high luminance contrasts", arguing that occupants could have higher glare tolerance, increasing the visual comfort variance. In fact, Frontczak and Wargocki [14] found from a survey analysis that visual comfort perception could be modified by age and type of job or activity executed. According to Bitsios et al. [34] findings, older people's pupil behaviour tends to dilate slower than in the case of younger people but constricts faster. In addition, Goncharov and Dainty [80] presented the relevant effect of ageing on the anatomical structure of the human eye, in particular for the cornea, the lens' thickness and lens' anterior radius; thus, modifying the eye's light regulation capacity.

Finally, Pierson et al. [36] highlighted the degree of certainty of the effect of some of these factors on the occupants' visual perception:

- gender and optical correction influence to be most certainly null;
- age, self-glare assessment, iris pigmentation impact to be yet inconclusive;
- culture, somewhat likely;
- macular pigment optical density (age-dependent), cortical hyperexcitability and contrast sensitivity influence have been considered likely.

Consequently, to correct this variability based on their physical features the scientific community has proposed to monitor the natural body light perception regulatory system (i.e., the eye). Reinhart [81] highlighted the human eye as the perfect light sensor and system actuator. As a personalized, naturally and highly evolved system, monitoring it could further calibrate visual comfort assessments (analogue to the body temperature in case of thermal comfort). For instance, the following approaches could be used:

- analysing the way the pupil size varies with respect to a certain lighting provision [82–85];
- monitoring pupil response when exposed to different lighting and work task type [86];
- screening the degree of eye opening [79];
- monitoring the view direction distribution and history under certain lighting provision [87];
- combining blinking, gaze direction, and pupil size variations records under a certain lighting environment [88]; and,
- analysing the frequency and extent of the facial muscles movement when exposed to expected discomfort glare [89].

The above-stated facts motivated a new approach towards visual comfort level evaluation, by coupling the building occupants' involuntary (i.e., body) response to the building management systems (BMS). The actual paradigm in which BMS systems are operated is contrasted on the basis of the averaged surveyed indoor physical parameters (see Figure 3). Therefore, hoping to reduce the gap between modelled and operating buildings' performance, by linking the occupant visual response to the BMS and using the data for more comprehensive understanding of the actual visual comfort. Consequently, increasing the satisfaction of occupants with the indoor luminous environment while also improving building energy efficiency, potentially reducing the risk of Sick Building Syndrome occurrence [90] and enhancing the livability of indoor spaces.

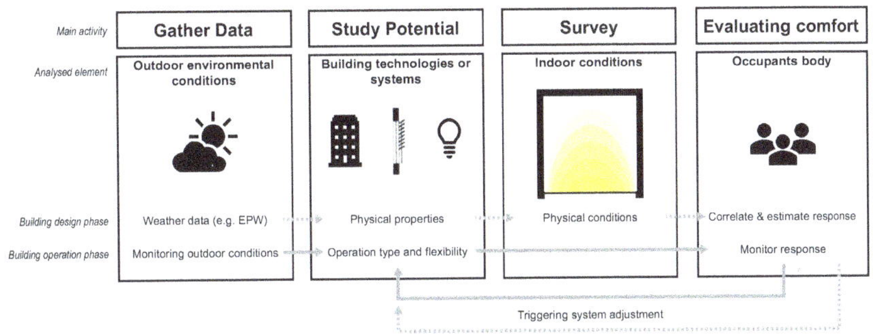

Figure 3. Description of the new concept for building components operation, shifting from monitoring the indoor physical conditions into monitoring the hosted occupant's body's physiological condition.

5.1. Eye Response to Light Variations

From the gathered literature, only a few metrics were found to directly assess the human eye response to the lighting environment, or indirectly account for the possibility of eye adaptation. These metrics are mostly based on the eye kinematics and only slightly on the quantity of light (either E or L). The identified metrics are summarized below:

- **Pupil diameter (d):** studied and condensed satisfactorily by De Groot and Gebhard [82]. An index is calculated based on the relationship between the pupil size and the luminance emitted towards it. It is grounded on a compendium of studies performed on human subjects for the medical research field of optometry. Equation (5) presents the most simplified form. Meanwhile, Equation (6) presented a greater correlation accuracy, especially when the corrected form of retinal illuminance (E_R) is used (Equation (7)) instead of the non-adjusted one (Equation (8)).

$$d = 4.9 - 3 \times \tan h[0.4(\log L + 0.5)] \tag{5}$$

$$\log E_R = \log L + 1.8614 - 0.000986 \times (\log L + 6.5) \tag{6}$$

$$\log E_R = 10 \times r^2 \times L \tag{7}$$

$$\log E_R = 10 \times r^2 \times L \times \left(1 - 0.0425 \times r^2 + 0.00067 \times r^4\right), \tag{8}$$

where:

E_R: retinal illuminance [trolands]
d: pupil diameter [mm]
r: pupil radius [mm]
L: luminance of the visual field [miliamberts]

- **Normalized pupil size (*NPS*):** Choi and Zhu [85] presented an analysis on how using a normalized pupil size value correlated to qualitative visual sensation communicated by the tested human subjects. The pupil size was obtained using a dedicated pupilometer, and the NPS was computed using Equation (7).

$$NPS = \frac{Ps_i - Ps_{neutral}}{Ps_{neutral}} \times 100, \tag{9}$$

where:

Ps_i: pupil size at the current state [mm]
$Ps_{neutral}$: pupil size at a neutral lighting state [mm]

- **Task-evoked Pupillary Response (*TEPR*):** Klinger et al. [86] studied the pupil diameter variations when different task and lighting conditions were imposed to the human subjects, measured by head-mounted eye trackers and remote eye trackers using image processing (i.e., counting the pixels composing the pupil).
- **Pupil diameter-unified formula (D_U):** Watson and Yellott [91] presented a review on the previous research carried out to correlate the pupil response to lighting stimulus, highlighting the equation's accuracy, and their potential adjustments. Moreover, it proposed to combine different effects that tend to deviate the pupils' diameter calculation (i.e., L, age, field size and the monocular effect) to construct Equation (10) (applicable for occupants aged between 20–83).

$$\begin{gathered} D_U = D_{SD} + A \\ D_{SD} = 7.75 - 5.75\left(\frac{(La/846)^{0.41}}{(La/846)^{0.41} + 2}\right) \\ A = (y - y_o) \times S \\ S = 0.021323 - 0.0095623 \times D_{SD}, \end{gathered} \tag{10}$$

where:

L: luminance [cd/m^2]
a: area [deg^2]
y: age of the human subject
y_o: minimum/reference/benchmark age

- **Degree of Eye Opening (*DEO*):** Yamin Garreton et al. [79] presented an analysis on how to use a normalized eye (or eyelid) height value. The pupil size was obtained using an analysis of images recording the state of the eyelid in correspondence with the maximum eye height (Equation (11)).

$$DEO = \frac{h_i}{h_{max}} \tag{11}$$

where:

h_i: eye/eyelid height [mm] or [pixels]
h_{max}: maximum eye/eyelid height [mm] or [pixels]

- **Gaze driven illuminance (E_{vg}):** Sarey Khanie et al. [92] exposed an analysis of how much would it change to compute the E_v considering accurately occupant's gaze direction, aided by the use of Radiance images and evalglare tool [63,66].

Referring to the occupants' eye condition instead of the physical light quantity in the room would implicitly include the perceived lighting within the field of view and all the personal factors affecting the occupant perception. Using a correlation amongst the two would further bridge the gap between designed and monitored building performance. However, no guidelines have embraced any of the metrics described above or, incorporated them into an obligatory assessment method of the lighting design in buildings as an initial attempt to promote such an approach. A summary of the advantages and limitations of the above presented metrics have been condensed into Table 4.

Table 4. Critical analysis summary of visual perception metrics based on eye conditions, highlighting their benefits and drawbacks.

Metric	Advantages	Disadvantages
d	Accounts for directionality and human adaptation, highly responsive and reliable, light at eye known.	There is no established correlation with the occurrence of glare, no human variability, only one location, hard to measure and model.
NPS	Accounts for directionality and human adaptation, refers to the saturation level of adaptability.	There is no established correlation with the occurrence of glare nor illuminance, only one location, hard to measure (especially for neutral condition) and model.
$TEPR$	Measures pupil kinematics with high precision.	Highly intrusive measurement, no correlation with lighting environment performed, only pupil response recorded.
D_U	Accounts for directionality, human adaptation and variability. Highly responsive and reliable. Light at eye known.	There is no established correlation with the occurrence of glare, only referred to one location and occupant, not applicable for young occupants, hard to monitor and model.
DEO	Accounts for directionality and human adaptation, refers to saturation level of adaptability (not only pupil), correlation with DGP.	Scaling issues, only actions for eye-lid kinematics considered, no human variability considered, intrusive measurement, one location, hard to model.
E_{vg}	Accounts for directionality and human adaptation, indirectly refers to adaptability (not only pupil), correlation with DGP. Light at eye known.	No human variability considered, intrusive measurement, one location, hard to model (requires known gaze behaviour).

6. Emerging Occupant-Centred Visual Comfort Methods for Operating Building Lighting-Related Components

Boestra et al. [13] analysed the Health Optimisation Protocol for Energy-efficient Buildings database (HOPE, https://hope.epfl.ch/partners/partners-intro.htm (accessed on 23 December 2019)) that gathers post-occupancy comfort surveys on 60 office buildings with more than 6000 surveyed participants. Concluding that occupants frequently reported dissatisfaction with the lack or limited options for personal system control to adjust their surroundings to their demands and needs. The communication of occupants' intentions towards the BMS shall be personal and, if possible, highly responsive to the singular occupant or a group of occupants with similar needs.

Ensuring that the occupants' intended interaction correlates to the indoor environment's unpleasant conditions is vital for the building to acknowledge both the occupants' intentions and the indoor environment (physical) state. The use of sensors and automatized actuators has the potential to increase building operation performance. The availability of granular and real-time data enhances the response rate of any system towards more proficient results.

The industry has already demonstrated these informed control benefits, emphasizing the utility of sensors for acquiring critical variables' trends, process monitoring and control, but also sharing highlights on the advantages and weaknesses of current data-processing methodologies. Kadlec et al. [93] executed a review about available approaches to develop soft-sensors and to acquire and process data to build and train prediction models. Furthermore, they presented applications (e.g., on-line predictions, fault detection) and strategies on how to deal with common issues that arise when implementing them (e.g., missing data, outliers, diverse sampling rates and measurement delays). In addition, it is not only the implementation and monitoring of specific data that drives performance boost, it is also enhanced by: the collaborative interaction among systems or machines, the possibility of acknowledging and interpreting new experienced conditions for adaptation (executing "informed" decisions) [94]. However, it is challenging to achieve a smooth functioning under a collaborative data acquisition and processing; it requires first, that sensors' noise is identified and filtered accurately [95], and then effective algorithms that are able to prioritize and organize the effects of integrated actuators properly.

For example, Poli et al. [96] proposed and tested the use of sensors immersed within glazing components for monitoring indoor conditions and gathering occupants' response to tune automatic blind control to regulate light influx. Rinaldi et al. [97] further analysed such components, testing a predictive model to later anticipate the occupants' intentions to regulate the window-blind states. Likewise, Karjalainen [98] presented a general conceptual work-flow to deal concurrently with heating, cooling, ventilation, lighting and automated blinds operation; by allowing occupants to communicate their interaction desires and by collecting data inputs of the weather forecast.

In more detail, recent research has concentrated on studying comfort from the singular occupant perspective. Occupant-based metrics have been proposed, as O'Brien et al. [99] did by normalizing results with occupancy density. This approach shift provides the advantage of understanding the real impact and needs of occupants' micro-conditions by having sensors and actuators close to the building occupants' workstations (in case of office buildings) and remotely controlled actuators from mobile or website applications (in case of residential buildings). From the studied actuators, their operation was found to be more in line with occupants' preferences (i.e., accurate response), daylighting proxies (parameters) were monitored at the occupants' work plane or from the occupant's view perspective.

6.1. Monitoring Occupants' Work Plane

When monitoring the occupants' work plane, the collected strategies were found to be mainly focused on surveying E_h intensities and contrast (U_o). These strategies are easily applied on simulation-based design, likewise replicated in real-case scenarios. For instance:

- Jia et al. [100] presented a platform-based design framework for BMS considering data acquired from a camera, illuminance, temperature, CO_2, relative humidity (RH) and passive infra-red (PIR) sensors, and occupants' requests via mobile applications; managing decisions throughout the use of fuzzy logic to run HVAC and lighting appliances.
- Konstantakopoulos et al. [101] proposed the automation of shared lighting appliances through game theory for processing the building occupants' vote on lighting preferences.
- Reinhart [102] presented an approach of a light-switching and blind-operation model within a two occupants office, in which using simulation methods approach resulted in a 20% lighting energy reduction achieved by monitoring at every desk: occupants presence, light and radiation intensity, light appliances condition (ON/OFF) and predicting the possibility of turning them ON upon arrival [103], or during the day [22], and of turning them OFF [104].
- Gunay et al. [105] tested and studied the work-plane illuminance set-points of a control strategy for light-switching and blind operation on single office space through simulations with EnergyPlus with dynamically updated operation thresholds (modified by the probability models of user-interaction [22,103,104]). As a result, energy use reductions between 25–35% were achieved.
- Cheng et al. [106] studied the personal occupant visual comfort in parallel with energy savings (weighting cooling vs. lighting energy needs) and slat inclination angles of venetian blinds, monitoring lighting conditions with illuminance sensors on the desk and occupants interactions within a recreated office space facing east. The acquired datas was used to train and adjust the thresholds through a Q-learning process, leading to an energy saving potential up to 10%.
- Van De Meugheuvel et al. [107] proposed a different approach by studying ceiling-mounted sensor configuration through simulations in DIALux [108]. Using multiple intelligent dimmable luminaries equipped with occupancy and light sensors. These were calibrated at night to reach E_h = 500 lux, to maintain desired lighting conditions working as stand-alone systems or as an integrated system network (for optimized luminaries activation also see Rubinstein et al., Caicedo and Pandharipande [109,110]).
- Jin et al. [111] studied the Indoor Environmental Quality (IEQ) by continuously monitoring an office area by a moving punctual reading (mounting sensors on a robot, including E_h). Doing so, it is possible to interpolate and map the total area conditions

with decent accuracy, given the surveyed values obtained while moving the robot throughout a predefined path.

Nevertheless, comparing the previously mentioned methods and more frequently utilized metrics, most of them analysed the light intensity on the work-plane, neglecting the light intensity and contrast perceived at eye level. Doing this is rather challenging; nonetheless, other methodologies have been elaborated based on the field of view to assess visual perception with higher accuracy, giving more importance to the light intensity falling on the human eye, its distribution around the space, and the personal occupant preferences (see Figure 4).

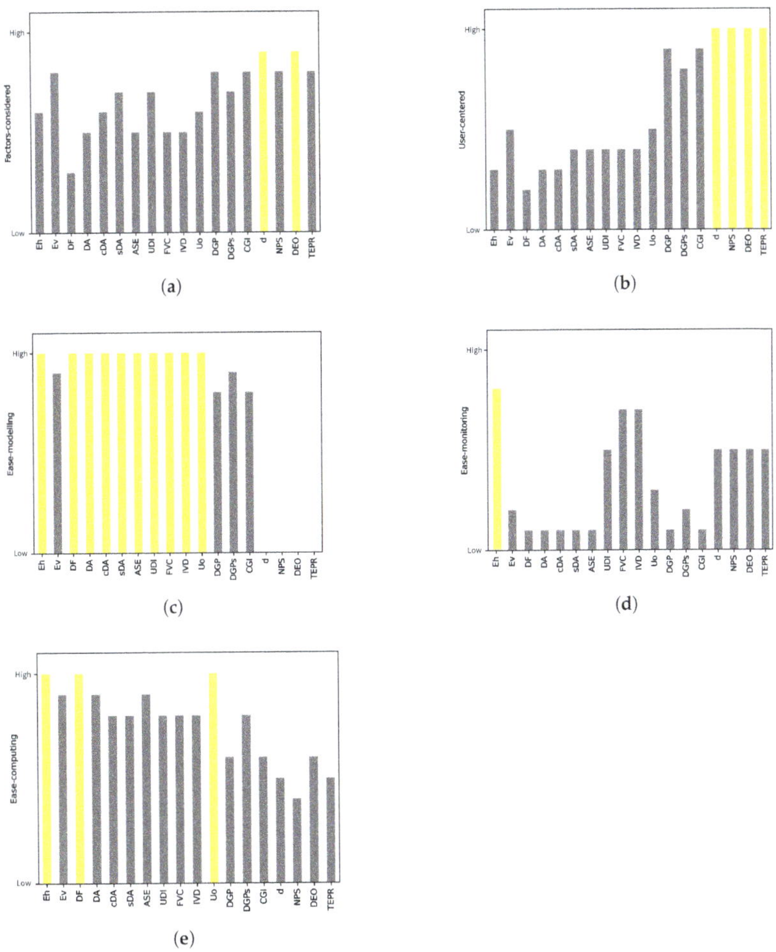

Figure 4. Direct comparison of the most utilized metrics under 5 different criteria, providing a subjective and experience-based rating, and highlighting in yellow the highest ranked in: (**a**) completeness of relevant factors considered (**b**) degree of occupant-centred approach (**c**) ease of modelling in the design phase (**d**) ease of monitoring in the operation phase (**e**) ease of computing.

6.2. Monitoring Occupants' Field of View

Recalling the information gathered in Tables 2–4 and the comparison made in Figure 4, personalized and, if possible, image-based analysis seems to be the most comprehensive and convenient way of assessing visual comfort. In this way, task lighting sufficiency, contrast and glare risk can be evaluated simultaneously. For this purpose, different methods have been proposed and hereby summarized; however, difficulties arise considering the limitations of replicating simulation-based analysis within real scenarios.

- Image-based—Although these methods demand higher computation time, they are the ones capable of capturing different aspects that, as discussed previously, can alter the perception of the surrounding luminous environment.
  - Guerry et al. [112] proposed an innovative image-based methodology on the evaluation of contrast able to take into consideration visual impairments in health. The proposed method screens the produced images, identifying disturbing surfaces or elements within the field of view, based on a set threshold of contrast of luminance between two juxtaposed surfaces.
  - Based on the work of Glenn et al. [113] for the automotive sector, aiming to reduce night accidents due to wrong street lighting disposition and intensity; Zatari et al. [114] studied the possibility of automatizing this methodology for estimating glare risk, L and E_v in real-time by employing Charge Coupled Device (CCD) cameras mounted on a vehicle.
  - Based on the results obtained by Wienold and Christoffersen [63], Konis [115] presented an actual office space survey on 14 participants, coupling HDR images analysis (captured with CCD cameras oriented to the most frequent occupants' field of view) with polling station results on visual comfort rating; in order to test the accuracy in terms of glare rating estimation, with horizontal and E_v based computations.
  - Using CCD cameras as a data acquisition system, it is possible to make, calibrate, and compute E_v and L from HDR images. These can be later used to monitor the interior conditions of an environment, as done by Parsae et al. [116] and Inanici [117]. The latter proposed the calculation by using the RGB values of an image and the D_{65} reference illuminant, which after the proper calibrations, and accounting for the camera settings (e.g., exposure, focal length), obtained errors below 8% when testing the methodology under different light sources. On the other hand, Moeck [118] proposed a geometrical-based procedure to understand the illuminance value coming from a certain surface using CCD cameras and HDR images, acknowledging not only the camera settings, but also the distance between the objects and the camera. Under controlled conditions, the error obtained was below 5%.
  - Goovaerts et al. [119] tested the computation of *DGP* from HDR images created from low-resolution camera pictures while monitoring E_h, to establish a venetian and roller blind control algorithm aiming to avoid the occurrence of visual discomfort. The *DGP* was initially calculated, and then the strategy was tested on those users which interacted with the shading systems to increase daylight influx indoors. However, the E_h was found to be underestimating the light intensity adaptation, and the *DGP* to be overestimating the glare risk rating.
  - Motamed et al. [120] used two HDR ceiling vision sensors for monitoring work-plane illuminance (ceiling mounted) and *DGP* (oriented towards occupants' visual display terminal (VDT)). This was done to lay down an advanced control algorithm for external roller blinds, which could in real-time, through fuzzy logic use, review the task illuminance compliance and glare risk from DGP values. By using this control logic to operate the external blinds confronted with a reference Eh at work-plane control, although all 30 subjects reported only slight variations on visual perception between control logics, energy needs from electrical appliances were reduced up to 31% compared to operate the roller blinds from monitored E_h at the work plane.
- Wearable—The use of wearable devices for indoor monitoring was initially proposed by researchers assessing thermal comfort, later embraced by the research stream in visual comfort. They have resulted in very useful input for the building management

systems as they communicate a more localized and personalized measurement of environmental conditions. However, from the literature analysis, the collected strategies hereby presented are only applicable for assessing buildings during operation.

- Similar to the methodology employed by Sarey Khanie et al. [92] to define E_{vg}, Schneider et al. [121] developed a prototype of a head-mounted camera, for studying human gaze-behaviour. Using a video-oculography device (VOG) and a camera motion device, synchronously measure binocular eye positions at up to 600 Hz, it is possible to acknowledge where and what the user is seeing or looking at. These could ease any gaze driven analysis for visual comfort, but the degree of intrusiveness is yet too high to consider it for continuous monitoring.
- Yamin Garreton et al. [79] proposed Equation (11) for computing glare risk rating as DEO. Using an eye-tracker, they computed this metric and tested it on 20 subjects within a recreated office space. Based on the eye-lid position, it displayed a decent correlation with DGP and E_v, under a wide range of lighting conditions (both low intensity/diffuse and high intensity/clear sky conditions). However, it still represents a highly intrusive method for continuous monitoring of the indoor environment lighting conditions and occupant perception.
- Choi and Zhu [85] investigated the potential of tracking the human pupil size to estimate visual sensation in office workplace environments. Profiting from its physiological capacity, as it reacts instantly with the visual environment variations. The tests were carried out on 20 subjects within a recreated office space, under different artificial lighting conditions and constant RH (avoiding dry-eye irritation effect). Pupil behaviour was monitored with a mobile pupilometer and the light intensity at the work-plane (E_h) with an illuminance meter. Although the intrusiveness is lower than in previously exposed studies, it is yet too restrictive to be applied as monitored visual comfort metric in operating buildings (requires users to wear the pupilometer constantly), and no shading effect from the eyelids was considered.

Thus, the above-presented new methodologies and the metrics condensed in Table 4, widen the possibilities to account for real occupants' lighting related needs or preference variability and bridge the gap between modelled and actual building performance. Visually comfortable ranges could be fine-tuned based on the actual reaction of the building occupants coupled with satisfaction surveys, including demographic factors that have been proven to affect the occupants' perception of the luminous environment. Furthermore, operating buildings could incorporate occupant-need responsive systems through clever monitoring networks incorporating some kind of occupant response tracking system.

7. Discussion

As mentioned in Section 3, acknowledging and incorporating the effect of occupants on building performance can contribute to narrowing down the performance gap between planned and actual building performance. Specially considering that based on building occupants demographic and physical features, occupants could be more susceptible to discomfort and, therefore, more inclined to intervene in building performance. Nevertheless, including this aspect in the definition of comfortable visual conditions for building design or operation is rather challenging. This variability is still unconsidered in current building guidelines, standards and methodologies for the assessment of the indoor luminous environment. Contrary to thermal comfort assessment, only general considerations proposing higher illuminance intensities have been made for visually impaired occupants [50,122].

Standards have been passive on including complex assessments for visual comfort, and established detailed approaches are mainly related to those proposed by certification procedures (e.g., LEED, BREAM). Nonetheless, the challenge remains on the complexity of associating a spatial and time-based approach to the sensitivity of the single building occupant. Moreover, when a diverse occupancy is considered, in complex environments, or in operating buildings, there are few possibilities to collect in parallel all their perceptions and needs to be able to satisfy all of them contemporarily.

The challenge presented above requires different but integrated solutions. One potential solution could be by starting to complement qualitative survey data with objective physiological response on the ratings given to different luminous environments. This would provide more accurate definitions of the comfortable visual environment; it will also enrich the knowledge and provide further certainties on the factor's affection visual perception. Then, such definitions can be diversified by testing similar conditions for representative subjects of each of the physiological factors that can affect their perception. Having these information enables the fine-tuning of existing metrics, or the proposal of new ones, to accurately estimate visual comfort in a single point for different space uses and demographic contexts. Thus, different simulation workflows can be optimized to maintain admissible computation time to perform time-based and spatial visual comfort assessments utilizing such metrics. Finally, non-intrusive monitoring strategies can be designed to run in operating buildings, targeting directly or indirectly the parameters defined to compute the developed or adjusted metrics.

The use of the body response as a proxy (i.e., eye response) to monitor visual stimulus can be one holistic solution. In fact, research has already been done on its relationship with environmental parameters (E_R, E_p and L) and in some of the factors that could impact the perception of the luminous environment (excluding those related to the emotional dimension). Using the body response is personalized for each occupant and it can be monitored on an operational building (taking care of the level of intrusion and privacy).

Limitations

Numerous visual comfort assessment methods and strategies have been collected, reviewed, and compared, finding as a potential solution the need for a more diversified and personalized metric, or number of metrics, and a more occupant-centred assessment. However, the collection of the literature was conducted by targeting diffused and innovative methods of visual comfort assessment (on an expert criteria basis). Additionally, literature that contained visual comfort analysis was put aside if a higher focus was given to thermal comfort or overall building energy use analysis. Thus, the literature collection procedure could have resulted in being unintentionally biased. Future updates to this work could be hampered as a generic query was utilized and variable filters were imposed. Nevertheless, the latest findings support that the course that visual comfort research is taking is aligned with what has been presented.

8. Conclusions and Further Developments

The present analysis concentrated on: (i) scrutinizing and comparing visual comfort assessment works and established procedures, to deduct advanced methods to characterize, monitor and rate the luminous environment. The most used metrics to rate visual comfort were compared and their strengths and weaknesses were presented, based on the identified need to capture a more accurate visual perception; (ii) Different methods and approaches to monitor the luminous environment and estimate the visual comfort were scrutinized for building design and operation. Their shortcomings were identified and potential improvements were proposed, in order to incorporate important missing influential factors such as the ones gathered by Persont et al. [36]. Consequently, tracking the eye response and correlating this response with illuminance was identified as a promising approach to enhance the accuracy to estimate visual comfort, to improve lighting design in buildings and thus, to cover the gap between expected and actual building performance.

However, further research is necessary following what has been done with the metrics presented in Section 5.1. In this regard, the actual eye response to light variations of both healthy and visually impaired occupants should be studied to attain more accurate visual comfort ratings. Obtained data could then be used to adjust the thresholds of the existing metrics (or, if deemed necessary, proposing new ones) to account for such physiological visual perception differences. As a result, better limit thresholds for building systems operation could be defined, while more appropriate occupancy definition and better monitoring of indoor lighting environments could be executed.

A proposal for a novel approach to evaluating a luminous indoor environment based on human-centric matrices could be defined. In general, it could be summarised as having a more detailed definition of the occupant inhabiting the building, better anticipating their preferences for luminous indoor conditions, and consequentially reducing unexpected building performance due to visual discomfort. In order to achieve the stated goals:

- In the instance of existing buildings for monitoring:
  - Transferring and integrating novel and growing technologies such as computer vision, image processing, scene understanding, and deep learning can further enable monitoring and operation strategies. They can help identify occupancy type (including occupant characteristics) that can be feed to the BMS for operational profiling. They can potentially decrease the level of intrusiveness for monitoring body response. Or, making components responsive, with data-training models able to capture occupant interactions and predict occupant needs [123].
  - Lean sensorial network and internet of things implementation have great potential to better interpret and communicate occupancy, overall indoor conditions and personalized task area conditions [124,125]. If these are fully integrated into a flexible plant system, personalized or localized building responses can be activated for providing, overall, better indoor environmental quality.
  - Alternatively, qualitative surveys and post occupancy evaluations can be utilized in real-time for requesting to the building specific operation adjustments. Comparable to the upgrade of a traditional manual control enhanced by digital technologies [126].
- In the instance of both existing and new buildings for design:
  - Data acquired in occupancy studies could serve better for understanding building occupant preferences in tendency in occupation, lighting appliances and blinds activation schedules, and space occupation (complementing missing information according to the specific LOD) (i.e., occupant modelling) [10]. Then, these data can be used for a more accurate simulation output with realistic occupant behaviour [127].
  - Diversified post-occupancy surveys on perception of the luminous environment (or visual comfort rating) could be used for training algorithms and defining prediction models resulting in average virtual occupant typology. Or, with such databases, occupants' preferences profiling can be carried out to adjust traditional or new visual comfort performance metrics [128].
  - Furthermore, utilizing personalized and complex visual comfort performance metrics would yield the analysis more complete. These could include demographics and spatial dependent factors by fine-tuning the correlating between body response (in this case the eye for visual comfort, using d and/or DEO) with environmental parameters (i.e., E_v, E_R or E_p, and DGP).
  - To assure that such calculations are done within applicable computational time, annual and spatial simulations could be: (i) performed using daylight coefficients, or cubic illuminance, instead of rendered-image-based analysis [72]; (ii) they can be structured in a way that these can be parametric so they can be run in parallel through cloud computing services [74]; (iii) Or, as a mid-term solution, simulations can be decomposed in such a way that many smaller and more specialized cores can be used to process the computational task (delivering higher computational performance) in GPUs [75].

In conclusion, this work is foreseen as a catalyzer to modernize the traditional way of assessing visual comfort, estimating visual perception and rating the lighting provision. Transiting from monitoring/studying the environment onto monitoring/studying the occupant response or occupant interactions itself. Motivating the inclusion of the occupant demographic features could render the analysis completer and more accurate. Thus, boosting the capacity of designer to predict the actual performance of buildings and to reduce the performance gap issues stemming from the occupants visual dissatisfaction.

Author Contributions: Conceptualization, J.D.B.C. and T.P.; methodology, J.D.B.C. and T.P.; validation, J.D.B.C., T.P., M.K. and G.L.; formal analysis, J.D.B.C.; investigation, J.D.B.C., T.P., M.K., G.L., A.G.M. and A.S.; resources, J.D.B.C., T.P., A.G.M. and A.S.; data curation, J.D.B.C., T.P., M.K., G.L., A.G.M. and A.S.; writing—original draft preparation, J.D.B.C.; writing—review and editing, J.D.B.C., T.P., M.K., G.L., A.G.M. and A.S.; visualization, J.D.B.C. and T.P.; supervision, T.P., M.K., G.L. and A.G.M.; funding acquisition, J.D.B.C., T.P., M.K., G.L., A.G.M. and A.S. All authors have read and agreed to the published version of the manuscript.

Funding: This research was funded by ABC Department—Politecnico di Milano (finanziamento delle attività per la valorizzazione della ricerca e TM DABC). The author M. K. acknowledges the financial support from the Slovenian Research Agency (research core funding No. P2-0158). The author G.L. acknowledges the financial support from the Norwegian Research Council (research project FRIPRO-FRINATEK no. 324243 HELIOS).

Institutional Review Board Statement: Not applicable.

Informed Consent Statement: Not applicable.

Data Availability Statement: Not applicable.

Conflicts of Interest: The authors declare no conflict of interest.

Abbreviations

The following abbreviations are used in this manuscript:

ASE	Annual Sun Exposure
BEM	building energy modelling
BGI	British Glare Index
BMS	Building Management System
CCD	Charge Coupled Device
cDA	continuous Daylight Autonomy
CGI	International Commission on Illumination Glare Index
D_U	Pupil diameter unified formula
DA	Daylight Autonomy
DEO	Degree of Eye Opening
DF	Daylight Factor
DGP	Daylight Glare Probability
DGP_s	simplified Daylight Glare Probability
E	Illuminance
E_h	Horizontal illuminance
E_p	Pupil illuminance
E_R	Retinal illuminance
E_v	Vertical illuminance
E_{vg}	Gaze driven illuminance
EN	European Committee for Standardization (CEN) standard
FVC	Frequency of Visual Comfort
HDR	High Dynamic Range
HVAC	Heating, Ventilation and Air Conditioning
IEQ	Indoor Air Quality

IVD	Intensity of Visual Discomfort
KPI	Key Performance Indicator
L	Luminance
NPS	Normalized pupil size
RGB	Red, Blue and Green colour space
RH	Relative humidity
sDA	spatial DA
TEPR	Task-evoked Pupillary Response
U_o	Daylight Uniformity
UDI	Useful Daylight Illuminance
VDT	Video Display Terminal

References

1. Matz, C.J.; Stieb, D.M.; Davis, K.; Egyed, M.; Rose, A.; Chou, B.; Brion, O. Effects of Age, Season, Gender and Urban-Rural Status on Time-Activity: Canadian Human Activity Pattern Survey 2 (CHAPS 2). *Int. J. Environ. Res. Public Health* **2014**, *11*, 2108–2124. [CrossRef] [PubMed]
2. Klepeis, N.E.; Nelson, W.C.; Ott, W.R.; Robinson, J.P.; Tsang, A.M.; Switzer, P.; Behar, J.V.; Hern, S.C.; Engelmann, W.H. The National Human Activity Pattern Survey (NHAPS): A resource for assessing exposure to environmental pollutants. *J. Expo. Sci. Environ. Epidemiol.* **2001**, *11*, 231–252. [CrossRef] [PubMed]
3. ANSI; ASHRAE. *ANSI/ASHRAE Standard 55–2017 : Thermal Environmental Conditions for Human Occupancy*; Technical Report; ANSI: New York, NY, USA; ASHRAE: Atlanta, GA, USA, 2017.
4. *ISO 7730:2005*; Ergonomics of the Thermal Environment—Analytical Determination and Interpretation of Thermal Comfort Using Calculation of the PMV and PPD Indices and Local Thermal Comfort Criteria. ISO: Bristol, UK, 2005.
5. UN Environment Programme. SDG11:Sustainable Cities and Communities Make Cities and Human Settlements Inclusive, Safe, Resilient and Sustainable. Available online: https://www.unenvironment.org/explore-topics/sustainable-development-goals/why-do-sustainable-development-goals-matter/goal-11 (accessed on 7 March 2018).
6. Directive EPBD Recast. Directive 2010/31/EU of the European Parliament and of the Council of 19 May 2010 on the energy performance of buildings (recast). *Off. J. Eur. Union* **2010**, *18*, 2010.
7. Directive EE. Directive 2012/27/EU of the European Parliament and of the Council of 25 October 2012 on energy efficiency, amending Directives 2009/125/EC and 2010/30/EU and repealing Directives 2004/8/EC and 2006/32. *Off. J. Eur. Union* **2012**, *315*, 1–56.
8. Erhorn, H.; Erhorn-Kluttig, H. Terms and definitions for high performance buildings. In *Detailed Report for the Concerted Action, Energy Performance of Buildings*; European Union: Luxembourg, 2011.
9. Masoso, O.T.; Grobler, L.J. The dark side of occupants' behaviour on building energy use. *Energy Build.* **2010**, *42*, 173–177. [CrossRef]
10. Happle, G.; Fonseca, J.A.; Schlueter, A. Impacts of diversity in commercial building occupancy profiles on district energy demand and supply. *Appl. Energy* **2020**, *277*, 115594. [CrossRef]
11. Hong, T.; Yan, D.; D'Oca, S.; Chen, C.F. Ten questions concerning occupant behavior in buildings: The big picture. *Build. Environ.* **2016**, *114*, 518–530. [CrossRef]
12. Arens, E.; Humphreys, M.A.; de Dear, R.; Zhang, H. Are 'class A' temperature requirements realistic or desirable? *Build. Environ.* **2010**, *45*, 4–10. [CrossRef]
13. Boerstra, A.; Beuker, T.; Loomans, M.; Hensen, J. Impact of available and perceived control on comfort and health in European offices. *Archit. Sci. Rev.* **2013**, *56*, 30–41. [CrossRef]
14. Frontczak, M.; Wargocki, P. Literature survey on how different factors influence human comfort in indoor environments. *Build. Environ.* **2011**, *46*, 922–937. [CrossRef]
15. Kamaruzzaman, S.; Egbu, C.; Zawawi, E.; Ali, A.; Che-Ani, A. The effect of indoor environmental quality on occupants' perception of performance: A case study of refurbished historic buildings in Malaysia. *Energy Build.* **2011**, *43*, 407–413. [CrossRef]
16. Dogbeh, A.; Jomaas, G.; Bjarlov, S.P.; Toftum, J. Field study of the indoor environment in a Danish prison. *Build. Environ.* **2014**, *88*, 20–26. [CrossRef]
17. Mirrahimi, S.; Mohamed, M.F.; Haw, L.C.; Ibrahim, N.L.N.; Yusoff, W.F.M.; Aflaki, A. The effect of building envelope on the thermal comfort and energy saving for high-rise buildings in hot-humid climate. *Renew. Sustain. Energy Rev.* **2016**, *53*, 1508–1519. [CrossRef]
18. Suratkon, A.; Chan, C.M.; Jusoh, S. Indicators for measuring satisfaction towards design quality of buildings. *Int. J. Geomate* **2016**, *11*, 2348–2355. [CrossRef]
19. Konstantoglou, M.; Tsangrassoulis, A. Dynamic operation of daylighting and shading systems: A literature review. *Renew. Sustain. Energy Rev.* **2016**, *60*, 268–283. [CrossRef]
20. Da Silva, P.C.; Leal, V.; Andersen, M. Influence of shading control patterns on the energy assessment of office spaces. *Energy Build.* **2012**, *50*, 35–48. [CrossRef]

21. Gunay, H.B.; O'Brien, W.; Beausoleil-Morrison, I.; Gilani, S. Development and implementation of an adaptive lighting and blinds control algorithm. *Build. Environ.* **2017**, *113*, 185–199. [CrossRef]
22. Reinhart, C.F.; Voss, K. Monitoring manual control of electric lighting and blinds. *Light. Res. Technol.* **2003**, *35*, 258–259. [CrossRef]
23. Bourgeois, D.; Reinhart, C.; Macdonald, I. Adding advanced behavioural models in whole building energy simulation: A study on the total energy impact of manual and automated lighting control. *Energy Build.* **2006**, *38*, 814–823. [CrossRef]
24. Escuyer, S.; Fontoynont, M. Lighting controls: A field study of office workers' reactions. *Light. Res. Technol.* **2001**, *33*, 77–94. [CrossRef]
25. Wang, S.; Chen, B.; Suo, J.; Zhao, J.R. Impact of building morphology and outdoor environment on light and thermal environment in campus buildings in cold region during winter. *Build. Environ.* **2021**, *204*, 108074. [CrossRef]
26. Mahmoudzadeh, P.; Afacan, Y.; Adi, M.N. Analyzing occupants' control over lighting systems in office settings using immersive virtual environments. *Build. Environ.* **2021**, *196*, 107823. [CrossRef]
27. da Silva, P.C.; Leal, V.; Andersen, M. Occupants interaction with electric lighting and shading systems in real single-occupied offices: Results from a monitoring campaign. *Build. Environ.* **2013**, *64*, 152–168. [CrossRef]
28. Yun, G.Y.; Kim, H.; Kim, J.T. Effects of occupancy and lighting use patterns on lighting energy consumption. *Energy Build.* **2012**, *46*, 152–158. [CrossRef]
29. Rea, M.; Dillon, R.; Levy, A. The effectiveness of light switch reminders in reducing light usage. *Light. Res. Amd Technol.* **1987**, *19*, 81–85. [CrossRef]
30. Azar, E.; Menassa, C.C. Agent-Based Modeling of Occupants and Their Impact on Energy Use in Commercial Buildings. *J. Comput. Civ. Eng.* **2012**, *26*, 506–518. [CrossRef]
31. Panda, S. Circadian physiology of metabolism. *Science* **2016**, *354*, 1008–1015. [CrossRef] [PubMed]
32. Turner, P.L.; Mainster, M.A. Circadian photoreception: Ageing and the eye's important role in systemic health. *Br. J. Ophthalmol.* **2008**, *92*, 1439–1444. [CrossRef] [PubMed]
33. Winn, B.; Whitaker, D.; Elliott, D.B.; Phillips, N.J. Factors affecting light-adapted pupil size in normal human subjects. *Investig. Ophthalmol. Vis. Sci.* **1994**, *35*, 1132–1137.
34. Bitsios, P.; Prettyman, R.; Szabadi, E. Changes in Autonomic Function with Age: A Study of Pupillary Kinetics in Healthy Young and Old People. *Age Ageing* **1996**, *25*, 432–438. [CrossRef] [PubMed]
35. Kumnick, L.S. Aging and the efficiency of the pupillary mechanism. *J. Gerontol.* **1956**, *11*, 160–164. [CrossRef] [PubMed]
36. Pierson, C.; Wienold, J.; Bodart, M. Review of Factors Influencing Discomfort Glare Perception from Daylight. *Leukos* **2018**, *14*, 111–148. [CrossRef]
37. Tsushima, S.; ichi Tanabe, S.; Utsumi, K. Workers' awareness and indoor environmental quality in electricity-saving offices. *Build. Environ.* **2015**, *88*, 10–19. [CrossRef]
38. Ahmadi-Karvigh, S.; Ghahramani, A.; Becerik-Gerber, B.; Soibelman, L. One size does not fit all: Understanding user preferences for building automation systems. *Energy Build.* **2017**, *145*, 163–173. [CrossRef]
39. Liang, X.; Hong, T.; Shen, G.Q. Occupancy data analytics and prediction: A case study. *Build. Environ.* **2016**, *102*, 179–192. [CrossRef]
40. Wilke, U.; Haldi, F.; Scartezzini, J.L.; Robinson, D. A bottom-up stochastic model to predict building occupants' time-dependent activities. *Build. Environ.* **2013**, *60*, 254–264. [CrossRef]
41. Adamopoulou, A.A.; Tryferidis, A.M.; Tzovaras, D.K. A context-aware method for building occupancy prediction. *Energy Build.* **2016**, *110*, 229–244. [CrossRef]
42. Mahdavi, A.; Tahmasebi, F. Predicting people's presence in buildings: An empirically based model performance analysis. *Energy Build.* **2015**, *86*, 349–355. [CrossRef]
43. D'Oca, S.; Hong, T. Occupancy schedules learning process through a data mining framework. *Energy Build.* **2015**, *88*, 395–408. [CrossRef]
44. Yang, J.; Santamouris, M.; Lee, S.E.; Deb, C. Energy performance model development and occupancy number identification of institutional buildings. *Energy Build.* **2016**, *123*, 192–204. [CrossRef]
45. Stoppel, C.M.; Leite, F. Integrating probabilistic methods for describing occupant presence with building energy simulation models. *Energy Build.* **2014**, *68*, 99–107. [CrossRef]
46. Langevin, J.; Gurian, P.L.; Wen, J. Tracking the human–building interaction: A longitudinal field study of occupant behavior in air-conditioned offices. *J. Environ. Psychol.* **2015**, *42*, 94–115. [CrossRef]
47. Yan, D.; Hong, T.; Dong, B.; Mahdavi, A.; D'Oca, S.; Gaetani, I.; Feng, X. IEA EBC Annex 66: Definition and simulation of occupant behavior in buildings. *Energy Build.* **2017**, *156*, 258–270. [CrossRef]
48. Wagner, A.; O'Brien, L. *Occupant Behaviour-Centric building Design and Operation EBC Annex 79 (Draft)*; Technical Report May; International Energy Agency: Paris, France, 2018.
49. Andrews, C.J.; Yi, D.; Krogmann, U.; Senick, J.A.; Wener, R.E. Designing buildings for real occupants: An agent-based approach. *IEEE Trans. Syst. Man, Cybern. Part A Syst. Hum.* **2011**, *41*, 1077–1091. [CrossRef]
50. CIE. *Lighting Needs for the Partially Sighted*; Technical Report; International Comission on Illumination (CIE): Vienna, Austria, 1997.
51. Al Horr, Y.; Arif, M.; Kaushik, A.; Mazroei, A.; Katafygiotou, M.; Elsarrag, E. Occupant Product. *Off. A Indoor Environ. Qual. Rev. Lit.* **2016**, *105*, 369–389. [CrossRef]
52. *EN ISO 8996*; Ergonomics of the Thermal Environment—Determination of Metabolic Rate. ISO: Bristol, UK, 2004.

53. Bourne, R.R.; Flaxman, S.R.; Braithwaite, T.; Cicinelli, M.V.; Das, A.; Jonas, J.B.; Keeffe, J.; Kempen, J.; Leasher, J.; Limburg, H.; et al. Magnitude, temporal trends, and projections of the global prevalence of blindness and distance and near vision impairment: A systematic review and meta-analysis. *Lancet Glob. Health* **2017**, *5*, e888–e897. [CrossRef]
54. Ko, W.H.; Kent, M.G.; Schiavon, S.; Levitt, B.; Betti, G. A Window View Quality Assessment Framework. *Leukos* **2021**, 1–26. [CrossRef]
55. Walsh, J.W.T. The Early Years of Illuminating Engineering in Great Britain. *Light. Res. Technol.* **1951**, *16*, 49–60. [CrossRef]
56. *BS 8206-2;* Lighting for Buildings—Part 2: Code of Practice for Daylighting. British Standard Institution (BSI): London, UK, 2008.
57. *European Standard EN 17037;* Daylight in Buildings. British Standard Institution (BSI): London, UK, 2018.
58. Reinhart, C.F.; Walkenhorst, O. Validation of dynamic RADIANCE-based daylight simulations for a test office with external blinds. *Energy Build.* **2001**, *33*, 683–697. [CrossRef]
59. *IES LM-83-12;* Approved Method: IES Spatial Daylight Autonomy (sDA) and Annual Sunlight Exposure (ASE). IES: Tokyo, Japan, 2012.
60. Nabil, A.; Mardaljevic, J. Useful daylight illuminances: A replacement for daylight factors. *Energy Build.* **2006**, *38*, 905–913. [CrossRef]
61. Sicurella, F.; Evola, G.; Wurtz, E. A statistical approach for the evaluation of thermal and visual comfort in free-running buildings. *Energy Build.* **2012**, *47*, 402–410. [CrossRef]
62. *EN 12464-1;* Light and Lighting—Lighting of Work Places Part 1: Indoor Work Places. International Standard Organization: Geneva, Switzerland, 2011; pp. 1–57.
63. Wienold, J.; Christoffersen, J. Evaluation methods and development of a new glare prediction model for daylight environments with the use of CCD cameras. *Energy Build.* **2006**, *38*, 743–757. [CrossRef]
64. Wienold, J. Dynamic simulation of blind control strategies for visual comfort and energy balance analysis. In *Building Simulation;* IBPSA: Beijing, China, 2007; pp. 1197–1204.
65. Hviid, C.A.; Nielsen, T.R.; Svendsen, S. Simple tool to evaluate the impact of daylight on building energy consumption. *Sol. Energy* **2008**, *82*, 787–798. [CrossRef]
66. Ward, G.; Shakespeare, R. *Rendering with Radiance: The Art and Science of Lighting Visualization;* SciX Open Publishing: San Francisco, CA, USA, 1998. [CrossRef]
67. Einhorn, H. Discomfort glare: A formula to bridge differences. *Light. Res. Technol.* **1979**, *11*, 90–94. [CrossRef]
68. Kirimtat, A.; Koyunbaba, B.K.; Chatzikonstantinou, I.; Sariyildiz, S. Review of simulation modeling for shading devices in buildings. *Renew. Sustain. Energy Rev.* **2016**, *53*, 23–49. [CrossRef]
69. Directive EE. Commission Delegated Regulation (EU) 2020/2155 of 14 October 2020 supplementing Directive (EU) 2010/31/EU of the European Parliament and of the Council by establishing an optional common European Union scheme for rating the smart readiness of buildings. *Off. J. Eur. Union* **2020**, *431*, pp. 9–24.
70. Liu, Y.; Colburn, A.; Inanici, M. Computing long-term daylighting simulations from high dynamic range imagery using deep neural networks. In *Building Performance Analysis Conference and SimBuild;* Zillow Group, University of Washington: Seattle, WA, USA, 2018; pp. 119–126.
71. Giovannini, L.; Favoino, F.; Lo Verso, V.; Pellegrino, A.; Serra, V. A Simplified Approach for the Annual and Spatial Evaluation of the Comfort Classes of Daylight Glare Using Vertical Illuminances. *Buildings* **2018**, *8*, 171. [CrossRef]
72. Jones, N.L.; Reinhart, C.F. Fast daylight coefficient calculation using graphics hardware. In Proceedings of the BS2015: 14th International Conference of the International Building Performance Simulation Association, Hyderabad, India, 7–9 December 2015; pp. 1237–1244.
73. Ladybug-Tools. Cross-Platform Python Libraries and Plugins for Grasshopper and Dynamo. Available online: https://www.ladybug.tools/ (accessed on 9 March 2022).
74. Ladybug-Tools. Pollination—Platform for Environmental Simulation. Available online: https://www.pollination.cloud/ (accessed on 9 March 2022).
75. Jones, N.L.; Reinhart, C.F. Physically based global illumination calculation using graphics hardware. In Proceedings of the eSim 2014: The Canadian Conference on Building Simulation, Ottawa, ON, Canada, 7–10 May 2014; pp. 474–487.
76. Cuttle, C. Cubic illumination. *Int. J. Light. Res. Technol.* **1997**, *29*, 1–14. [CrossRef]
77. Cuttle, C. Research note: A practical approach to cubic illuminance measurement. *Light. Res. Technol.* **2014**, *46*, 31–34. [CrossRef]
78. Raynham, P.; Unwin, J.; Guan, L. A new metric to predict perceived adequacy of illumination. *Light. Res. Technol.* **2019**, *51*, 642–648. [CrossRef]
79. Yamin Garretón, J.; Rodriguez, R.; Ruiz, A.; Pattini, A. Degree of eye opening: A new discomfort glare indicator. *Build. Environ.* **2015**, *88*, 142–150. [CrossRef]
80. Goncharov, A.V.; Dainty, C. Wide-field schematic eye models with gradient-index lens. *J. Opt. Soc. Am. A* **2007**, *24*, 2157. [CrossRef] [PubMed]
81. Reinhart, C.F. *Daylighting Handbook I: Fundamentals, Designing with the Sun;* Christoph Reinhart: Cambridge, MA, USA, 2014. [CrossRef]
82. De Groot, S.G.; Gebhard, J.W. Pupil size as determined by adapting luminance. *J. Opt. Soc. Am.* **1952**, *42*, 492–495. [CrossRef] [PubMed]
83. Fugate, J.M.; Fry, G.A. Relation of changes in pupil size to visual discomfort. *Illum. Eng.* **1956**, *51*, 537.
84. Fry, G.A.; King, V.M. The Pupillary Response and Discomfort Glare. *J. Illum. Eng. Soc.* **1975**, *4*, 307–324. [CrossRef]

85. Choi, J.H.; Zhu, R. Investigation of the potential use of human eye pupil sizes to estimate visual sensations in the workplace environment. *Build. Environ.* **2015**, *88*, 73–81. [CrossRef]
86. Klingner, J.; Kumar, R.; Hanrahan, P. Measuring the task-evoked pupillary response with a remote eye tracker. In Proceedings of the 2008 Symposium on Eye Tracking Research & Applications-ETRA '08, Savannah, GA, USA, 26–28 March 2008; ACM Press: New York, NY, USA, 2008; Volume 1, p. 69. [CrossRef]
87. Sarey, K.M.; Andersen, M.; Hart, B.M.; Stoll, J.; Einhäuser, W. *Integration of Eye-Tracking Methods in Visual Comfort Assessments*; CISBAT 11; EPFL: Lausanne, Switzerland, 2011; pp. 14–15.
88. Hamedani, Z.; Solgi, E.; Hine, T.; Skates, H. Revealing the relationships between luminous environment characteristics and physiological, ocular and performance measures: An experimental study. *Build. Environ.* **2020**, *172*, 106702. [CrossRef]
89. Berman, S.; Bullimore, M.; Jacobs, R.; Bailey, I.; Gandhi, N. An Objective Measure of Discomfort Glare. *J. Illum. Eng. Soc.* **1994**, *23*, 40–49. [CrossRef]
90. Al horr, Y.; Arif, M.; Katafygiotou, M.; Mazroei, A.; Kaushik, A.; Elsarrag, E. Impact of indoor environmental quality on occupant well-being and comfort: A review of the literature. *Int. J. Sustain. Built Environ.* **2016**, *5*, 1–11. [CrossRef]
91. Watson, A.B.; Yellott, J.I. A unified formula for light-adapted pupil size. *J. Vis.* **2012**, *12*, 12. [CrossRef] [PubMed]
92. Sarey Khanie, M.; Stoll, J.; Einhäuser, W.; Wienold, J.; Andersen, M. Gaze and discomfort glare, Part 1: Development of a gaze-driven photometry. *Light. Res. Technol.* **2017**, *49*, 845–865. [CrossRef]
93. Kadlec, P.; Gabrys, B.; Strandt, S. Data-driven Soft Sensors in the process industry. *Comput. Chem. Eng.* **2009**, *33*, 795–814. [CrossRef]
94. Lee, J.; Kao, H.A.; Yang, S. Service innovation and smart analytics for Industry 4.0 and big data environment. *Procedia CIRP* **2014**, *16*, 3–8. [CrossRef]
95. Wu, S.; Clements-Croome, D. Understanding the indoor environment through mining sensory data—A case study. *Energy Build.* **2007**, *39*, 1183–1191. [CrossRef]
96. Poli, T.; Mainini, A.G.; Speroni, A.; Cadena, J.D.B.; Moretti, N. The Effect of Real-Time Sensing of a Window on Energy Efficiency, Comfort, Health and User Behavior. In *Digital Transformation of the Design, Construction and Management Processes of the Built Environment*; Springer: Berlin/Heidelberg, Germany, 2020; pp. 291–296.
97. Rinaldi, S.; Bellagente, P.; Ciribini, A.L.C.; Tagliabue, L.C.; Poli, T.; Mainini, A.G.; Speroni, A.; Blanco Cadena, J.D.; Lupica Spagnolo, S. A Cognitive-Driven Building Renovation for Improving Energy Efficiency: The Experience of the ELISIR Project. *Electronics* **2020**, *9*, 666. [CrossRef]
98. Karjalainen, S. Should we design buildings that are less sensitive to occupant behaviour? A simulation study of effects of behaviour and design on office energy consumption. *Energy Effic.* **2016**, *9*, 1257–1270. [CrossRef]
99. O'Brien, W.; Gaetani, I.; Carlucci, S.; Hoes, P.J.; Hensen, J.L. On occupant-centric building performance metrics. *Build. Environ.* **2017**, *122*, 373–385. [CrossRef]
100. Jia, R.; Jin, B.; Jin, M.; Zhou, Y.; Konstantakopoulos, I.C.; Zou, H.; Kim, J.; Li, D.; Gu, W.; Arghandeh, R.; et al. Design Automation for Smart Building Systems. *Proc. IEEE* **2018**, *106*, 1680–1699. [CrossRef]
101. Konstantakopoulos, I.C.; Ratliff, L.J.; Jin, M.; Sastry, S.S.; Spanos, C.J. A Robust Utility Learning Framework via Inverse Optimization. *IEEE Trans. Control Syst. Technol.* **2018**, *26*, 954–970. [CrossRef]
102. Reinhart, C.F. Lightswitch-2002: A model for manual and automated control of electric lighting and blinds. *Sol. Energy* **2004**, *77*, 15–28. [CrossRef]
103. Hunt, D. Predicting artificial lighting use—A method based upon observed patterns of behaviour. *Light. Res. Technol.* **1980**, *12*, 7–14. [CrossRef]
104. Pigg, S.; Eilers, M.; Reed, J. Behavioral aspects of lighting and occupancy sensors in private offices: A case study of a university office building. In *ACEEE 1996 Summer Study on Energy Efficiency in Buildings*; American Council for an Energy Efficient Economy (ACEEE): Asilomar, CA, USA, 1996; pp. 161–170.
105. Gunay, H.B.; O'Brien, W.; Beausoleil-Morrison, I.; Huchuk, B. On adaptive occupant-learning window blind and lighting controls. *Build. Res. Inf.* **2014**, *42*, 739–756. [CrossRef]
106. Cheng, Z.; Zhao, Q.; Wang, F.; Jiang, Y.; Xia, L.; Ding, J. Satisfaction based Q-learning for integrated lighting and blind control. *Energy Build.* **2016**, *127*, 43–55. [CrossRef]
107. Van De Meugheuvel, N.; Pandharipande, A.; Caicedo, D.; Van Den Hof, P.P.J. Distributed lighting control with daylight and occupancy adaptation. *Energy Build.* **2014**, *75*, 321–329. [CrossRef]
108. Fakra, A.H.; Boyer, H.; Maamari, F. Experimental validation for software DIALUX: Application in CIE test Cases for building daylighting simulation. In *International Conference on Building Energy and Environment (COBEE 2008)*; HAL: Bengaluru, India, 2008.
109. Rubinstein, F. Photoelectric control of equi-illumination lighting systems. *Energy Build.* **1984**, *6*, 141–150. [CrossRef]
110. Caicedo, D.; Pandharipande, A. Distributed Illumination Control With Local Sensing and Actuation in Networked Lighting Systems. *IEEE Sens. J.* **2013**, *13*, 1092–1104. [CrossRef]
111. Jin, M.; Liu, S.; Schiavon, S.; Spanos, C. Automated mobile sensing: Towards high-granularity agile indoor environmental quality monitoring. *Build. Environ.* **2017**, *127*, 268–276. [CrossRef]
112. Guerry, E.; Gălățanu, C.D.; Canale, L.; Zissis, G. Luminance contrast assessment for elderly visual comfort using imaging measurements. *Procedia Manuf.* **2019**, *32*, 474–479. [CrossRef]
113. Glenn, J.; Dodds, G.; Robinson, R. Calibration and use of camera-based systems for road lighting assessment. *Light. Res. Technol.* **2000**, *32*, 33–40. [CrossRef]

114. Zatari, A.; Dodds, G.; McMenemy, K.; Robinson, R. Glare, luminance, and illuminance measurements of road lighting using vehicle mounted CCD cameras. *LEUKOS J. Illum. Eng. Soc. N. Am.* **2004**, *1*, 85–106. [CrossRef]
115. Konis, K. Predicting visual comfort in side-lit open-plan core zones: Results of a field study pairing high dynamic range images with subjective responses. *Energy Build.* **2014**, *77*, 67–79. [CrossRef]
116. Parsaee, M.; Demers, C.M.; Potvin, A.; Lalonde, J.F.; Inanici, M.; Hébert, M. Biophilic photobiological adaptive envelopes for sub-Arctic buildings: Exploring impacts of window sizes and shading panels' color, reflectance, and configuration. *Sol. Energy* **2021**, *220*, 802–827. [CrossRef]
117. Inanici, M. Evaluation of high dynamic range photography as a luminance data acquisition system. *Light. Res. Technol.* **2006**, *38*, 123–136. [CrossRef]
118. Moeck, M.; Anaokar, S. Illuminance analysis from high dynamic range images. *LEUKOS J. Illum. Eng. Soc. N. Am.* **2006**, *2*, 211–228. [CrossRef]
119. Goovaerts, C.; Descamps, F.; Jacobs, V.A. Shading control strategy to avoid visual discomfort by using a low-cost camera: A field study of two cases. *Build. Environ.* **2017**, *125*, 26–38. [CrossRef]
120. Motamed, A.; Deschamps, L.; Scartezzini, J.L. On-site monitoring and subjective comfort assessment of a sun shadings and electric lighting controller based on novel High Dynamic Range vision sensors. *Energy Build.* **2017**, *149*, 58–72. [CrossRef]
121. Schneider, E.; Villgrattner, T.; Vockeroth, J.; Bartl, K.; Kohlbecher, S.; Bardins, S.; Ulbrich, H.; Brandt, T. EyeSeeCam: An Eye Movement-Driven Head Camera for the Examination of Natural Visual Exploration. *Ann. N. Y. Acad. Sci.* **2009**, *1164*, 461–467. [CrossRef]
122. *BS EN ISO 28803*; Ergonomics of the Physical Environment—Application of International Standards to People with Special Requirements. ISO: Geneva, Switzerland, 2012.
123. Blanco Cadena, J.D. Human Eye Kinematics for Adaptable Visual Comfort Assessment. Personalized Responsive Control Strategies to Integrate Building Envelope and Artificial Lighting. Ph.D. Thesis, Politecnico di Milano, Milan, Italy, 30 March 2020.
124. Amayri, M.; Arora, A.; Ploix, S.; Bandhyopadyay, S.; Ngo, Q.D.; Badarla, V.R. Estimating occupancy in heterogeneous sensor environment. *Energy Build.* **2016**, *129*, 46–58. [CrossRef]
125. Gilani, S.; O'Brien, W. Review of current methods, opportunities, and challenges for in-situ monitoring to support occupant modelling in office spaces. *J. Build. Perform. Simul.* **2016**, *1493*, 1–27. [CrossRef]
126. Luna-Navarro, A.; Allen, M.; Meizoso, M.; Overend, M. BIT—Building Impulse Toolkit: A novel digital toolkit for productive, healthy and resource efficient buildings. *J. Phys. Conf. Ser.* **2019**, *1343*, 012139. [CrossRef]
127. Hong, T.; Sun, H.; Chen, Y.; Taylor-Lange, S.C.; Yan, D. An occupant behavior modeling tool for co-simulation. *Energy Build.* **2016**, *117*, 272–281. [CrossRef]
128. Imperadori, M.; Poli, T.; Cadena, J.D.B.; Brunone, F.; Mainini, A.G. Comparison of Comfort Performance Criteria and Sensing Approach in Office Space: Analysis of the Impact on Shading Devices' Efficiency. In *Regeneration of the Built Environment from a Circular Economy Perspective*; Springer: Berlin/Heidelberg, Germany, 2020; p. 381.

applied sciences

MDPI

Article

Decision-Making Processes of Residents in Preservation, Thermal Comfort, and Energy Efficiency in Heritage Buildings: A Pilot Study in Mexico City

Krisangella Sofia Murillo Camacho [1,*], Kalliopi Fouseki [1] and Hector Altamirano Medina [2]

[1] UCL Institute for Sustainable Heritage, University College London, London WC1E 6BT, UK; kalliopi.fouseki@ucl.ac.uk
[2] UCL Institute for Environmental Design and Engineering, University College London, London WC1E 6BT, UK; h.altamirano-medina@ucl.ac.uk
* Correspondence: ucftkmu@ucl.ac.uk

Featured Application: To date, the 'values' associated with 'heritage buildings' have received little attention in designing technical interventions for energy efficiency. This may be due to a fear that modern interventions for improving energy performance clash with heritage conservation, especially conservation of original features. We argue that energy efficiency interventions and heritage conservation can co-exist if an in-depth understanding of people's heritage values and attitudes is obtained. This paper adds to our limited knowledge of residents' approaches to heritage conservation and energy efficiency. It does so by presenting the first study of its kind in Mexico's City Historic Centre (a World Heritage Site since 1987). The results reveal the type of heritage values that residents assigned to their buildings when seeking to achieve thermal comfort and energy efficiency in heritage buildings.

Abstract: With building construction representing one of the largest sectors responsible for the use of natural resources, retrofitting existing heritage buildings becomes a necessity, albeit a challenging one. The emergence of specific guidance on retrofitting heritage buildings has unveiled more than never the need to understand how residents negotiate, thermal comfort, energy efficiency, and heritage conservation decisions. The paper reports the complexity of the decision-making process of residents of heritage buildings in the Historic Centre of Mexico City regarding energy efficiency, intending to improve thermal comfort and reduce energy consumption while preserving heritage values. The study involved in-depth semi-structured interviews with users of heritage buildings that were thematically analysed, complemented by the monitoring of internal environmental conditions and system dynamics analysis. The results show that although the residents perceived the buildings' temperature as poor, passive thermal comfort actions (e.g., wearing more clothes and closing windows) were preferred against invasive retrofitting solutions for thermal comfort due to residents' resistance to a potential loss in the buildings' values and the high cost of changes. The degree of change necessary for maintenance, renovation, and actions for improving the thermal comfort of a heritage building is related to values and to their preservation for future generations. The users' changes were limited to small-scale interventions in floors and ceilings while avoiding touching what they consider essential to preserve and protect (i.e., social and cultural values). Integrating the user into the decision-making process would enhance the long-term continuity and sustainability of retrofitting policies and guidelines, thus avoiding losing heritage-built stock.

Keywords: heritage values; user; sustainability; thermal comfort; energy efficiency; heritage buildings; decision-making; preservation; Mexico City

Citation: Murillo Camacho, K.S.; Fouseki, K.; Altamirano Medina, H. Decision-Making Processes of Residents in Preservation, Thermal Comfort, and Energy Efficiency in Heritage Buildings: A Pilot Study in Mexico City. *Appl. Sci.* **2022**, *12*, 1486. https://doi.org/10.3390/app12031486

Academic Editors: Tiziana Poli, Andrea Giovanni Mainini, Gabriele Lobaccaro, Mitja Košir and Juan Diego Blanco Cadena

Received: 1 December 2021
Accepted: 27 January 2022
Published: 29 January 2022

1. Introduction

Historic buildings are increasingly considered in literature dealing with energy efficiency actions and renewable energy systems [1–9]. Recent projects such as Energy Efficiency for EU Historic Districts' Sustainability, Climate for Culture, and Efficient Energy for EU Cultural Heritage are some examples addressing the impacts of changing climate conditions on historic buildings. The projects propose energy-efficient retrofit solutions while respecting buildings' heritage values. Historic England has developed one of the most comprehensive guidelines that intends to adopt a holistic approach to historic buildings and energy efficiency. The guidance seeks to address every factor that affects energy use in the buildings and tries to balance saving energy, sustaining heritage significance, and ensuring comfort (a whole-building approach) [9]. However, although the emphasis is placed on heritage values, there is an assumption on what these 'values' are and how they should be retained [10]. These programmes rely mostly on research to integrate technology into the built heritage but do not explore users' 'heritage values' in depth. Qualitative research on heritage and energy efficiency began to show the connection between energy efficiency and values. For instance, Yarrow [11] offers perspectives that underline a social negotiation of the values by inhabitants of old buildings and heritage professionals and involves concepts of climate change and energy efficiency. Yarrow conducted ethnographical research of specialists, planners, and homeowners interested in renovating and retrofitting buildings of attributed historical value. Koukou and Fouseki [12] conducted a study in Greece that included residents' approaches to heritage conservation and energy efficiency in neoclassical buildings. Their research explored how residents' meanings and values regarding historic buildings drive or inhibit energy-efficiency interventions. Their study involved semi-structured interviews with residents that revealed a conflict between the urgency to improve thermal comfort during winter through installing a mechanical heating system (at the loss of the original characteristics of the building) and against interventions on the façade of the building. The study found that in most cases, the changes made by residents complied with current legislation and the architectural significance of the building as most buildings were listed.

Overall, we lack in-depth, qualitative studies on energy efficiency and heritage conservation. In this regard, Fouseki et al.'s [10] research have understood and integrated heritage values into decision-making to improve the energy performance of the heritage buildings' stock, which motivated this research. Fouseki et al. [10] show that decision making in thermal comfort improvement, energy efficiency, and heritage conservation is a sociocultural and dynamic practice. In the connection or discontinuity of elements (materials, competencies, resources, values, senses, and time), the decision-making process changes based on external factors surrounding the building (context, listed status, age, climate, and ownership status) [10].

In Mexico, Murillo et al. [13] reported on the heritage values attributed to historic buildings and how they change, drive, or prevent energy efficiency changes over time. Their study consisted of in depth semi-structured interviews complemented by monitoring indoors environmental conditions that revealed what users prioritise between energy efficiency interventions and heritage values. Their research shows a tension in the limitations on buildings with listed statuses, which restrict changes in use and prevent energy-efficiency interventions and highlighted that a fundamental requirement for developing effective energy policies, standards, and guidelines is understanding the meaning of heritage attributed by the users. More international approaches that contribute to decarbonising the built environment are needed.

Given the foregoing, this paper details the results of a qualitative study conducted in Mexico's City Historical Centre (a World Heritage Site) that includes residents' and buildings' thermal conditions to understand user decision-making processes for energy efficiency and thermal comfort. The study focused on the social and cultural values residents of listed and non-listed buildings attach to their buildings and which values they prioritise during energy-efficiency interventions. The work offers new research insights from Mexico

City, where heritage values, user behaviour, and thermal comfort data were obtained in a protected heritage site. The study's premise is that social meaning, spatial structures, heritage values, sustainability preservation, and energy efficiency are interconnected and reinforced over time, driving or preventing changes in energy efficiency. Moreover, tensions can arise between limitations on a listed building (that impede change) and the lack of interventions applied (which directly affects the structure). The study assumes that both values associated with the tangible characteristics of a building (e.g., architectural, historical, and aesthetic) and sentimental, symbolic values (e.g., family attachment) increase the overall value of a residence over time and determine which building characteristics residents are willing to change, compromise, or maintain as they strive to improve the building's energy performance.

Improving Energy Efficiency in Mexico City

In Mexico City, the need to retrofit existing residential buildings [14] has been highlighted in recent years to prevent buildings' abandonment and obsolescence due to indoor thermal problems. The retrofitting of existing residential buildings to improve energy efficiency or reduce green housing emissions is not applied either [14]. More studies regarding improving energy efficiency in Mexico have only focused on electricity consumption. Mexican government programmes to reduce electricity use nationwide include the Trust for the Thermal Insulation of Housing and the Program for Integral Systematic Savings. Both programmes—managed by the Secretary of Energy [15]—have achieved a national energy use reduction of 3410.72 GWh and 1,534,824 fewer tonnes of concentrated CO_2 in the atmosphere [14,16] obtained by 120,703 thermal insulation actions. Despite the progress and goals set regarding retrofitting existing buildings in the heritage sector, significant challenges remained unsolved.

2. Methods and Materials

2.1. Case Study

The Historic Centre of Mexico City was chosen as the case study. The urban area has approximately 1,500 listed buildings with historical and artistic value [17].

Declared a World Heritage Site in 1987 [18], the Historic area has faced challenges on the social, political, environmental, and economic fronts, given its geopolitical location and historical transformation of the social context. The social transformation began after the earthquake of 1985, which caused the area's depopulation. In 1987, with the World Heritage Site declaration, the local government started urban revitalisation management plans to repopulate and attract more inhabitants. Precisely, Heritage status imposes restrictions upon users concerning what they can and cannot modify. Therefore, it is intriguing to study how residents negotiate their heritage conservation and energy efficiency decisions in this specific context. Furthermore, social housing (planned for people who cannot afford to buy housing with the private sector) also makes this area compelling to examine.

2.2. Sample Selection

Through system dynamics [19,20], social data (related to resident attitudes regarding heritage values, thermal comfort, and energy efficiency) were collected, analysed, and synthesised, alongside environmental (relative humidity and temperature) and building condition data (materials and maintenance). As stated by Levi-Strauss [21], due to the complexity of social phenomena, the data collection is guided by successively evolving interpretations made during the study and by the researcher interpretations. To this end, a number of participants and apartments is based on an in depth case-oriented analysis essential to qualitative research [22]. Additionally, qualitative samples are selected by their capacity to provide rich information relevant to the study. As a result, it sets an 'information-rich' case [23] and with deep understanding. The sociotechnical approach employed consists of qualitative and quantitative methodologies, summarised in Table 1. The procedure of our approach starts firstly with a formulation of objectives clarifying the research—next with a description of analysis and theoretical framework required for qualitative and quantitative methodologies. Afterwards, the experimental procedure collects and processes the data (with tools and instruments), confirming (or not) our initial objectives. The results interpretation in data analysis leads to indicators (relevant for research objectives) and finally to the results.

Table 1. Sociotechnical approach.

Stages-Process				
	Objectives			
Clarification of research	User Heritage values	Perception of the conditions of the buildings	Perceived thermal comfort Energy efficiency interventions Heritage conservation processes	Dynamic interplay related to heritage values, conditions of the buildings and thermal comfort
Description of study and Theoretical framework	**Methodology**			
	Qualitative		Quantitative	Mixed
	Case study selection			
	Selection of participants and buildings			
	Data collection			
Experimental procedure	Interview with users (Social data)		Building physical conditions survey Installation of monitors into the properties (Environmental data)	
	Tools and instruments			
	Semi-structured interviews photo-elicitation	Building visual inspection with camera	Tiny-tags monitors Electricity bills Thermal camera and photo viewer Microsoft Excel	Vensim NVivo
	Data analysis			
Interpretation of the results	Thematic analysis through NVivo Coding and categorisation of cause-effect relationships Causal loop visualisation in Vensim Exploration of monitoring data in Excel and contrasted with users perception			Causal loop diagrams with System Dynamics
	Indicators			
Identification of relevant characteristics	Time Values Preferences Expectations	Deterioration Interventions	Temperature Relative humidity Cost Consumption of electricity	Cathegories and code groups
Results	**Evaluation and conclusions**			

The final sample of five buildings for this study was considered suitable for an in-depth case study to test methods and tools and gain the residents' approval to install the monitors in the properties. We identified buildings that have preserved their original typology and belonged to the Heritage Site protection list of Mexico City Council (a total of 134 housing buildings making up the total housing unit). The typology of housing building in the Historic Centre (known as vecindades) has a significant connotation on the architectural characteristics dating from the colonial period. Inside, the rooms are organised around a rectangular patio or corridor that serves as the central circulation and, simultaneously, a source of ventilation and lighting (Figure 1). A layout with current internal uses of social housing is shown in Figure 1. The buildings have mixed uses consisting of housing and commerce.

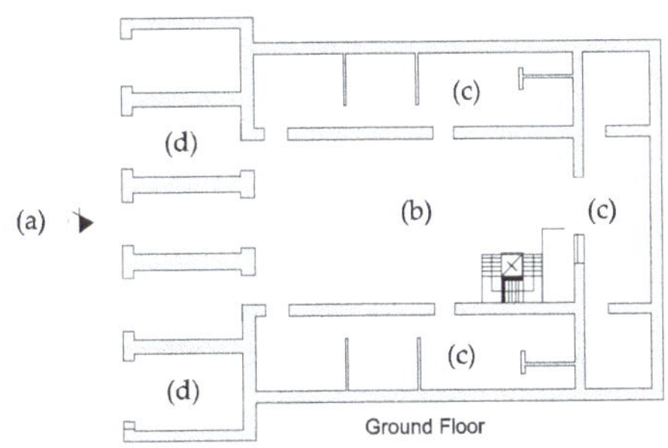

Figure 1. Building layout produced during data collection in the Historic Centre: (**a**) main entrance to the building; (**b**) patio or corridor; (**c**) apartments; (**d**) commerce.

The final sample includes one listed monument (assets linked to the nation's history), one non-listed but protected apartment building, and three listed apartment buildings (within the area of historical monuments and with architectural value). Their architectural styles are dated from the 16th to 19th century (primary baroque or colonial style), which influenced each building's materials (tepetate, masonry, and brick) (Figure 2).

(a) (b) (c) (d) (e)

Figure 2. Building architectural styles: (**a–d**) colonial style—typical colonial house found in cities, analogous to the houses of pre-Hispanic Mexico. These houses were built inwards, usually with two stories and simple façades, surrounded by walled gardens; (**e**) modern style—rationalist language with a series of horizontal windows on its façade. The modulation in the proportions is an important element of the early stage that characterised the architectural design.

2.3. Data Collection

Data collection was conducted over one month during the winter (from December 2019 to January 2020). Participants were recruited for the study fulfilling at least one of the following two criteria: they must have lived in the area or in the building since 1987 (the Declaration of World Heritage), or their construction must belong to the government official heritage catalogue, which provides more information about the building (historical files, interventions, renovations, ownership, and previous uses of the building). Semi-structured interviews were conducted with five users (four owners and one tenant), and written informed consent was obtained complemented with photo-elicitation (i.e., interviewers were shown energy-efficiency options and prompted to comment on them) [24] to gain deeper insight into the values assigned. The semi-structured questionnaire design was divided into four sections: the first part focussed on building conditions; the second concerned understanding the attributed values, combined with photo-elicitation; the third part was designed to explore attitudes towards thermal comfort and energy efficiency, and the fourth explored user actions and interventions undertaken in the building. Each resident and apartment was assigned with a unique code of identification, including the country and number of participants. For instance, the first one was MX-U1 (see Table 2). For environmental data collection, the tiny tag data monitors were placed during the visit where the user felt more dissatisfaction with perceived thermal comfort (the data loggers are battery-powered instruments with electronic sensors that use data logger software to download the data from the monitors). The installation was in living rooms and bedrooms.

Table 2. Building main characteristics and numbers, including resident code.

Building Number and Resident Code	Orientation of the Facade	Location of the Apartment	Total M2	Total Apartments	Number of Floors	Space Monitored	Total M2	Walls Width
Building 1 MX-U1	north	Ground floor	1923	6	2	Bedroom Living room	20 20	0.60 mts
Buiding 2 MX-U2	south	First floor	400	4	2	Bedroom Living room	13 24	0.60 mts
Building 3 MX-U3	north	Fifth floor	544	15	5	Bed room Living room	15 20	0.30 mts
Building 4 MX-U4	north	First floor	926	24	2	Living room	12	0.60 mts
Building 5 MX-U5	north	First floor	2046	120	2	Bedroom Living room	13 20	0.60 mts

The thermal imaging exercise helped explain materials' thermal properties to the residents (when pointing at an object or area, the thermal camera sensor allowed the user to view the infrared spectrum. The warmer regions were shown as red, orange, and yellow on the camera's colour screen. In contrast, the colder parts showed up as purple and blue). During the data collection stage, collaboration with local authorities was also essential, including a participatory workshop on energy efficiency and heritage values with 12 neighbours, 3 stakeholders and 4 professionals in the Trust of the Historic Centre (Fideicomiso del Centro Historico). The course aimed to introduce the present pilot study and exchange ideas for proposals related to the Historic Centre regarding energy efficiency, conservation, and heritage values. This study is the first of two research studies carried out in the heritage site. The second study was conducted in 11 buildings and 11 residents during the winter of 2020–2021; the results are part of another publication.

2.4. Data Processing

The interviews were transcribed and translated from Spanish to English. Afterwards, they were thematically analysed using NVivo software, which enabled coding and the identification of cause-and-effect relationships among factors that affect a specific intervention (or lack thereof). The approach followed principles of grounded theory allowing the data to drive the hypothesis [21]. The interviews were first coded through an open coding process, identifying all factors. The codes were then reclassified through axial coding into 251 final codes related to the main aim and objectives and different subnodes from the decision-making process regarding energy-efficiency actions to address heritage building conditions and user values over time. The coding process and analysis used were based on Fouseki et al.'s [10] research. During the coding process, cause-and-effect relationships between nodes were discovered [10] and recorded in Excel. Table 3 shows the 12 categories of codes groups. The objective of the identification was to show the system elements that affected the dynamic interaction between values, thermal comfort, and energy-efficiency actions and practices. These categories interconnect, reinforce, and balance the system over time.

Table 3. Categories and code groups.

Category	Code	Category	Code
Time	Time living in the property, changes, age of the building	Materials	Façades, walls, humidity, porosity
Needs	Thermal comfort in winter and summer, perception of thermal comfort	Practice	Qualified interventions
Feelings	Satisfaction, guardianship, family attachment	Cost	Rent, cost of changes
Value	Sentimental, aesthetics, historic, symbolic, originality	Risk	Earthquakes, humidity, lack of maintenance
Place/Space	Urban context, size of the rooms	Ownership status	Owner, tenant
Actions	Type of maintenance, thermal comfort actions, ventilation, preservation	Building status	Listed, not listed

2.5. Mapping Data through Systems Dynamics

The relationships were mapped using Vensim software for systems dynamics analyses [25,26] to illustrate cause-and-effect relationships identified through the feedback thinking [25,26] and that are part of the system of preservation, thermal comfort, and energy efficiency. Systems dynamics is grounded in the theory of nonlinear dynamics, feedback, and interconnected loops and helps capture the complex causal structure of the system in a formal model that can be simulated and validated against real-world observations. A causal loop can grow or decline, allowing identification of the gap between the current objective and the desired one [25,26].

The five interviews were analysed using the same process, resulting in five diagrams that portray the dynamic system between heritage buildings, values, thermal comfort, and energy efficiency (decision-making process). To summarise the 251 interconnected variables identified during the interviews, an aggregate version of the five interviews (and five apartments) is shown in Figure 3. The symbols positive (+) or negative (−) are used to indicate reinforcing (growing) (R symbol) or balancing (B symbol) relationships. A reinforcing relationship between the two variables indicates continuous growth. On the other hand, an equilibrium is intended to be achieved by balancing relationships. For instance, a reinforcing relationship between original features and aesthetic values means that the more original the façades, the higher the aesthetic value assigned by the user. On the other hand, a balancing interrelationship was noted between the deterioration of the physical condition and replacing old materials with modern ones.

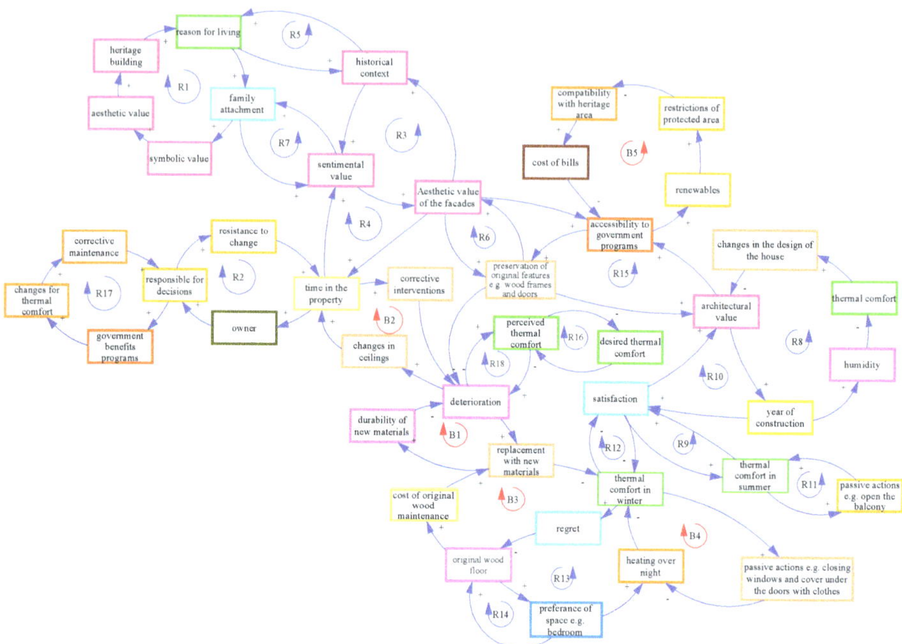

Figure 3. Causal loop diagram produced with Vensim showing relationships between values, interventions, and thermal comfort. Colour scheme: yellow, time; green, needs; light blue, feelings; pink, value; dark blue, place; orange, actions and practice; purple, materials; brown, cost; grey, risks; dark green, type of ownership.

3. Results

3.1. Heritage Values Assigned to Historic Buildings

Residents' values to their buildings and apartments varied according to their lifestyles. The everyday life of the inhabitants of the Historic Centre is characterised by commerce, tourism, architecture, and the political powers that are concentrated in the area. The inhabitants combine work activities (formal and informal) with those of the home. Most have different educational levels and have lived in the Historic Centre for more than 35 years, witnessing this space's social and urban transformations. In the same way, the Historic Centre represents a labour resource for many of them, who have had a commerce or workshop there.The first value associated with the house was the historical value, followed by sentimental and aesthetic value: *"Well, for the year it was made and for its architecture, its balconies, its corridors because it is very beautiful. I like everything. I like it a lot"* (MX-U2). The existing relationship between the preservation of the original features and aesthetic value is shown in R6 (see Figure 3). When mapped on Vensim (R7 in Figure 3), the family attachment to the property is a significant relationship; the owner has a motive for living in the building that becomes stronger over time due to the values appreciating and their willingness to preserve and protect. The residents endeavoured to maintain the building in fidelity to its original structure because of its architectural and sentimental value (R3 in Figure 3), in part due to family ties (R7 in Figure 3). For instance, respondent MX-U2 said, *"Well, the greatest meaning is sentimental because I have lived here (for many years with my family) . . . because it is the original, the architecture and the shape of the windows, the doors, I like it a lot, we do not touch it (we maintain it)"*. Over time, the reason for living in the apartment and family attachment reinforces the owner's relationship with its architectural, aesthetic, and symbolic value. As stated by MX-U2, *"Well, it has a symbolic value and to me, they are very beautiful, as well as are the architectural features that characterise the construction and the architecture*" (Figure 4).

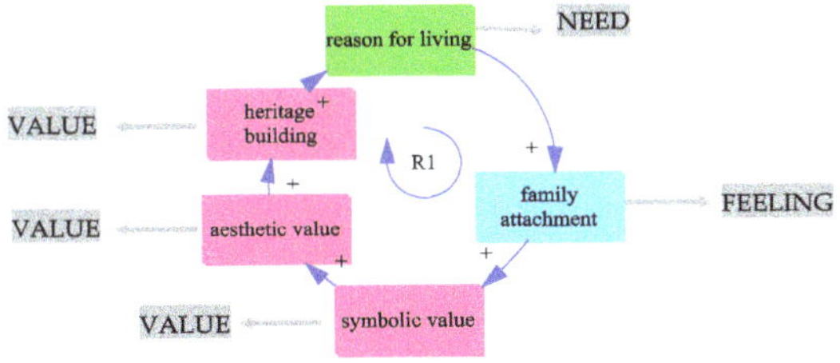

Figure 4. Reinforcing (R1) causal loop of relationships between values, need, and feeling.

Aesthetic value was directly related to architectural qualities such as space, form, use, dimensions, and structure. Originality was important in terms of the materials and features of the apartment: "*The architecture and the shape of everything, the windows, the doors, and all of that, I like a lot. We do not touch it*" (MX-U1). The residents felt responsible for what happens inside the building. They mentioned that all users needed heritage knowledge to make good decisions: "People must become *aware of how to treat a listed building*" (MX-U5). The architectural value, reinforced by the historic context, is related as a reason for living in the building: "*Because of the location, it is centric and for the benefits that we have in the Centre of the city, we have everything and then it is something that interests us*" *(MX-U3)* (see Figure 5).

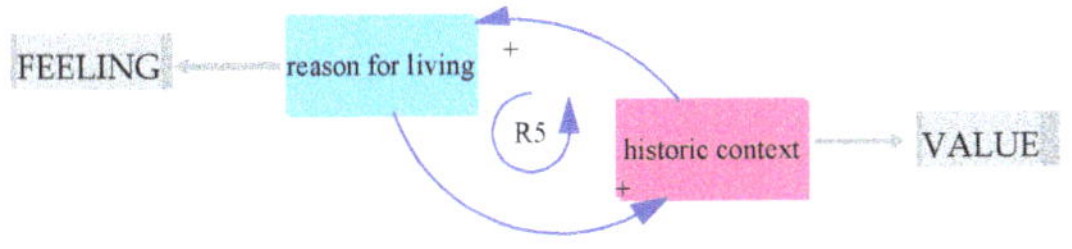

Figure 5. Reinforcing causal loop of values and feeling.

3.2. User Perceptions of the Condition of the Building over Time

This part was intended to understand the energy-efficiency interventions and their impact on the environmental conditions provided by the buildings: the users' perceived thermal comfort and the presumptions about building performance, which could be confirmed (or not) with environmental data, related to what was happening inside the building. It was found that humidity, deterioration, and use are interconnected and increase the need to maintain in good quality the originality of materials and façades. Humidity deteriorating the original façades' materials in all buildings raised the cost of changes due to their aesthetic value, and old structures required specialised interventions and maintenance. The older the building, the more susceptible it is (because of the original materials). MX-U5 said, "*Yes, it's because the use of buildings has changed over time. For example, (original materials) need a lot of maintenance*". However, there is regret regarding the changes made (unsupervised), as reported by MX-U2: "*Well, the floor [previous wooden floor] was cosier, warmer inside the apartment, so I disagree that it was removed. But it had to be removed because it needed a lot of maintenance and had deteriorated, and it costs a lot. The changes were made because they had deteriorated, and the new material was supposed to last longer*". While exterior windows were changed due to humidity and deterioration, interior modifications for comfort were further driven by new materials' durability over time (which implies less maintenance). However, the users' actions for preserving the original materials are related to the supervised government programmes for interventions despite costs, damage, and time (see Figure 6). Respondent MX-04 said, "*The roofs needed waterproof paint and, on the façade, some painting. The local government approved carrying out the interventions on the façade*". MX-U1 stated, "*Fideicomiso Trust arranged everything for us, painted the façade of the house, all of this because the whole façade was severely damaged; the government gave us support with the material, and with money . . . They are very beautiful façades that have been preserved very well, everything with the supervision of the National Institute of Anthropology and History"*.

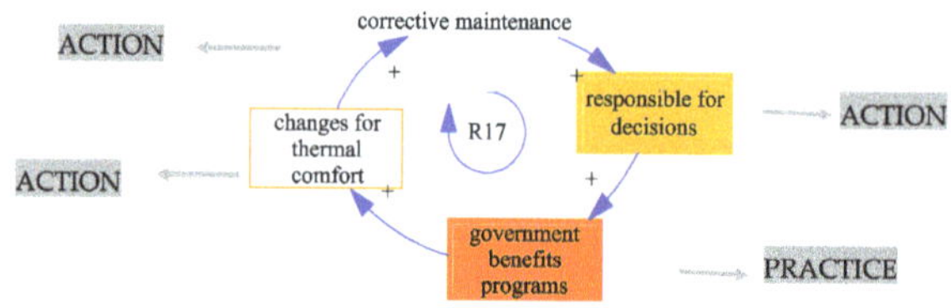

Figure 6. Reinforcing causal loop of relationships (R17) between actions and practice.

Despite the special treatment that the original materials of a historic building require, residents felt satisfied with the maintenance over time and have resisted introducing additional major interventions for comfort on façades. Such resistance is related to the architectural value and elevated cost of interventions for historic buildings. Costs *"are high because all the interventions have to be done by expert people, people who know what material is and they 'll do it right"* (MX-U2). The residents considered heritage; they were aware of and willing to preserve the value. *"Well, it's pretty. The building is beautiful, well preserved; it looks very beautiful. We cooperate with INAH (National Institute of Anthropology and History social housing programme) to preserve our building; we've shared the ownership with INAH for 80 years . . . Well, it's maintained, it's presentable. We're trying to keep it that way. So, it's a benefit to present the building in good condition. The government gives us the benefit of housing because we are cooperating with the building"* (MX-U3). The users also considered the economic value of future interventions: *"Because the government doesn't give maintenance (listed private buildings), the user must do it and look for the best price"* (MX-U3).

3.3. Interrelationship between Thermal Comfort, Energy Efficiency, and Heritage Conservation

This section is intended to describe thermal comfort through the residents of heritage buildings, their attitudes, and what is happening inside the building. It was possible to compare users' desired and perceived thermal comfort with the actual interior temperature (see Table 4). From the interviews, there was clearly a difference between the users' desired and perceived temperature described as freezing, cold or unbearable. For instance, the ideal internal temperature for some residents was 28 °C or over: *"I say 28 to 30 °C [would be a suitable temperature for me]"* (MX-U1). *"28 °C"*, seconded MX-U5. From monitoring, indoor temperature and external temperature showed no significant difference. However, when contrasted with users' perceived temperature, the performance is experienced as bad (Table 4) (Building 4 bedroom and living room share the same space).

Table 4. Buildings average temperature monitored, perceived, and desired temperature by the user.

Building Number	Average Indoor Temperature Living Room °C	Average Indoor Temperature Bedroom °C	Average External Temperature °C	User Desired Temperature °C	User Perceived Temperature
Building 1	18.91	16.25	19	28–30	freezing
Buiding 2	19.37	19.57	19	28	cold
Building 3	18.88	19.69	19	28	cold
Building 4	18.91	18.91	19	24	cold
Buildng 5	17.85	17.77	19	28	unbearable

According to the International Organization for Standardization, thermal comfort is a mental condition related to the satisfaction of the thermal environment [27,28]. In the apartments studied, thermal comfort is vital for residents, as it can affect their health and quality of life. It is shown (Table 4) that users have their own criteria for thermal comfort. Thermal comfort is also related to a space and its characteristics, where they feel pleasant. For instance, MX-U4 stated, *"In my bedroom, because I like to rest there [and] because I have health problems, then I can watch TV and rest"*. MX-U2 said, *"In the living room, I like it more because we have the windows, natural light, and I see people (outside) come and go and the atmosphere is very warm"*. Perceived thermal comfort during winter and summer is influenced by the building architecture and materials, as reported by respondent MX-U1: *"During winter, it's a little bit cold. I always bring my coat. In the summer, it's a delight. It's a delight to get inside from the street and find natural air conditioning. Thanks to the height of ceilings and thanks to the walls that keep a delightful temperature, it's very nice"*. Users also related internal conditions (temperature) with specific building materials: *"I recognise that the building is very well designed, [including] the spaces. The building is solid; it has masonry walls. Both function and form are very well designed. The spaces are wide, large, spacious ... What I do not like is that it's very cold because of the concrete that was used to change the old structure that had deteriorated from the old building"* (MX-U5). The architectural characteristics appreciated by the user show that although the structure reinforces the building against earthquakes, it is also more susceptible to humidity and deterioration. The older the building, the more susceptible it is (because of the original materials).

The climate in Mexico City is temperate subhumid in most (87%) of its territory. During monitoring, the external temperature was a maximum of 24 °C, a minimum of 14 °C, and an average of 19 °C. Figure 7 shows temperatures in the bedrooms and living rooms monitored. From the figure, we can see that the lowest temperature occurred in Building 1 with, 16.25 °C in the bedroom and 18.91 °C in the living room. In contrast, the highest temperature was found in Building 2, with 19.57 °C in the bedroom and 19.37 °C in the living room.

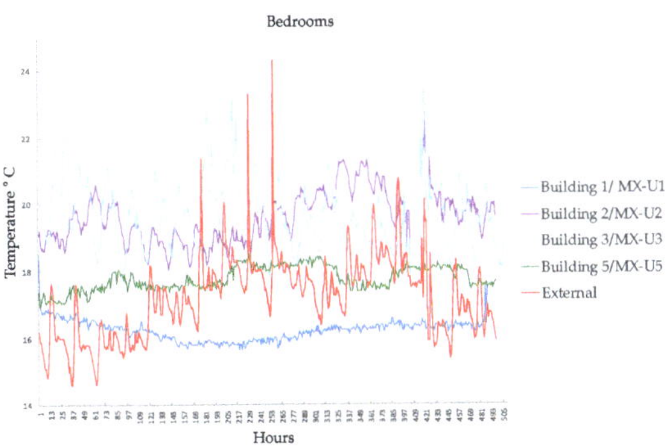

(a)

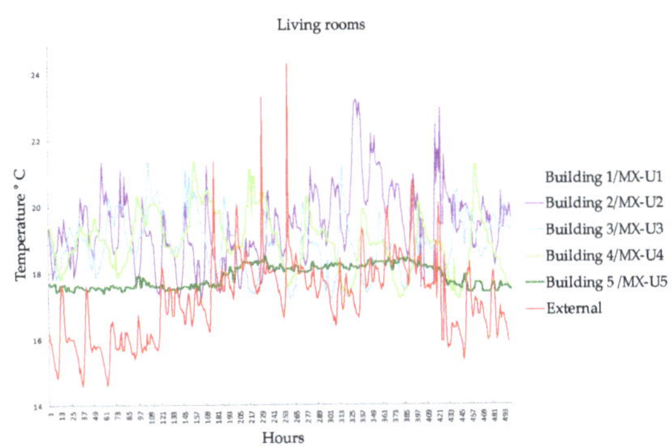

(b)

Figure 7. Temperature per hour in bedrooms and living rooms in the five buildings; (**a**) indoor temperature of the bedrooms contrasted with external temperature; (**b**) indoor temperature of the living rooms contrasted with external temperature.

Residents control the indoor temperature with natural ventilation and by performing passive activities such as opening and closing windows or, as last resort, with a portable heater during nights. The passive actions performed gave them the opportunity to understand their needs for thermal comfort and to help them adapt to the apartment thermal conditions. As reported by MX-U3, *"Well, today it's a little bit cold, but during winter, I close them (the windows and doors), and the inside temperature remains warmer because the roof retains the interior heat"*.Monitoring was also performed for relative humidity. The data for relative humidity was between acceptable ranges of 40% and 70% [28,29]. User dissatisfaction regarding the buildings' indoor thermal comfort was related to deterioration caused by humidity. MX-U1 noted, *"If you look, the temperature is something that impacts; it's very cold. It's very cold here because of the height; we have high ceilings and then the materials. It feels humid"*. MX-U4 commented, *"We have humidity in all the bedrooms because the material is so porous that I feel that it filters, and I have to repair regularly because the paint disappears quickly"*. Within the building, the deterioration and humidity were expressed on the surfaces of walls, ceilings, and façades (because of the materials' natural properties), and the subsoil characteristics could be contributing. During the Spanish conquest, the Mexico City Historic Centre was constructed on a lake. What is of interest here is that despite the relatively stable humidity levels, the perception of how humid the houses felt was fundamentally different from the data. This may be explained by the limited environmental data, as they reflect only a short (but cold) period in a year. It may also relate to the overall perception (or misconception) that "old" buildings are cold and humid without necessarily being so. The misalignment between "actual" and "perceived" humidity is an area for further exploration.

3.4. Interaction between Heritage Values, Perceived Buildings Conditions and Thermal Comfort

Residents' perceptions of thermal comfort and thermal building conditions interact with the actions to preserve the assigned value while improving thermal comfort. Perceived thermal comfort was affected by floor level, façade orientation, ventilation, room size, natural light, and humidity. The lack of direct solar radiation has to do with apartment orientation, location, and interior layout. Two buildings (first-floor apartment and fifth-floor apartment) were compared to show differences in perceived thermal comfort. Figure 8 shows building 3 and 5 temperatures in bedrooms and living rooms. Although monitoring showed not much variation from the external temperature, in building 3 (fifth-floor apartment), we can see fluctuations in temperature that achieved higher temperatures than building 5. In both buildings (north orientation façade but different floor location), the perception of indoor temperature was bad, but one of them was perceived as unbearable.

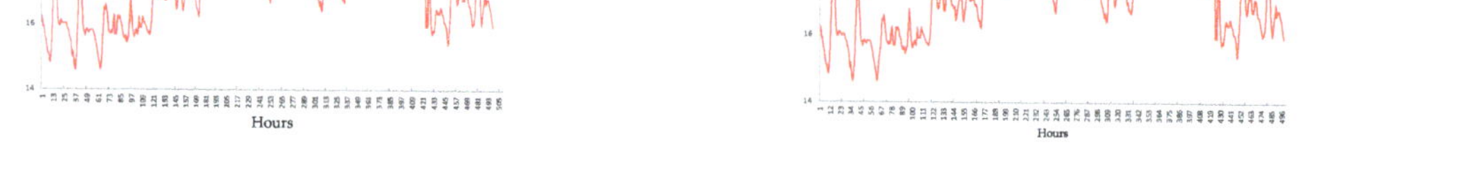

(a) (b)

Figure 8. Monitoring of indoor temperature building 3 and 5; (**a**) buildings 3 and 5: bedroom indoor temperature and external temperature; (**b**) buildings 3 and 5: living room internal and external temperatures.

As reported by the user of building 3 MX-U3, "In the summer, I feel very hot. It's very hot because the sun always enters everywhere. It is super hot, but during this season, it's okay". The thickness of the walls and the windows' height and dimensions also influenced user perceptions of the space's temperature. For example, the thermal perception was cold in a first-floor north orientation façade (Building 5) compared to the same orientation façade on the highest floor (Building 3), where it was perceived as warm. MX-U5 stated, "How comfortable do I feel? Very uncomfortable; it's very cold. The truth is that it's unbearable during winter ... because my apartment is on the first floor. I feel that it's affected in that sense. I don't have enough natural light either". The ventilation and size of the rooms could increase user perceptions of thermal comfort. Although there was a consensus that internally it was cold, depending on the floor the residents occupied, indoor temperature was perceived differently: "I feel warmer in the living room and my bedroom because the sun hits them up, and the others are in the back (so they don´t get much sun) ... I live on the fifth floor and particularly like where I live on because it's the top last floor" (MX-U3). In addition, time reinforced thermal comfort; for instance, the more time residents had lived in the apartment, the greater their awareness of changes (during different seasons) affecting its thermal properties. Over time, they improved comfort and conditioning spaces for their lifestyles and needs but preserved the originality of the building.

The relationships between thermal comfort, actions, and changes by the residents can be observed in Figure 9. The narrative of the diagram starts in R12 (reinforcing loop), which is at the centre of the diagram where it is shown the relation between poor thermal comfort in winter. The less thermal comfort in winter, the less satisfied the user was. Therefore, thermal comfort in winter is balanced with passive actions and less use of heating overnight (B4 balancing loop): *"Well, we have a heater, which we use in the evenings for a while, then we have a warmer environment, a little bit, not a big deal. Yes, if I'm freezing, I turn it on. Look at this heater, but close to me. I install it near the bedpost, connect it, leave it here, then it heats us a bit, and it gives us peace"* (MX-U1).

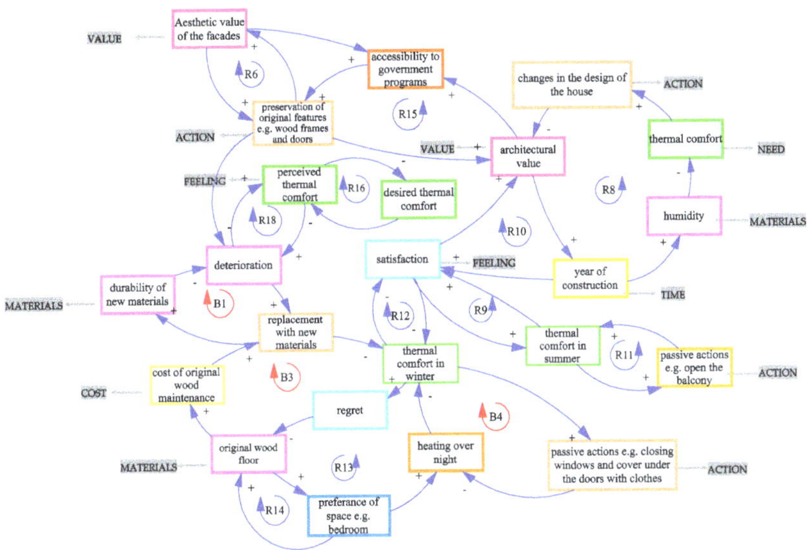

Figure 9. Causal loop diagram created in Vensim (part of the aggregate version in Figure 3 showing relationships between values, interventions, and thermal comfort, and changes by the residents) Colour scheme: yellow, time; green, needs; light blue, feelings; pink, value; dark blue, place; orange, actions and practice; purple, materials; brown, cost; grey, risks; dark green, type of ownership.

Because of the perceived poor thermal comfort in winter, some changed the original wood floors, but they regret the changes made. As stated by MX-U2: *"Well, what I didn't like was to change the floor (ceramic tile) because the wooden floor (previous floor), that was cosier, it made you feel warmer inside the apartment, so I disagree that it was removed"*. Still, the changes were made because the cost of maintenance and the replacement occurred with materials that did not always improve internal conditions in winter, creating a balancing loop (B3) (Figure 9). The desired thermal comfort reinforces the perceived thermal comfort (R16) and because of materials' deterioration, the bad perception of thermal comfort creates reinforcement (R18). There is a relationship between the aesthetic of the façade and preservation of original features in R6. The preservation simultaneously has a relationship between architectural value, materials, thermal comfort, and actions in R8. The value was a driver for preserving and conserving, but, at the same time, the residents resisted changing for thermal comfort. Passive actions were preferred during the winter and summer (R11 and B4).

Despite the discomfort perceived inside the buildings, the users did not want to implement intrusive changes for thermal comfort in interiors for three fundamental reasons. The first reason was an attachment to the building and its value (sentimental, architectural, and historical). Second, status for the protection of the historic area was an additional barrier to interventions. The third reason relates to preservation for generations to come. If residents felt thermally uncomfortable inside, they would reflect on making minimal improvements, such as replacing the materials in floors to wood (thought to have better thermal properties) and in ceilings (thermal insulation).

4. Discussion

Research on the values that society assigns to heritage has shown that it is a complex concept based on intellectual, cultural, historical, and psychological references and that it varies according to the person and time [10]. The present authors used a sociotechnical approach to comprehensively evaluate the following parameters as part of a whole system: historic buildings, users, values, thermal comfort, and energy efficiency. The findings present different understandings of the type and degree of change necessary for the maintenance (corrective or preventive), renovation, preservation, and conservation of the value and actions for the thermal comfort of a heritage building. The values assigned to the buildings were related to elements that residents were willing to change or maintain. User-assigned values need a focussed assessment in accordance with buildings' heritage meanings and cultural heritage, which are more than objects of aesthetic, architectural, and historic value [10,30]. Thus, instead of a standard approach to buildings, guidelines are needed to integrate users. In the case of Mexico City (World Heritage City), residents' attachment to buildings is strong. Values are related to the physical attributes of the buildings, but there is also family attachment. This relationship between value and feelings is explained by the preservation of original materials. The users sustain the sentimental value associated with their own memories, reinforcing the will to maintain the building for future generations. Mapping the impact of this interaction on the buildings revealed a holistic path that shows the reinforcement of values over time (aesthetic, symbolic, sentimental, and historical) and the balance created (actions, feeling, and practice) in the decision-making process. In heritage buildings, the conservation process involves the interconnection of practices, actions, needs, values, feelings, time, and materials that reinforce and balance the system.

In Mexico, heritage preservation of original features becomes stronger with the values appreciated and a willingness to preserve and protect over time. The process is affected by the restrictions on the heritage area that inhibit residents from making changes for thermal comfort. The users thus used their own resources to adapt to the space, temperature, and humidity (e.g., wearing more clothes, closing windows, and blocking cold air from outside and dividing areas). These passive activities and low-cost solutions improved a sense of thermal comfort and thereby are likely to improve the historic building's energy performance, reducing the use of natural resources while preserving its value. In this case, they would deliberate on making minimal improvements in the future, such as changing the floors and ceilings' materials. Due to the buildings' original features, aesthetic value, architectural value, and cost of interventions, the residents resisted significant changes.

This study illustrates the findings of a rather unique setting. It is likely, as has been shown to some extent, that there are distinct attitudes in each context. It is therefore imperative to acknowledge that a one-fit 'whole house' approach is not feasible. As was stated by Fouseki et al. [10], efforts for a correct assessment and choice of energy-efficiency measures—such as a whole-house approach [9]—must consider residents' decisions on heritage conservation, thermal comfort, and energy efficiency. The final objective must balance users' needs and values comprehensively and effectively. Understanding the complexity of user decision-making processes for preservation, thermal comfort, and energy efficiency in heritage buildings is fundamental for long-term, practical applications in government management plans, where actions regarding sustainability in heritage buildings in Mexico are needed [31]. After all, the ultimate managers of the historic residential buildings are the residents themselves. Therefore, it is essential to develop, with them, management strategies for sustainable living while preserving the cultural values with which such buildings are embedded.

5. Conclusions

International guidelines and policies on energy efficiency of historic buildings suggest that values of aesthetics and authenticity often determine the correct type of energy-efficiency intervention. However, in practice, users' values are often neglected; consequently, interventions can lead to the potential loss of value appreciated by users. 'Value' is a critical element in the user's decision-making process when there is a negotiation involving their thermal comfort requirement, energy bills cost reduction, and heritage conservation. For instance, residents may initially prioritise the aesthetics of original windows over the need for thermal comfort because of the potential loss of value. In contrast, the cost and the need for energy-efficiency changes may change over time [32]. In many listed buildings and protected areas, there is a set of values that the introduction of energy-efficiency interventions should not compromise. This concern becomes even more challenging in social housing characterised as a heritage site, which was examined as part of this paper, and which showed that residents might have their own value systems that may or may not align with official 'value systems'(for instance included in international guidelines and conventions). Thus, it is important to capture the complete sets of values so that regulations on the energy efficiency of historic buildings can better be contextualised. Mexico City has been through many urban management plans without significant progress. While one area has been conserved and renovated, another has been prioritised for commercial uses, exacerbating depopulation in housing buildings and resulting in abandoned, deteriorated buildings. Heritage management challenges should be addressed in future studies where questions on sustainability should also be included. This research adds to the knowledge base on this subject by using an example where no studies have been reported in Latin America. Therefore, examining Mexico offers a new understanding of the subject matter. The sociotechnical approach evaluated and explored the parameters of values, energy efficiency, and users alongside building components (including environmental conditions). The analysis, combined with systems dynamics [10], unveiled the complex and, to some extent, dynamic interrelationships of the parameters contributing to users' decision-making processes. The residents adapt to the conditions of the building and prefer passive actions that produce thermal comfort and low-cost solutions for them. In the same way, the dynamic and complex relationships identified in the paper illustrate that heritage values, preservation, thermal comfort, and energy-efficiency interplay when residents make decisions involving energy and thermal comfort interventions. By preserving the unique materials, occupants sustain the sentimental value associated with personal and family memories. When they conduct maintenance and interventions for conservation, they maintain the building and its authenticity for the generations to come. With the active participation of the users, the risk that deterioration and abandonment cause to the heritage site on a larger scale would be prevented. Mexico City has shown challenges for Historic Centres where user values and sustainable goals must be included. We advocate for a participatory approach, such as this one, in guidelines to sustainable design in the environment of buildings with historic value.

Author Contributions: Conceptualisation, fieldwork in residences of the Historic Centre of Mexico City, formal analysis, and writing—original draft preparation, K.S.M.C.; fieldwork supervision, K.F. and H.A.M.; writing—review and editing, K.S.M.C., K.F. and H.A.M. All authors have read and agreed to the published version of the manuscript.

Funding: This research was funded by SENER-CONACyT, Mexico, and the EPSRC Centre for Doctoral Training in Science and Engineering in Arts, Heritage and Archaeology (EP/L016036/1).

Institutional Review Board Statement: The study was conducted in accordance with the Declaration of Helsinki and approved by the UCL Research Ethics Committee of UNIVERSITY COLLEGE LONDON, THE BARTLETT SCHOOL OF ENVIRONMENT, ENERGY AND RESOURCES data protection Legal Services (protocol code No Z6364106/2018/12/09 social research, approved 21 December 2018). for studies involving humans.

Informed Consent Statement: Informed consent was obtained from all subjects involved in the study.

Acknowledgments: The authors would like to thank SENER-CONACyT, Mexico, and the EPSRC Centre for Doctoral Training in Science and Engineering in Arts, Heritage and Archaeology for supporting the research in Mexico City. A very special thanks to the participants who agreed to install monitors in their homes and kindly spend their time on the study.

Conflicts of Interest: The authors declare no conflict of interest.

References

1. Fouseki, K.; Cassar, M. Editorial: Energy Efficiency in Heritage Buildings: Future Challenges and Research Needs. *Hist. Environ.* **2014**, *5*, 95–100. Available online: https://www.tandfonline.com/doi/full/10.1179/1756750514Z.00000000058 (accessed on 28 November 2021). [CrossRef]
2. Berg, F.; Flyen, A.; Godbolt, A.; Broström, T. User-Driven Energy Efficiency in Historic Buildings: A Review. *J. Cult. Herit.* **2017**, *28*, 188–195. Available online: https://www.sciencedirect.com/science/article/pii/S129620741730362X (accessed on 10 November 2021). [CrossRef]
3. Cassar, M. Energy Reduction and the Conservation of Cultural Heritage: A Review of Past, Present, and Forthcoming Initiatives. *Int. Preserv. News* **2011**, *55*, 6–9.
4. TC346. EN-16883. *Conservation of Cultural Heritage—Guidelines for Improving the Energy Performance of Historic Buildings*; Comité Europeen de Normalisation: Brussels, Belgium, 2017.
5. Newton, D.; Fouseki, K. Heritage Values as a Driver or Obstacle for Energy Efficiency in Victorian and Edwardian Buildings. In Proceedings of the 3rd International Conference on Energy Efficiency in Historic Buildings (EEHB2018), Visby, Sweden, 26–27 September 2018; pp. 530–538.
6. EECHB. Energy Efficiency and Comfort of Historic Buildings. Proceedings, Brussels, Belgium. Efficiency in Historic Buildings: A Review. *J. Cult. Herit.* **2016**, *28*, 188–195.
7. Webb, A.L. Energy Retrofits in Historic and Traditional Buildings: A Review of Problems and Methods. *Renew. Sustain. Energy Rev.* **2017**, *77*, 748–759. [CrossRef]
8. Martínez-Molina, A.; Tort-Ausina, I.; Cho, S.; Vivancos, J.-L. Energy Efficiency and Thermal Comfort in Historic Buildings: A Review. *Renew. Sustain. Energy Rev.* **2016**, *61*, 70–85. [CrossRef]
9. Historic England. Energy Efficiency and Historic Buildings: How to Improve Energy Efficiency. 2018. Available online: https://historicengland.org.uk/images-books/publications/eehb-how-to-improve-energy-efficiency/heag094-how-to-improve-energy-efficiency/ (accessed on 10 November 2021).
10. Fouseki, K.; Newton, D.; Camacho, K.S.M.; Nandi, S.; Koukou, T. Energy Efficiency, Thermal Comfort, and Heritage Conservation in Residential Historic Buildings as Dynamic and Systemic Socio-Cultural Practices. *Atmosphere* **2020**, *11*, 604. [CrossRef]
11. Yarrow, T. Negotiating Heritage and Energy Conservation: An Ethnography of Domestic Renovation. *Hist. Environ. Policy Pract.* **2016**, *7*, 340–351. [CrossRef]
12. Koukou, T.; Fouseki, K. Heritage Values and Thermal Comfort in Neoclassical Residential Buildings of Athens, Greece: Tension or Co-Existence? In Proceedings of the 3rd International Conference on Energy Efficiency in Historic Buildings (EEHB2018), Visby, Sweden, 26–27 September 2018; pp. 463–471.
13. Murillo, K.S.; Fouseki, K.; Altamirano, H. A sociotechnical approach to users' heritage values and decision-making processes for energy efficiency and thermal comfort in heritage buildings: A pilot study in Mexico City. *IOP Conf. Ser. Earth Environ. Sci.* **2021**, *863*, 012031. [CrossRef]
14. Medrano-Gómez, L.E.; Izquierdo, A.E. Social Housing Retrofit: Improving Energy Efficiency and Thermal Comfort for the Housing Stock Recovery in Mexico. *Energy Procedia* **2017**, *121*, 41–48. [CrossRef]
15. Secretary of Energy. Secretary Energy Funds—Strategy 2014–2018. Available online: https://www.gob.mx/sener/acciones-y-programas/fondos-sectoriales-de-energia (accessed on 28 November 2021).

16. Ma, Z.; Cooper, P.; Daly, D.; Ledo, L. Existing Building Retrofits: Methodology and State-of-the-Art. *Energy Build.* **2012**, *55*, 889–902. [CrossRef]
17. Carlos Slim Trust, Centro Histórico. *10 Years of Revitalization*; Historic Centre Trust A.C.Fundación del Centro Histórico de la Ciudad de México: Mexico City, Mexico, 2011; ISBN 9786079507572.
18. UNESCO Convention for the Protection of the World Cultural and Natural Heritage, United Nations Educational, Scientific and Cultural Organization, 1972–2014. Available online: http://unesdoc.unesco.org/images/0011/001140/114044s.pdf#page=139 (accessed on 10 November 2021).
19. Forrester, J.W. Lessons from System Dynamics Modelling. *Syst. Dyn. Rev.* **1987**, *3*, 136–149. [CrossRef]
20. Eker, S.; Zimmermann, N. Using Textual Data in System Dynamics Model Conceptualization. *Systems* **2016**, *4*, 28. [CrossRef]
21. Strauss, A.L. *Qualitative Analysis for Social Scientists*; Cambridge University Press: Cambridge, UK, 1987; pp. 1–39. ISBN 9780511557842.
22. Collins, R.J.; Fielder, H.J. Beckstrand's concept of practice theory: A critique. *Kango Kenkyu Jpn. J. Nurs. Res.* **1985**, *18*, 300–306. [CrossRef] [PubMed]
23. Patton, M.Q. *Qualitative Evaluation and Research Methods*, 2nd ed.; Thousand Oaks, CA, USA, 1990; Available online: https://psycnet.apa.org/record/1990-97369-000 (accessed on 31 November 2021).
24. Banks, M.; Zeitlyn, D. *Visual Methods in Social Research*; SAGE: Thousand Oaks, CA, USA, 2015; ISBN 978-1446269756.
25. Sterman, J. *Business Dynamics: Systems Thinking and Modelling for a Complex World*; McGraw-Hill Education: New York, NY, USA, 2000; ISBN 978-0071179898.
26. Randers, J. *Elements of the System Dynamics Method*; Wright Allen Press: Cambridge, MA, USA, 1980; ISBN 978-0915299393.
27. ASHRAE. *Standard 55–92: Thermal Environmental Condition for Human Occupancy*; American Society of Heating Refrigerating and Air-Conditioning Engineers: Atlanta, GA, USA, 2010; p. 4.
28. Arundel, A.; Sterling, E.; Biggin, J.; Sterling, I. Indirect Health Effects of Relative Humidity in Indoor Environments. *Environ. Health Perspect.* **1986**, *65*, 351–361. [CrossRef]
29. International Organization for Standardization. *ISO 7730:2005. Ergonomics of the Thermal Environment—Analytical Determination and Interpretation of Thermal Comfort Using Calculation of the PMV and PPD Indices and Local Thermal Comfort Criteria*; International Organization for Standardization: Geneva, Switzerland, 2005; p. 1.
30. Smith, L. *Uses of Heritage*; Routledge: London, UK, 2006; ISBN 978-0415318310.
31. United Nations. Goal 13: Take Urgent Action to Combat Climate Change and Its Impacts. Available online: https://www.un.org/sustainabledevelopment/climate-change/ (accessed on 10 November 2021).
32. Fouseki, K.; Bobrova, Y. Understanding the Change of Heritage Values over Time and Its Impact on Energy Efficiency Decision-Making at Residential Historic Buildings through System Dynamics. In Proceedings of the 3rd International Conference on Energy Efficiency in Historic Buildings (EEHB2018), Visby, Sweden, 26–27 September 2018; pp. 11–21.

applied sciences

MDPI

Article

Use of Low-Cost Devices for the Control and Monitoring of CO_2 Concentration in Existing Buildings after the COVID Era

Andrés Pastor-Fernández [1,*], Alberto Cerezo-Narváez [1], Paz Montero-Gutiérrez [1], Pablo Ballesteros-Pérez [2] and Manuel Otero-Mateo [1]

[1] School of Engineering, University of Cadiz, Av. University of Cadiz 10, 11519 Puerto Real, Spain; alberto.cerezo@uca.es (A.C.-N.); paz.monterogutierrez@mail.uca.es (P.M.-G.); manuel.otero@uca.es (M.O.-M.)

[2] Project Management, Innovation and Sustainability Research Centre (PRINS), Universitat Politècnica de València, Camino de Vera s/n, 46022 Valencia, Spain; pabbalpe@dpi.upv.es

* Correspondence: andres.pastor@uca.es

Abstract: In the COVID-19 era, a direct relationship has been consolidated between the concentration of the pollutant carbon dioxide (CO_2) and indoor disease transmission. For reducing its spread, recommendations have been established among which air renewal is a key element to improve indoor air quality (IAQ). In this study, a low-cost CO_2 measurement device was designed, developed, assembled, prototyped, and openly programmed so that the IAQ can be monitored remotely. In addition, this clonic device was calibrated for correct data acquisition. In parallel, computational fluid dynamics (CFD) modeling analysis was used to study the indoor air flows to eliminate non-representative singular measurement points, providing possible locations. The results in four scenarios (cross ventilation, outdoor ventilation, indoor ventilation, and no ventilation) showed that the measurements provided by the clonic device are comparable to those obtained by laboratory instruments, with an average error of less than 3%. These data collected wirelessly for interpretation were evaluated on an Internet of Things (IoT) platform in real time or deferred. As a result, remaining lifespan of buildings can be exploited interconnecting IAQ devices with other systems (as HVAC systems) in an IoT environment. This can transform them into smart buildings, adding value to their refurbishment and modernization.

Keywords: indoor air quality (IAQ); smart building; sustainable development goals (SDGs); COVID-19; CO_2; refurbishment

Citation: Pastor-Fernández, A.; Cerezo-Narváez, A.; Montero-Gutiérrez, P.; Ballesteros-Pérez, P.; Otero-Mateo, M. Use of Low-Cost Devices for the Control and Monitoring of CO_2 Concentration in Existing Buildings after the COVID Era. *Appl. Sci.* **2022**, *12*, 3927. https://doi.org/10.3390/app12083927

Academic Editors: Tiziana Poli, Andrea Giovanni Mainini, Mitja Košir, Juan Diego Blanco Cadena and Gabriele Lobaccaro

Received: 30 November 2021
Accepted: 11 April 2022
Published: 13 April 2022

1. Introduction

The World Health Organisation (WHO) declared on 11 March 2020 that COVID-19 disease had become pandemic [1]. Since then, the WHO has been constantly reporting on the outbreak and providing global guidelines for its control [2]. This has been a global challenge in air quality (AQ) research [3–5], especially on its prevention modes, and detection and protection systems [6,7]. For example, research has been performed on barriers to transmission, infection, and vaccines [6,8]. Many researchers are still working on these and other issues. Among them, the importance of air renewal to achieve adequate IAQ and avoid health risks can be highlighted.

The WHO identified direct contact with people suffering from the disease as an obvious source of contagion. Therefore, the social distance was delimited, the use of masks was made compulsory, and other types of actions were taken such as eye protection and use of biohazard clothing for professionals directly exposed (for example, hospital workers) [9,10]. In addition, general applicable options that are not dependent on the decision of the individual were also sought during this pandemic. For instance, closing public spaces, restricting seating capacity in places of assembly [11], and increasing indoor air renewal in public buildings [12,13] stand out. Capolongo et al. [14] summarised a series of recommendations in a decalogue related to health strategies and pandemic challenges, which can be a great reference for addressing the future times ahead.

On the other hand, two means of transmission of this type of disease were established: surface and airborne. To combat the first mode, surfaces that have been in contact with secretions or particles from sick people must be treated to inactivate the virus [15]. In addition, cleanliness of hands and face have been increased to reduce the likelihood of infection [16]. Concerning airborne transmission, the spread can occur through "droplets" [17,18], which are larger than 100 micrometres in size, and aerosols, which are smaller than 10 micrometres in size. Research has proven that contagion can involve aerosols from distances of up to 10 m [11,17]. Therefore, how to reduce aerosol concentration in indoor spaces has been explored [19,20]. In this context, aerosol monitoring can be performed by means of characterisation systems [21].

Apart from being infected with the COVID-19 disease, exposure to air pollutants can affect different organs of people [22], damaging the respiratory, nervous, urinary, and digestive systems, among others. International standards and regulations [23–29] establish the following parameters for determining air quality (AQ):

- Physical parameters: Wet bulb temperature (T) and relative humidity (RH) to assess thermal comfort as well as ambient particulate matter concentration ($PM_{2.5}$ y PM_{10}).
- Chemical parameters: Carbon dioxide (CO_2), carbon monoxide (CO), formaldehyde (HCHO), ozone (O_3) and volatile organic compounds (TVOCs).
- Biological parameters: Bacteria and fungi.
- Air renewals (based on metabolic CO_2 concentration).

It can be noted that these parameters are mostly assessed in outdoor environments using stations with specific instrumentation. The European Environment Agency (EEA) and the European Commission provide information on AQ [30] throughout Europe [31]. Clean air is a mixture of gases [32]. The percentage of CO_2 corresponds to 0.035% among the different elements. However, excess CO_2 is considered to be one of the main sources of environmental pollution. Unfortunately, measurement of CO_2 concentration is not available at all stations.

To reduce harmful emissions in outdoor and indoor environments [33], many countries have established strict policies to improve AQ and comfort. In the current context, they are also serving to identify areas with a potential danger of contagion in the COVID-19 pandemic. In indoor environments, CO_2 displaces oxygen and exposes users to the effects of hypoxia [34,35]. As proposed by Novakova and Kraus [36], the ideal indoor concentration can be established close to 350 parts per million (ppm) in environments with special requirements and 500 ppm in normal activity environments. Nevertheless, other reference values are provided in specific standards and regulations because it is considered an indicator that relates AQ and comfort level to CO_2 concentration [37,38]. It can be noted that in the context of COVID-19, CO_2 concentration [39–42] from occupant respiration has been used as an indicator of IAQ. As a baseline, the limit values set by the US Environmental Protection Agency (EPA) for CO_2 concentrations (in ppm) and the corresponding AQ are shown in Table 1. In the same vein, the Technical report "CEN/TR 16798-2" [43] indicates that a value of 400 ppm can be assumed as the average outdoor concentration.

Table 1. AQ related to the pollutant CO_2.

CO_2 (ppm)	AQ
340–600	Good
601–1000	Moderate
1001–1500	Unhealthy
1501–5000	Hazardous

In living spaces, the main focus of CO_2 increase is of metabolic origin, which is related to the type of activity [44,45]. As a result, the more effort a person makes, the more exhalations are produced, and the more CO_2 is emitted. The WHO has classified activities according to the "intensity of physical activity", relating them to the metabolic equivalent unit (MET). Table 2 provides a classification of activities with the range of physical intensities in METs. In this regard, Soares et al. conducted research [46] using low-cost devices, concluding that an increase in activity results in an increase in MET concentration and exhaled metabolic CO_2.

Table 2. Activity classification according to metabolic equivalent level (MET).

Classification	Description	MET
Soft	Activities that do not require excessive effort and are normally at rest (sleeping, watching TV, driving, eating, ...).	<3
Moderate	Activities that accelerate the heart rate in a sensitive way (housework, dancing, construction work and painting, ...).	3–6
Intensive	Activities that require greater effort and cause rapid or accelerated breathing, involving activities that increase the heart rate (aerobics, sports, or moving weights of more than 20 kg, ...).	>6

COVID-19 is currently a major global concern. Peng and Jimenez confirmed that CO_2 co-exhaled with aerosols containing SARS-CoV-2 by COVID-19 can be used as an indicator of indoor SARS-CoV-2 concentrations [47]. Other researchers have pointed out that group immunity is expected by the end of 2024 [48] thanks to the vaccination process, but broader strategies must be put in place that can respond to future diseases. Therefore, buildings should aim to act smarter [49], renovating indoor air to achieve better quality. These activities are in accordance with the sustainable development goals (SDGs) of the 2030 Agenda [50]. Although the need for AQ control in public, residential, and business buildings has been analysed in many studies [36,51,52], the price of the systems required for this purpose has been a barrier to implementation. This applies mainly to low-income households [53] and socially disadvantaged environments [54].

The Center for Disease Control and Prevention (CDC) has published extensive research on aerosol concentrations in different situations. In unventilated shared-use spaces, CO_2 concentration is high [36] and infections are higher. This context is unchanged even if a social distance greater than two metres or six feet is maintained [11,55,56]. Although there are other methods to reduce pollutants such as filtration with HEPA filters [57], ultraviolet radiation treatments, or disinfection with ozone [58], they were not considered because they are beyond the scope of this research. In this context, ventilation has become a fundamental solution [59,60] for the reduction in indoor pollutant concentrations and for the improvement of IAQ [61]. In some cases, the way to achieve air renewal is by means of natural ventilation (cross, outdoor, indoor) [33,62,63]. In other cases, a forced ventilation system must be used [64]. In all cases, pollutant concentrations are reduced if air is renewed. These lead to a decrease in CO_2 concentration in indoor spaces [65]. It can be noted that the current trend for effective AQ control is to promote natural ventilation through strategies combining large external openings [66] with intelligent ventilation [67], which also allows optimizing energy consumption by increasing its efficiency [68,69].

Several studies have analysed methods and models to determine the air renewal required to achieve acceptable IAQ parameters [70–73]. In all of them, the amount of exhaled CO_2 has emerged as a critical factor in establishing the number of air renewals. In addition, different standards indicate the number of air renewals recommended in the diverse living spaces of buildings according to their use and occupancy. Among them, those set by ASHRAE [74,75] and the national legislation in Spain [76], which comes from the Directive 2010/31/EU of the European Parliament and of the Council of 19 May 2010 on the energy performance of buildings, stand out.

Monitoring IAQ, and more specifically CO_2 concentration in buildings, can be realised using professional (commercial) equipment or clonic devices. These clonic systems are increasingly being used due to their reliability and prestige in the electronics market. Villanueva et al. [77] carried out a research focused on commercial equipment and clonic devices used in Spain for the measurement of CO_2 during the COVID-19 pandemic. As a noteworthy result of this research, it must be noted that the price of commercial instruments on the market ranges from EUR 75 for the cheapest to EUR 400 for the most expensive, although they include other functionalities to measure relative humidity, temperature, and particle size. In addition, they also evaluated clonic devices. They showed that the measurement error varies over a spectrum ranging from 9% to 15% in environments with 500 ppm CO_2 and from 7% to 12% when concentrations are closed to 700 ppm CO_2.

On the one hand, professional equipment are electronic devices designed and commercialised by manufacturers intended for a professional audience (laboratories and health and safety services), which can monitor any physical or chemical parameter of AQ. They can be classified according to their portability, autonomy, and number of parameters monitored. In addition, these systems have the advantage that their probes and sensors are calibrated in metrological laboratories with controlled atmospheres. They also allow data to be stored and sent instantly to other devices. However, the high weight of some of them, due to the high number of measuring sensors, is a disadvantage for their portability.

On the other hand, clonic measurement devices are composed of electronic components and open-sourced software. The scalable configuration of clonic devices makes it possible to customize and increase the number of features of professional equipment. To begin with, a motherboard must be selected. Raspberry Pi and Arduino are two of the manufacturers offering more competitiveness in this field. According to studies reviewed, 37.5% of clonic devices are based on Arduino microcontrollers and 35% on Raspberry Pi ones [78]. The use of a motherboard allows the connection of peripherals that help in capturing information on physical and chemical parameters of IAQ. Specifically, between 67% and 70% of the sensors that monitor IAQ are intended to measure CO_2 [78,79]. Measurement range, sensitivity, and response time are their key sensors features. In addition to the two elements described, a communication unit must be included. This enables data storage, processing, and further analysis to be performed [79].

The interconnection of clonic devices with other systems favors the use of the Internet of Things (IoT) and data storage in the cloud. These enabling technologies allow wirelessly collected data to be evaluated for interpretation, in real time or deferred, for a real monitoring and control of the IAQ [79,80]. This has the potential to transform an existing building to be renovated into a smart building [81]. Several researches have studied the application of IoT for IAQ control [82,83]. They stress the advantages of system integration with connections to smartphone applications, ease of installation and scalability as an element to achieve cost improvements [84]. In this regard, Marquez et al. [85] developed research on an IoT solution that monitors CO_2 concentration in smart buildings. In this way, the IAQ can be controlled and therefore, the health of the occupants be improved.

It can be noted that clonic devices present a series of limitations. These are related to the maintenance and calibration of the sensors assembled, the communication protocols, and their power consumption [86]. Fortunately, with the aim of creating devices that help to promote healthy, sustainable, and smart cities, considerable progress has been made in recent years dealing with these issues [79].

The first objective of this research was to design, develop, assemble, and prototype a low-cost clonic device to provide buildings [87] with smart functionalities in the IAQ field [88]. As indoor comfort is currently taking a back seat to the priority of maintaining safe CO_2 levels to protect from COVID-19, research focused on the use of CO_2 sensors. Another purpose addressed was the choice of the best location to place the clonic device. To this end, computational fluid dynamics (CFD) simulations [89] were carried out in different scenarios (ventilation options) for a living space that was used as case study. It can be noted that, to the best of our knowledge, this has not been addressed before. Once the best locations for the placement and distribution of the instruments were determined, an experimental study was performed. Data were obtained with a calibrated equipment that was used as a reference standard and also with the low-cost prototype device. Comparisons were then made to check if measurements were acceptable for IAQ monitoring and control. In summary, the three main objectives and their relationship are shown in Figure 1:

- CFD analysis of the interior air flows of a living space to establish the location of measurement points.
- Exhaustive analysis of the solution to design and develop the prototype.
- Calibration of the cloning device for proper data acquisition.

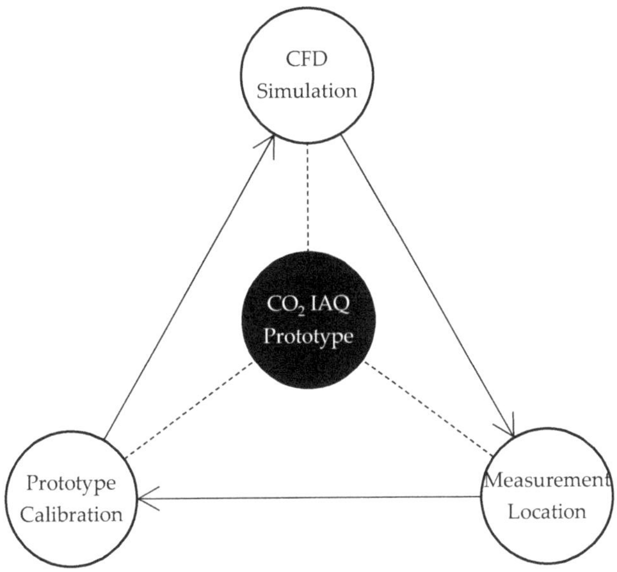

Figure 1. Objectives of the research.

The main novel contributions of this paper can be summarised in the following four elements:

- Complete design, development, and prototyping of a clonic device for CO_2 measurement as a control and monitoring parameter for IAQ. This parameter relates to air renewals for minimizing the spread of COVID. We showed all the physical components (sensor, microprocessor board and connections) and the programming code necessary for its operation.
- Description in detail the calibration protocol. In this paper, we used the KIMO HQ-210 commercial instrument as a reference standard for the clonic device. However, other instruments of similar characteristics with a calibration certificate from a laboratory legally accredited for this purpose could have been selected. In this context, we showed how to prepare a clonic device to properly measure CO_2 concentration.
- Process of all collected data in the cloud for further analysis and possible connection to other smart systems in the building. Therefore, we showed the applicability of the device to control and monitor CO_2 concentration and even act, if required, on HVAC systems.
- Definition of a methodology to obtain the best location of sensors for data collection. For this purpose, a CFD analysis was carried out in different scenarios of a case study. In this way, we improved the layout suggested by the different standards, which were limited to establishing separation values for the envelope (boundary) of the room without considering the air flows caused by the renewal of the interior air.

In summary, we showed for the first time, to the best of our knowledge, all the details of the components of the clonic device so that it can be cloned by anyone interested in the "do it yourself" philosophy. It also incorporated open source programming code, allowing other researchers to replicate the research and generalize this contribution to the scientific community and society. Furthermore, the use of the CFD methodology provided the most suitable points to determine the CO_2 concentration in the space to be studied, eliminating points that could generate distortions in the control and monitoring systems. In addition, using an open-source system allowed the connection with other smart building systems, which in the COVID-19 era can be considered a preventive measure to analyze IAQ.

2. Materials and Methods

This section describes the methodology used in the research, including its application in a case study. The methodology is primarily based on the use of CFD to determine the positioning of the CO_2 measurement system, for which simulations were carried out to identify those places where there is no air renewal. Next, it is necessary to design, configure, and develop a programmable clonic device.

The starting point was to select the living space where the prototype will be tested. A single-family house was taken in which a single bedroom was to be used as a study room. In this space, the calibration of the clonic device was carried out using a high-precision and high-cost calibration instrument. For this purpose, different assumptions and scenarios were defined, which are described in detail further in the next sections. Furthermore, this room is the space used to perform the CFD simulations. For obtaining the main results, the geometry was modeled in CATIA V5, and the different boundary conditions defined in ANSYS Discovery 2021 R2. These results allowed us to identify the best location of the programmable clonic device inside the room.

2.1. CFD Analysis

CFD is an area of engineering knowledge that belongs to the computer aided engineering (CAE) simulation programs. These analyses allow the numerical and visual simulation of flows, heat transfer, and even chemical reactions. These simulations and analyses are increasingly used to characterise living spaces, determining how the ventilation is operating and therefore the level of existing pollution. In short, CFD analysis helps to detect where the air is fouled. Thanks to this, preventive measures can be taken to avoid air quality risks [90]. These rises may be due to an increase in the number of people inside living spaces or a lack of ventilation in them [86,90]. In this study, CFD simulations helped to detect CO_2 concentration.

For this research, two different software packages were used: ANSYS Fluent 2021 R2 and ANSYS Discovery 2021 R2 (available at www.ansys.com). ANSYS Fluent was used for static and dynamic simulations. In addition, ANSYS Discovery was used only for dynamic simulations. The main features that had to be selected are:

- Dimension: 3D.
- Display options: Display mesh after reading.
- Options: Double precision.
- Processing Options: Serial.

To begin with the analysis, the geometry and boundary conditions had to be previously defined to establish the convergence of the solution [86]. These boundary conditions were related to CO_2 concentration and ventilation flow, helping to determine the location of the measurement instruments. The case study analyses the indoor atmosphere of a medium-sized room in a single-family house was simulated. In order to perform it, a series of conditions and restrictions had to be established:

- The building is located in Medina Sidonia, a village in the province of Cadiz (southern Spain). Its height above the sea is 337 m. In addition, its coordinates are as follows: longitude 5°55′37.81″ W and latitude 36°27′25.02″ N.

- The usual outdoor pollution levels are low, thanks to its location near Alcornocales Natural Park on the town periphery. In some previous measurements with the standard reference equipment, the CO_2 concentration ranged from 380 to 410 ppm under different conditions of traffic, industrial activity, and prevailing winds. It can be noted that although some of the values are even lower than those indicated by the technical report "CEN/TR 16798-2", a value of 400 ppm is assumed as the average outdoor concentration, as suggested by the report.
- The characteristic wind in this area comes from the east. As the main opening is in the southeast direction, the room received an adequate amount of fresh air.
- The room consists of a rectangular enclosure of average dimensions (3.9 × 2.4 × 2.5 m). The geometry allowed the simulations to be carried out easily. It can be noted that this is a factor that directly influences the meshing performed by ANSYS. In addition, the boundary conditions defined refer to the amount of air entering the room, which was considered constant in the defined scenarios.
- Only one person lives in the room. No additional CO_2 emissions were considered that could have an impact on the defined boundary conditions.
- The test room is not forced ventilated and has no HVAC systems to renew the air inside. Therefore, ventilation is natural, with air being renewed through the openings in the envelope. The room has an exterior sliding window (height 0.93 m and width 0.63 m by leaf) and an interior door (height 2.03 m and width 0.72 m) that communicates with the rest of the rooms in the building.
- We intended to locate a clonic device inside the room. For this purpose, it was essential to determine the areas where there is no air renewal, as well as the presence of vortices that may affect data collection.

Four scenarios were defined for the CFD analysis. This made four situations possible: closed window and door (no ventilation), closed window but opened door (internal ventilation which is an indirect ventilation through other spaces that are also being renovated), opened window but closed door (external ventilation) and opened window and door (cross ventilation). In each of them, it was essential to guarantee the continuity of the simulation. For this purpose, an air inlet and outlet opening had to be established.

- Scenario 1 (cross-ventilation: Air renewal with opened window and door). Air enters through the window. At the same time, air exits through the door.
- Scenario 2 (outdoor-ventilation: Air exchange with opened window and closed door). In this case, the door is closed. The air enters through the window and exists through the lower area of the door (due to its permeability). In this way, continuity is ensured.
- Scenario 3 (indoor-ventilation: Air renewal with opened door and closed window). Opposite case to the previous one.
- Scenario 4 (no-ventilation: door and window closed). No air is exchanged. In this case, only the concentration of CO_2 metabolically generated by the occupant is simulated. To ensure continuity, it is necessary to define their permeability.

The boundary conditions refer to the mass flow of air and CO_2 entering and exiting the room. For the analysis of the renewal of IAQ in the presence of a CO_2 emission source, the geometry was created using DesignModeler, defining the inlets and outlets for each of the gases considered. A turbulent flow was modelled, specifically governed by the equations of the k-omega SST turbulence model. The physico-chemical characteristics of the gases involved were taken from the ANSYS database, activating the "Species" model. The composition of exhaled air is 78.6% N_2, 1.8% H_2O, 15.6% O_2 and 4% CO_2. Accordingly, the mass flux of CO_2 emitted by a person breathing is small. For instance, the amount of air emitted by a person while performing activities at rest is approximately 0.0004 kg/s [91]. This emission was considered to be ideal and constant. Although a mesh with small elements would lead to a more precise solution, this increased the calculation time. This mesh was defined with small triangular elements which were between 0.05 and 0.1 m in size. In addition, the air inlet was considered to be the exterior window opened, and the air outlet was considered to be the interior door opened. To determine the mass flow, the following Equations (1) and (2) were used:

$$V_i = L_i \times W_i \times H_i \tag{1}$$

where:

V_i is the volume of the $scenario_i$
L_i is the length of the $scenario_i$
W_i is the width of the $scenario_i$
H_i is the heigth of the $scenario_i$

$$M_f = D_{air} \times V_f \tag{2}$$

where:

D_{air} is the density of the air at temperature of 25 °C
M_f is the Mass flow
V_f is the Volumetric flow

Applying the values from the case study:

$$V = 3.9 \times 2.4 \times 2.5 = 23.4 \text{ m}^3,$$
$$D_{air}\ (25\ ^\circ\text{C}) = 1.3 \text{ Kg/m}^3$$

In the first simulation, cross ventilation: A static analysis was performed in which the gas present in the indoor atmosphere was air. Similarly, for the inlet and outlet opened, the incoming and outgoing gas was also defined as air. For the inlet, the mass flow considered was 0.024 kg/s, which corresponds to three renewals per hour for the volumetric flow. The output of this simulation was defined by the "outflow", which allowed the law of conservation of mass to be satisfied: the entire mass flow of air entering through the window passes out the door. In addition, CO_2 was injected, simulating the constant exhalation of a person in a resting state.

In the second simulation, outdoor ventilation: The renewal airflow was performed through the window opened (one sliding leaf), the door being closed. A static analysis was carried out in which the gas present in the indoor atmosphere was air. Similarly, through the inlet opened, the incoming gas was air. The mass flow rate considered corresponded to one point five hourly renewals for the volumetric flow, then the mass flow was 0.012 kg/s. In this case, a slit at the bottom of the door (its clearance) was defined as the exit. This allowed for continuity within the room. The output of this simulation was defined by the "outflow", which allowed the law of conservation of mass to be satisfied: all the mass flow of air entering through the window opened passes out the door slit. Analogous to the previous simulation, CO_2 was injected, simulating the constant exhalation of a person in a resting state.

In the third simulation, indoor ventilation: The renewal air flow was performed through the door opened, the window being closed. A static analysis is carried out in which the gas present in the indoor atmosphere was air. Similarly, through the inlet opened, the incoming gas was air. The renewal air mass flow rate was 0.008 kg/s. In this case, a slit in the side window (its permeability) was defined as the exit. This allowed for continuity within the room. The output of this simulation was defined by the "outflow", which allowed the law of conservation of mass to be satisfied: all the mass flow of air entering through the door opened passes out the window slit. Analogous to previous simulations, CO_2 was injected, simulating the constant exhalation of a person in a resting state.

2.2. Professional Equipment

To select the professional equipment to be considered as a reference standard, a series of characteristics related to its weight, dimensions, and price were required. To begin with, the equipment must be certified by a recognised accredited agency. In addition, they must include human and battery autonomy, which allows measurements to be taken over long periods of time (greater than 50 h). A probe with infrared sensor for CO_2 measurement must be coupled. A wireless connection between the equipment and the probes available must be provided. The measurement modules must operate with different parameters and temperature ranges (from −20 °C to +80 °C). Finally, the professional equipment must measure in working environments with neutral gases. Among those available on the market, a KIMO HQ-210 was chosen. The equipment is supplied with a calibration certificate for the temperature, hygrometry, and CO_2 concentration probes (code NEM1700592 SCOH 112).

2.3. Low-Cost (Clonic) Device

Main advantages and disadvantages of the commercial equipment were considered as a reference to select the main components of the clonic device. In this regard, two elements were fundamental to create it: a sensor to monitor the concentration of CO_2 and an electronic motherboard that allows the processing of atmospheric data. In order to select the most suitable components, a series of criteria matrices were used to score the characteristics considered, which are detailed below.

Firstly, a CO_2 sensor was chosen, for which its main characteristics evaluated were:

- Measurement range: it should be high (0–5000 ppm).
- Accuracy: it should be less than ±0.1%
- Capability of monitoring: it should measure one IAQ gas. In this case, CO_2.
- Sensitivity: it must provide high sensitivity and speed of response (in minutes).
- Heating time.
- Price: it should be less than EUR 30.

Four different sensors were considered: Amphenol Advanced Sensors Telaire T6713, Sensirion SCD30, DFRobot SEN0219, and Amphenol SGX Sensortech MiCS-VZ-89TE. The result of the criteria matrix is shown in Appendix A. After performing the criteria matrix (see Table A1), the most suitable sensor was the SEN0219. This is an analog sensor capable of monitoring CO_2 in an exclusive way by using non-dispersive infrared (NDIR) principle to detect the existence of CO_2 in the air. It features a long life, high resolution, accuracy and sensitivity, as well as a medium response speed. In addition, it is easy to install on any microcontroller and its price is relatively low (between EUR 30 and EUR 60).

Secondly, the selected motherboards included the Raspberry Pi Model B+ microcomputer and the Arduino MEGA 2560 R3 and ELEGOO MEGA 2560 R3 microcontrollers. Due to the similarities between the Arduino and ELEGOO microcontrollers, ELEGOO was considered because of its lower price. Their main factors evaluated were:

- General characteristics.
- Software and hardware.
- Innovation.
- Interconnection.
- Price.

The result of the criteria matrix is shown in Appendix A. Although both boards were suitable and had the right features for mounting the programmable device (see Table A2), ELEGOO MEGA 2560 Rev3 had a more intuitive and user-friendly software interface. The C++-based programming language allows fast programming of the SEN0219 sensor, as it is designed for implementation on microcontrollers. Although Raspberry Pi has many advantages, this project could not make full use of them. Additionally, the programming language (mainly Python, which is one of the base programming languages of Raspberry

Pi) is complex and designed for expert programmers. After performing the criteria matrix (see Table A3), the most suitable motherboard was the ELEGOO MEGA 2560 Rev3.

2.4. Programming and Calibration of the Device

Once the motherboard and the sensor to be used were selected, they were assembled and programmed. The SEN0219 CO_2 sensor could be easily connected to the ELEGOO MEGA 2560 R3 microcontroller (see Appendix B, Figure A1). The software and programming language used was Arduino. The prototype is shown in Figure 2.

Figure 2. Clonic device assembled with SEN0219 and ELEGOO MEGA 2560 R3.

Data captured by the sensor in the Arduino Serial Monitor were stored in a Microsoft Excel spreadsheet. For this purpose, additional programming code was carried out in the Visual Studio Code editor. This allowed the data stored to be analysed from time to time. This parameter was defined in the programming code depending on the monitoring needs. Once the system was ready, the next step consisted of the calibration of the zero point (which is based on the idea that if the sensor is placed into a 100% concentration of an alternate gas, the sensor reads absolute zero). This process was performed according to the datasheet of the manufacturer of the sensor, by shorting the circuit between the HD and GND pins for 7 s at least, but previously ensuring that the sensor was stable for more than 20 min at fresh air (400 ppm ambient environment). However, other alternative methods could have been used such as exposing the sensor to a gas with a known CO_2 concentration or to a gas with no CO_2 present. After that, since it was intended as a low-cost device, it was not required to correct the input voltage and gain (offset voltage) unless anomalies were detected in the readouts. Finally, if measurements were not within the tolerance range, then an adjustment inside the calibration was required. This was done by linear regression.

The calibration line was included in the programming code (see Appendix C). It relates the CO_2 concentration (in ppm) with the voltage values (in mV) captured by the sensor. Both parameters were related by two constants (C_1 and C_2 in Equation (3)), which were provided by the DataSheet of the analog sensor. These constants needed to be calculated in order to modify the regression line and achieve the adjustment within the calibration of the programmable device. The calibration line consisted of the following form:

$$CO_{2i} = (C_1 \times V_i) - C_2 \quad (3)$$

where:

CO_{2i} is the concentration in ppm unit at "i" time.
V_i is the voltage in mV unit at "i" time.
C_1 is the first constant in linear regression line.
C_2 is the second constant in linear regression line.

The regression line obtained allowed the calibration process of the clonic device to be carried out. An analytical process of data collection is explained below. To achieve this, the data from the device was compared with a calibrated highly accurate and costly measuring instrument. To be precise, the KIMO HQ-210 equipment from KIMO Electronic Pvt. Ltd. (Mumbai, India), was selected. It consists of two probes capable of monitoring CO_2, temperature, and humidity. The error of this measuring instrument was reflected in the calibration certificate provided by the manufacturer. This error was also considered in the error of the clonic device calibration. Prototype calibration was also possible thanks to the collection of a lot of data from both the sensor device and the KIMO equipment. To obtain the calibration line, the voltage measured by the sensor was related to the exact and real concentration value provided by the reference standard. This line changed every time a new calibration was performed. To be accurate, three rounds of calibration were needed to achieve a maximum target error of less than 10%.

All the calibrations were performed in the same location during a 9 h measurement interval. This location was the same as the one considered in the CFD simulations. The SEN0219 sensor device and the KIMO equipment were placed side-by-side. Only the characteristics considered were changed for obtaining the minimum error for each calibration. Data collection by the sensor device and the KIMO equipment was carried out at regular intervals of 5 min (which provided 108 measurement points each). A shorter time between measurements was beneficial to obtain more data, helping to obtain a more accurate and successful calibration line. Additionally, this time could be linked to the resolution of the IAQ monitoring device. A longer time between data collection meant that changes in CO_2 concentration were neglected.

Calibrations compare the experimental data from the KIMO equipment with the prototype estimates. To validate these comparisons, different error rates are used. This allows to observe how well the prototype device reproduces the behavior of the reference standard equipment. Several indices for validating models are reported in the literature, which can be classified according to their dependence on the scale of the compared signals. To begin with, errors that depend on the scale [92] such as the mean absolute error (MAE) used by [93] which measures the average magnitude of the error between the measured data by the equipment and the data estimated by the device, the mean error (ME) used by [94] which measures whether the device measures whether the device overestimates or underestimates the measured data, and the root mean square error (RMSE), used by [95] which weights the forecasts that are farthest away from the measured value by the equipment. Additional examples include the errors that do not depend on the scale [96] such as the mean absolute percentage error (MAPE) used by [97] which measures the average percentage error of the estimates and the best fit index (FIT) used by [98] which compares the measured and estimated data with respect to the average of the measured data.

It can be noted that the coefficient of determination (R^2) only measures the proportion of the variance in the dependent variable that is predictable from the independent one [99]. In addition, R^2 is invariant for linear transformations of the distribution of the independent variables [100], so that an output value close to one does not always produce a good prediction regardless of the scale on which these variables are measured. In this context, indicators as MAPE are recommended when it is more important being sensitive to relative variations than to absolute variations [101]. However, although scale-independent errors are widely used in the literature, since they are used when comparing data with different scales, the weighting of the larger-scale error causes adjustments with larger deviation the higher the CO_2 concentration to penalize against others. For instance, RMSE gives a relatively high weight to large errors since the errors are squared before they are averaged, so that it is useful when large errors are particularly undesirable [102].

2.4.1. First Round of Calibration

This was the first contact with the sensor. The upward and downward trend of the concentration recorded by the sensor corresponded to those obtained by the KIMO equipment. Even so, there was a large error that must be resolved by calibration. Figure 3 shows the first calibration line. This error became larger in case of high CO_2 concentrations. This is shown in Figure 4. However, after the adjustment, the error between the prototype device and the KIMO equipment decreased significantly, as shown in Table 3. Therefore, a further calibration had to be performed, taking into account these considerations:

1. The atmosphere at the time of data acquisition must be as stable as possible. Sudden changes in temperature, humidity, or CO_2 must be avoided.
2. The clonic device and the KIMO equipment are placed elsewhere. Areas close to the window or door of the room must be avoided. This makes it possible to ensure the previous point considered.
3. The number of data collected (minimum 60) must be increased.

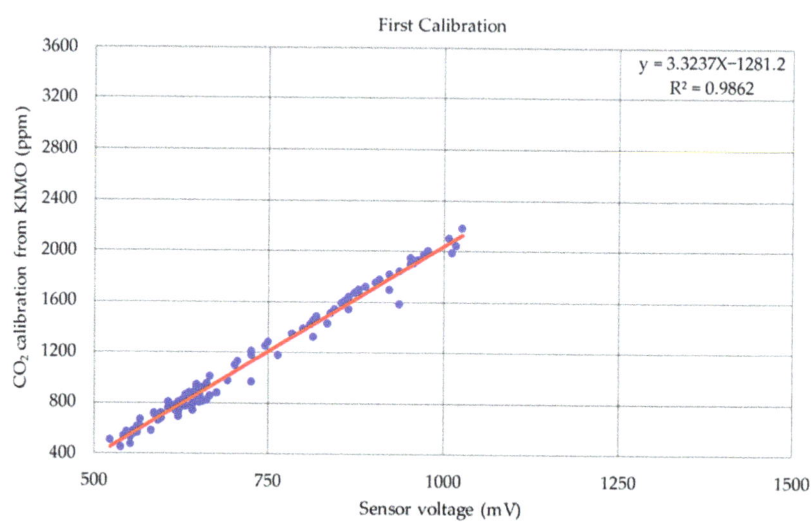

Figure 3. Linear regression of the first round of calibration.

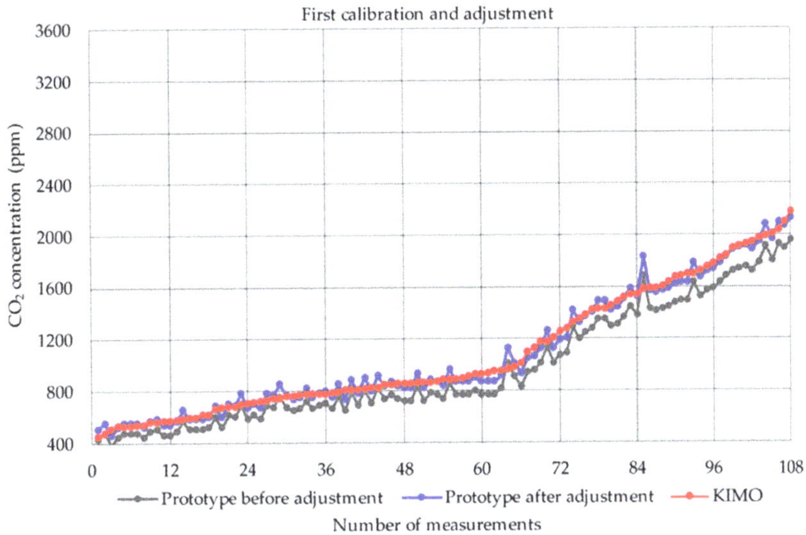

Figure 4. Measurements of the first calibration and adjustment.

Table 3. Statistical parameters related to the errors of the first calibration.

First Calibration	Datasheet	After Adjustment
	($y = 3.1249x - 1251.6$)	($y = 3.3237x - 1281.2$)
MAE	114.74 ppm	43.03 ppm
ME	111.85 ppm	0.05 ppm
RMSE	127.92 ppm	55.41 ppm
MAPE	11.10%	04.49%
FIT	72.32%	89.46%
R^2	98.62%	

2.4.2. Second Round of Calibration

Since the errors after the first calibration were greater than 10%, a second calibration was performed. Although with the adjustment the errors could be lowered, there was no guarantee that under other conditions this adjustment could be contradictory. With the adjustment after the first calibration, the second calibration was performed. In the subsequent data, there was still a significant difference between the values monitored by the two instruments. Figure 5 shows the second calibration line. This still resulted in an average error of more than 10%, as shown in Table 4. Despite this, progress was favourable (when adjusting) and the graphs began to look similar in low and medium CO_2 concentration, as shown in Figure 6. In addition, several conclusions could be drawn from this calibration:

- The occupant of the living space must be far away from both instruments (prototype and equipment). This avoids falsification of data by possible direct exhalation on them.
- Data collection must not be started until 15 min after a change in the atmosphere. These changes are caused by the opening of door and window. After 15 min, the atmosphere becomes stable again.

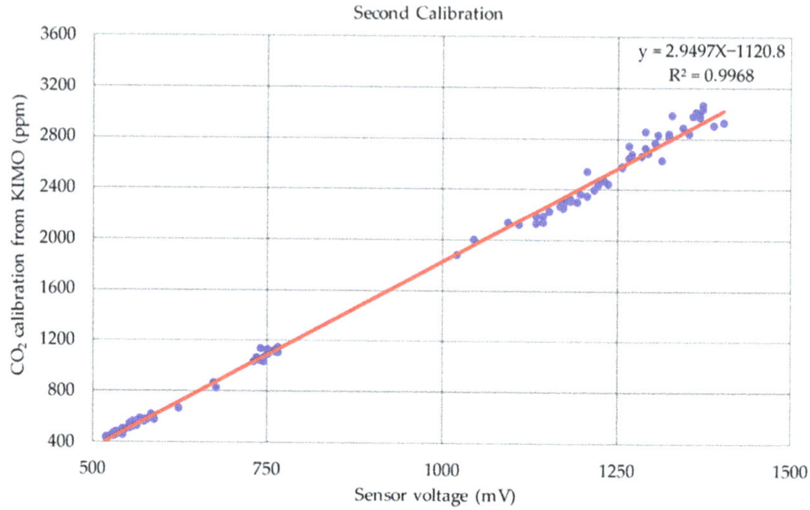

Figure 5. Linear regression of the second calibration.

Table 4. Statistical parameters related to the errors of the second calibration.

Second Calibration	Before Adjustment	After Adjustment
	$y = 3.3237x - 1281.2$	$2.9497x - 1120.8$
MAE	180.34 ppm	41.65 ppm
ME	−180.34 ppm	1.17 ppm
RMSE	225.37 ppm	55.17 ppm
MAPE	10.44%	2.97%
FIT	82.70%	95.56%
R^2	99.68%	

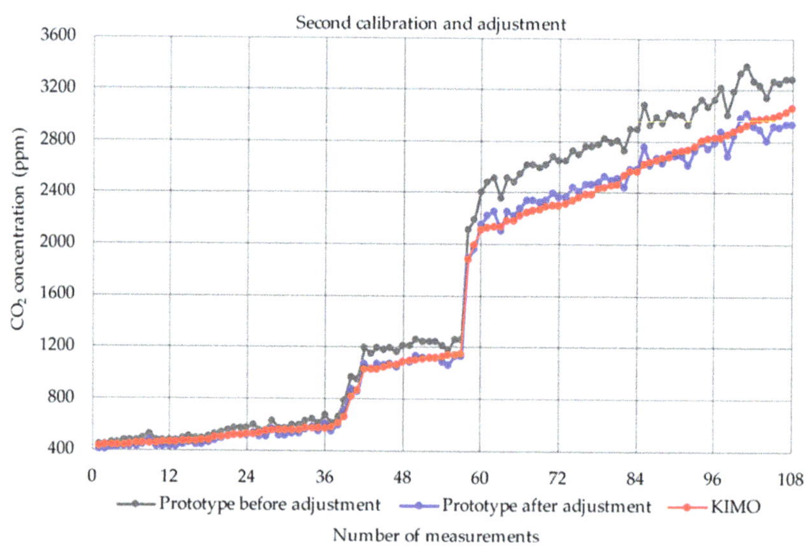

Figure 6. Measurements of the second calibration and adjustment.

2.4.3. Third Round of Calibration

All considerations from the two previous calibrations were made. With the adjustment after the second calibration, the third calibration was performed. The average error obtained was below 10% even before the adjustment, as shown in Table 5. Therefore, this calibration could already be considered suitable to start testing the different scenarios. Figure 7 shows the third regression line after calibration. As a result, the difference between the clonic device and the KIMO equipment was small, as shown in Figure 8. With this milestone, the calibration process was completed. However, to ensure that the clonic device measured constantly, regularly, and correctly, a preventive maintenance had to be scheduled. It is recommended to carry out calibrations every short period of time. This allows one to transform it into a predictive maintenance, since both absolute and relative error rates can be studied.

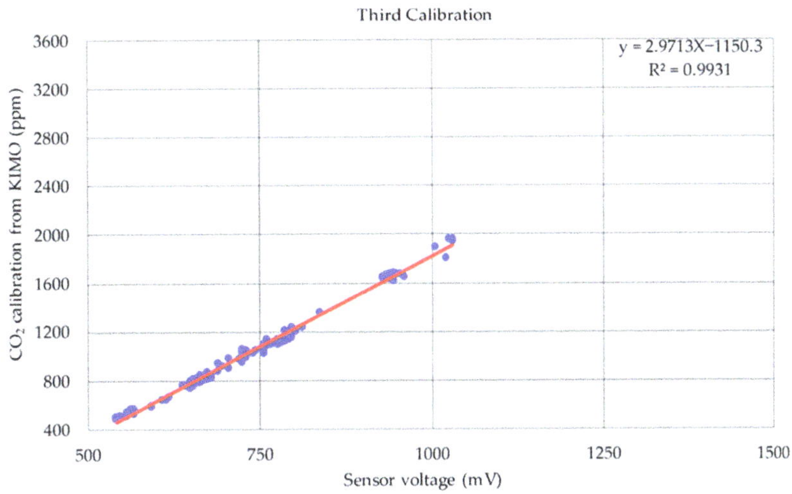

Figure 7. Regression line of the third calibration.

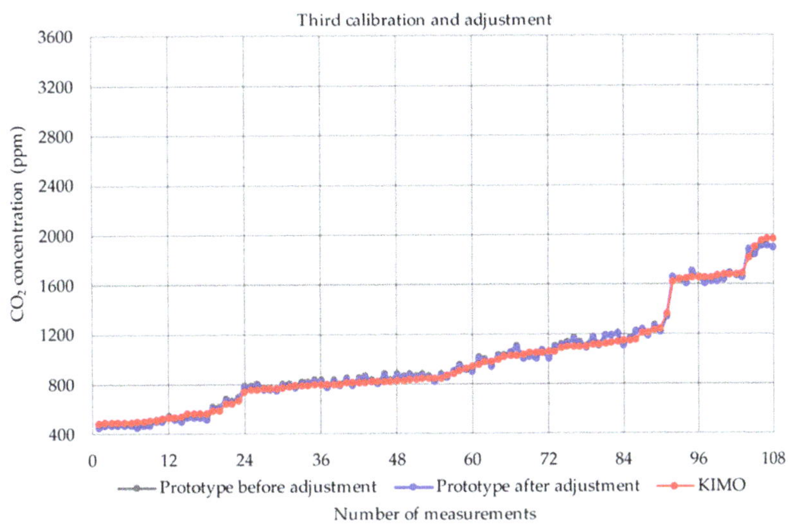

Figure 8. Measurements of the third calibration and adjustment.

Table 5. Statistical parameters related to the errors of the third calibration.

Third Calibration	Before Adjustment	After Adjustment
	2.9497x − 1120.8	**y = 2.9713x − 1150.3**
MAE	29.29 ppm	27.45 ppm
ME	−13.94 ppm	0.01 ppm
RMSE	35.17 ppm	32.17 ppm
MAPE	3.11%	3.03%
FIT	90.23%	90.96%
R^2	99.31%	

3. Results and Discussion

For presenting the results of the research and their discussion in an organised structure, the diagram presented in Figure 9 is followed. Then, the most relevant milestones involved in all phases of the research is summarised.

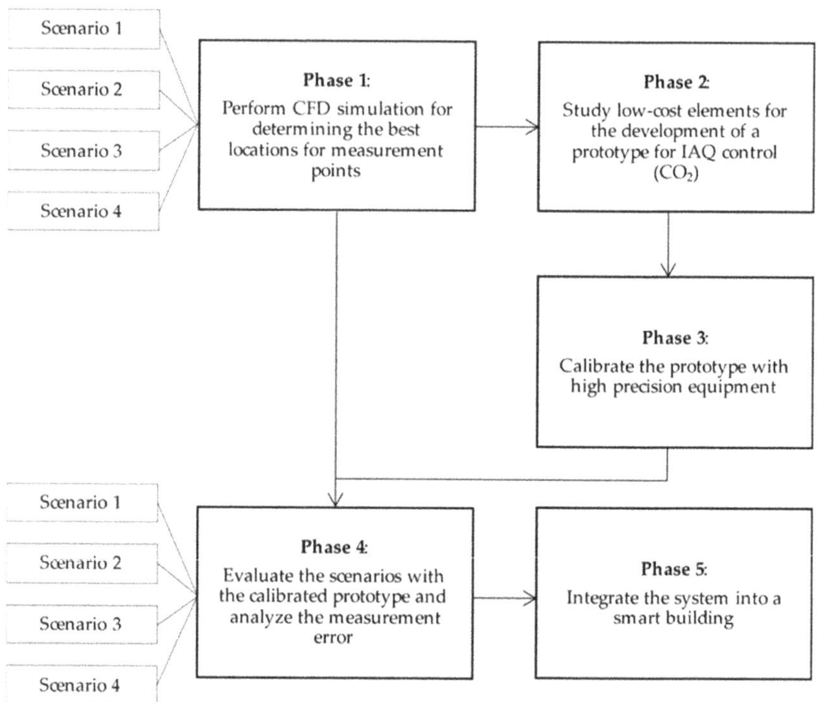

Figure 9. Research framework scheme.

3.1. *Phase 1: Perform CFD Simulations*

As indicated in the previous section, the study was performed in four scenarios. They were those corresponding to the same living space of 9.4 m^2 with a unique occupant in soft sorting activities (minor than 6 MET), but in the following conditions:

- Cross-ventilation: space with door and window opened.
- Outdoor-ventilation: space with door closed and window opened.
- Indoor-ventilation: space with door opened and window closed.
- No-ventilation: space with door and window closed.

The hygrometric comfort conditions were considered acceptable in all simulations, with temperature (T) ranging between 17 °C and 25 °C and relative humidity (RH) between 40% and 70%. The room faces southeast (taking the plane of the window as a reference). As the characteristic wind in this area comes from the east, ventilation with an open window was satisfactory. Figure 10 shows the working space employed in the research. The room was modelled with Ansys 2021 R2 version software.

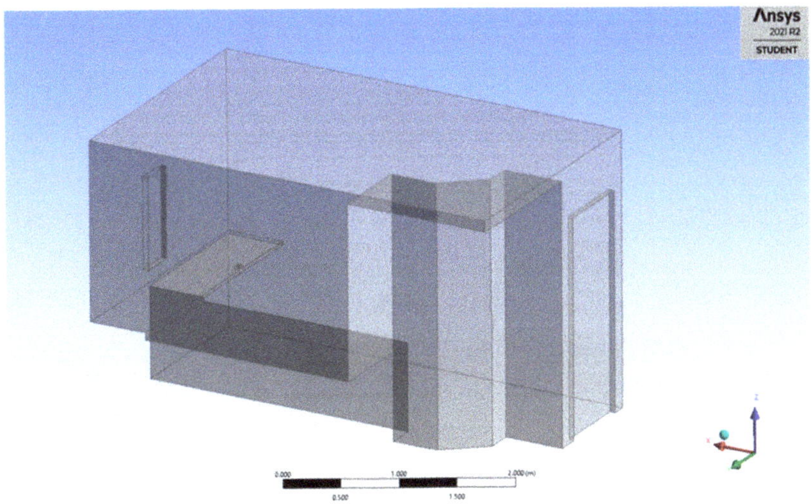

Figure 10. Working space for the research.

The main parameters considered in the simulations was the mass flow of incoming and outgoing air through the different openings and the concentration (or mass flow) of CO_2 from the metabolic exhalation of the occupant in the working space. Among four scenarios, cross-ventilation had the highest air renewal. This example could be extrapolated to public spaces that require a high air renewal rate to keep CO_2 levels low. Therefore, the location of measurement points was based on this scenario, as it was the most unfavourable. In this scenario, static and transitional tests were performed. For this purpose, an inlet air mass flow rate of 0.008 kg/s for each hourly renewal considered, depending on the scenario, was defined. Figure 11 shows the zones in the space where vortices are formed. It must be noted that these points were not suitable for the placement of the CO_2 sensors for IAQ control.

On the other hand, in the case of no-ventilation (Scenario 4), the same mass flow of air was considered and there was an increase of exhaled metabolic CO_2. The permeability of the door and window allowed for pressure balancing through minimal air renewal, which was considered. Figure 12 shows the simulation of this scenario for a period of 9 h. As can be observed, as time progressed, the CO_2 concentration in the working space became uniform, being higher in the areas close to the CO_2 emitter.

To decide a good sensor location in each of the four scenarios, the parameters for the mass flow of air from the concentration of CO_2 emitted by the people in the room measured and the ventilation were used. The amount of CO_2 reflected in all simulations performed was 0.0004 kg/s. However, the mass flow of air from the ventilation varied according to the scenario considered. Among all of them, the highest mass air flow came from the highest air renewal (Scenario 1, cross ventilation) with three renewals per hour, corresponding to a volumetric flow rate of 20 L/s per person. By studying the behavior of the turbulent flow by CFD, the vortices in the room were observed. These points were not suitable for sensor placement. On the contrary, in Scenario 4 (no ventilation), there was no air flow that caused vortices and the CO_2 concentration increased progressively, uniformly filling the volume tested. For this reason, the location of the sensors in Scenario 4 was the most demanding case for monitoring CO_2.

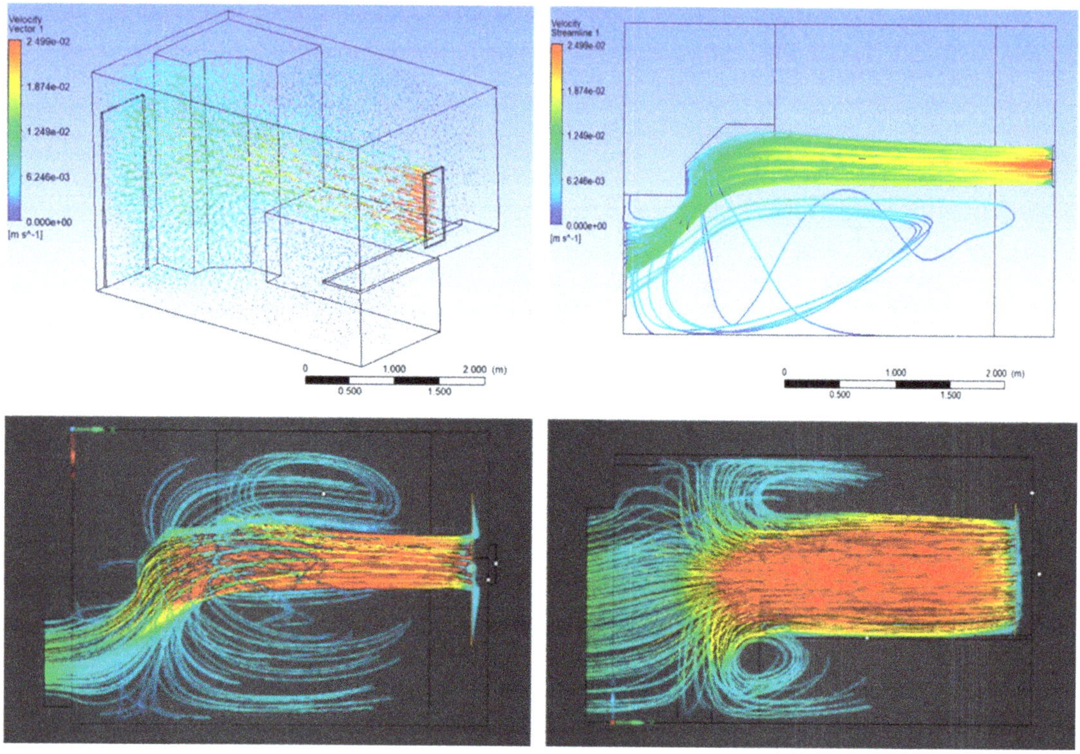

Figure 11. Process for determining vortices with ANSYS Discovery 2021 R2 Student Version.

Figure 13 shows the three points selected among suitable ones for the CO_2 concentration measurement. They were placed in positions that do not correspond to the vortices nor currents of air, avoiding distorting measurement points. The red colour shows areas with high renewal, so that placing the sensors in these areas would distort the average level in the space. Despite the statements of previous research (to the best of our knowledge) and current legislation [103,104] which establish that measurements should be taken at 1.2 metres above the ground, equivalent measurements can be taken at higher or lower points, as they are not modified either by renovations or by CO_2 concentration in the most unfavourable scenarios. They are as follows:

- Location 1: right side wall at a height of 1.2 m above floor and a separation of 2.5 m from the wall of the door.
- Location 2: exterior wall (corresponding to the window) at a height of 2 m and centered in the gap between the left wall and the window.
- Location 3: left side wall at a height of 2 m above the floor and a separation of 1.5 m from the window.

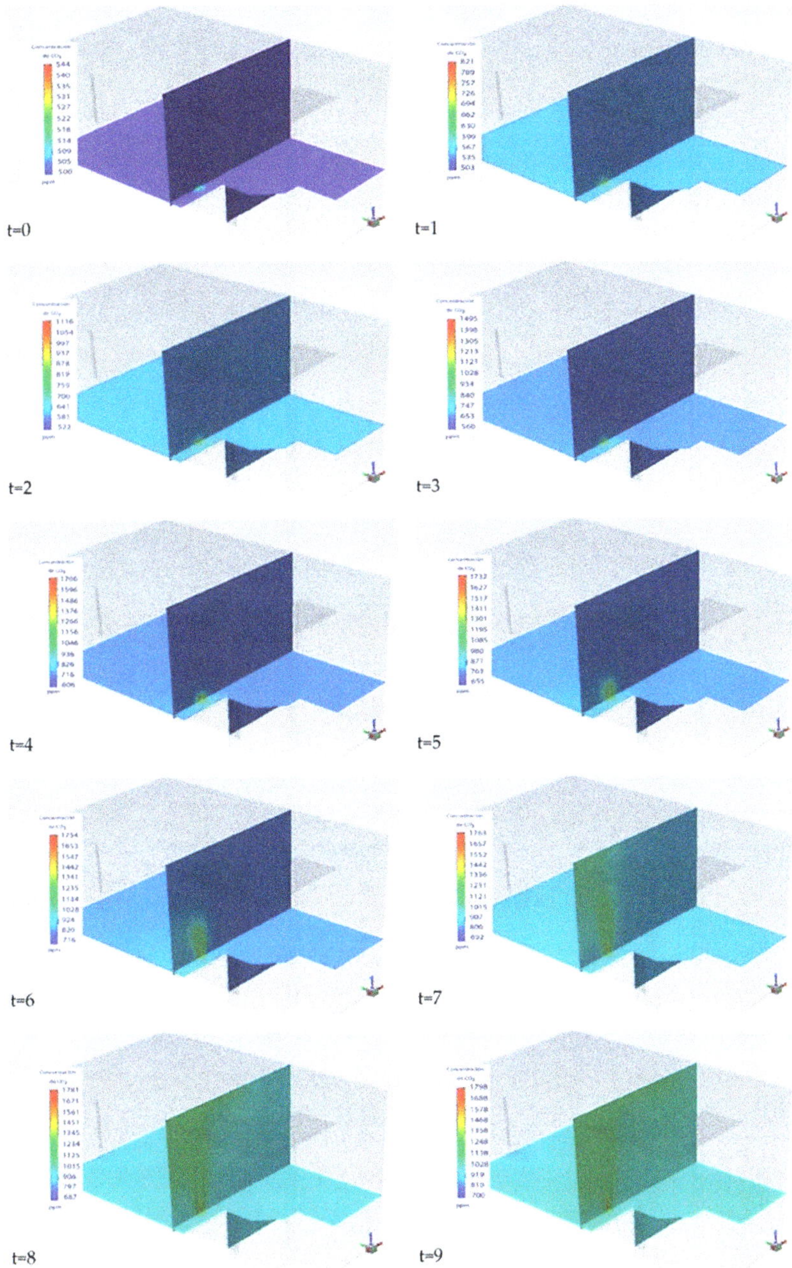

Figure 12. CO_2 concentration in a room with no-ventilation (closed window and closed door).

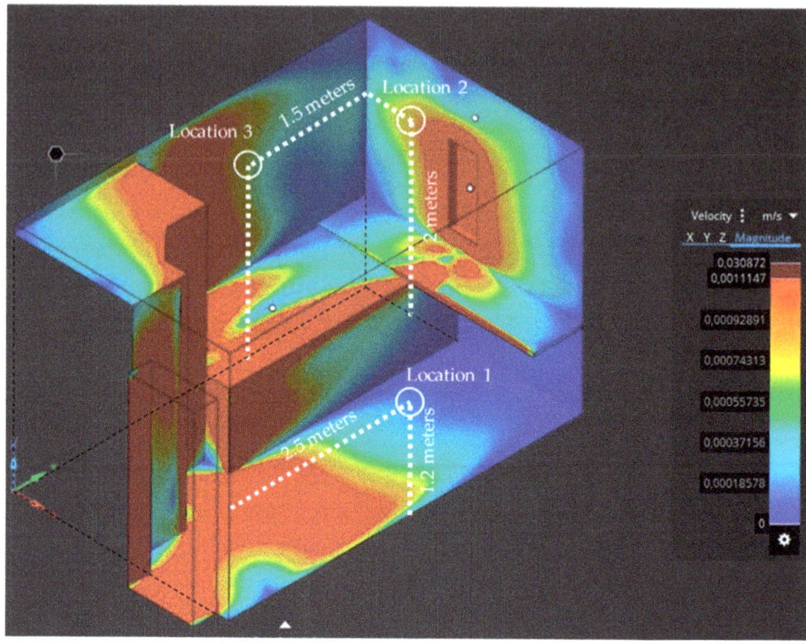

Figure 13. Location of three the points selected for the CO_2 concentration measurement.

As Location 1 complies with current legislation for all kind of activities that can be undertaken in the living space, this is the one selected for the series of measurements.

3.2. Phase 2: Study Low-Cost Elements for the Prototype

In this research, high-precision and high-cost instruments were used. However, other research with CO_2 measuring devices allowing CO_2 concentrations to be monitored less expensive were also analysed. The cheapest cost of these clonic devices, according to the latest research by Villanueva et al. [77], was EUR 75. The clonic device proposed has an open programming language, which is available in Appendix C. In addition, its price is approximately EUR 60, which is 20% cheaper. Therefore, users and construction companies can include it as an element for the transformation of buildings into smart buildings.

Regarding the parametric considerations, the error of the prototype compared to equipment calibrated was in the range between 0 and 7.49%. This error was much lower for Scenario 1 and comparable to commercial equipment in Scenario 4, according to research by Villanueva et al. [77]. The connection between prototype and cloud allows the collection of data and its further management, which is not possible with static devices that only provide a point measurement of the CO_2 concentration.

3.3. Phase 3: Calibrate the Prototype

As detailed in the previous section, the calibration of the prototype was performed through iterations with regression lines. In successive calibrations, better and better fits were obtained. The last calibration determined that the error between the measurements obtained by the calibrated equipment (KIMO) and the prototype sensor device was less than the principal milestone of the project: less than 10%. Absolute indicators as RMSE as well as relative ones as MAPE could be used to check this. Therefore, after this process, a reliable prototype was obtained whose performance is comparable to other commercial or even clonic devices on the market such as those studied by Villanueva et al. [77].

3.4. Phase 4: Evaluate the Scenarios and Determine the Measurement Error

In this phase, a comparative analysis of the measurements made by the prototype device compared to those from the calibrated instrument was carried out. This statistical analysis of the data obtained was performed for each of the four scenarios. It consisted of capturing data every 5 min over a period of 6 h. The data obtained for the measurements in the four scenarios are shown graphically.

Figure 14 shows the data for Scenario 1 (cross-ventilation). It can be observed that the data corresponding to the measurement with the prototype coincided with those of the equipment calibrated. The minimum difference between the recopilated data (error) could be caused by cross-ventilation because the atmosphere was not completely steady. In addition, the CO_2 concentration was observed to remain at healthy levels, close to the IAQ values that can be achieved in outdoor environments. The amount exhaled by the occupant of the space had hardly any effect on the total concentration when the ventilation was cross-ventilation.

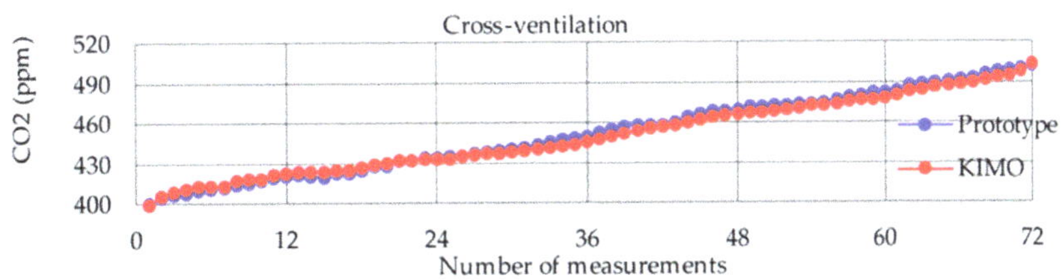

Figure 14. Data for Scenario 1.

Figure 15 shows the data for Scenario 2 (outdoor-ventilation). The data from the measurement with the prototype matched those of the equipment calibrated. The CO_2 concentration was close to health thresholds in a small range, as the window being open, the airflow renewed the amount of clean air, keeping the scene in a healthy condition for the occupant.

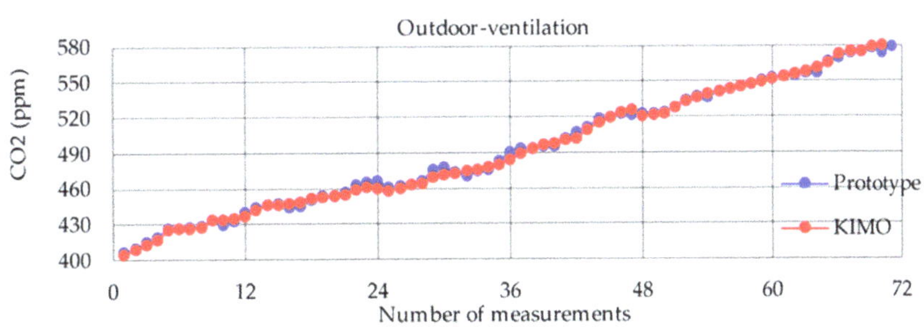

Figure 15. Data for Scenario 2.

Figure 16 shows the data for Scenario 3 (indoor-ventilation). The data from the measurement with the prototype match those of the equipment calibrated. Air renewal depends on the interior connection between the room and the rest of the building. Except for occasional measurements, the IAQ is considered to be adequate. In this case, in order to have more information, it would be necessary to perform the study of the entire building (including all the points where there could be air renewal).

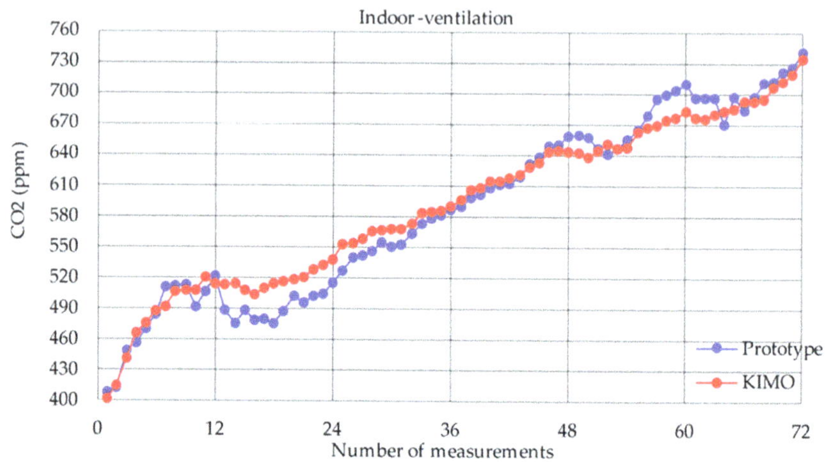

Figure 16. Data for Scenario 3.

Figure 17 shows the data for Scenario 4 (no-ventilation). The data from the measurement with the prototype matched those of the equipment calibrated. It can be observed how the amount of CO_2 increased to levels considered unhealthy. After approximately 50 min, concentrations close to 1000 ppm of CO_2 were obtained.

The instantaneous error of each measurement corresponded to the difference between the measurement obtained by the low-cost clonic device and the commercial instrument calibrated (KIMO). This measurement already incorporated the corrected measurement uncertainty of the KIMO instrument. Details of all measurements are included in the Supplementary Materials.

Simultaneous measurements were obtained with both the low-cost clonic device and the instrument calibrated (KIMO), which was employed as a reference standard. As stated before, these device measurements already included both the correction and the measurement uncertainty of the KIMO instrument. The values of the mean, minimum, maximum, and standard deviation values were studied. Table 6 shows these values:

Table 6. Comparative study of the sample with KIMO and prototype measurements.

Scenario	Device	Mean	Minimum	Maximum	Standard Deviation
Cross-ventilation	KIMO	450.88	400	504	27.00
	Prototype	452.49	401	503	28.98
Outdoor-ventilation	KIMO	494.07	405	585	51.55
	Prototype	494.45	407	582	50.66
Indoor-ventilation	KIMO	590.57	402	734	79.77
	Prototype	587.39	408	741	89.36
No-ventilation	KIMO	1087.69	401	1638	374.10
	Prototype	1091.24	420	1664	368.80

Note: The unit of measurement for CO_2 concentrations is ppm.

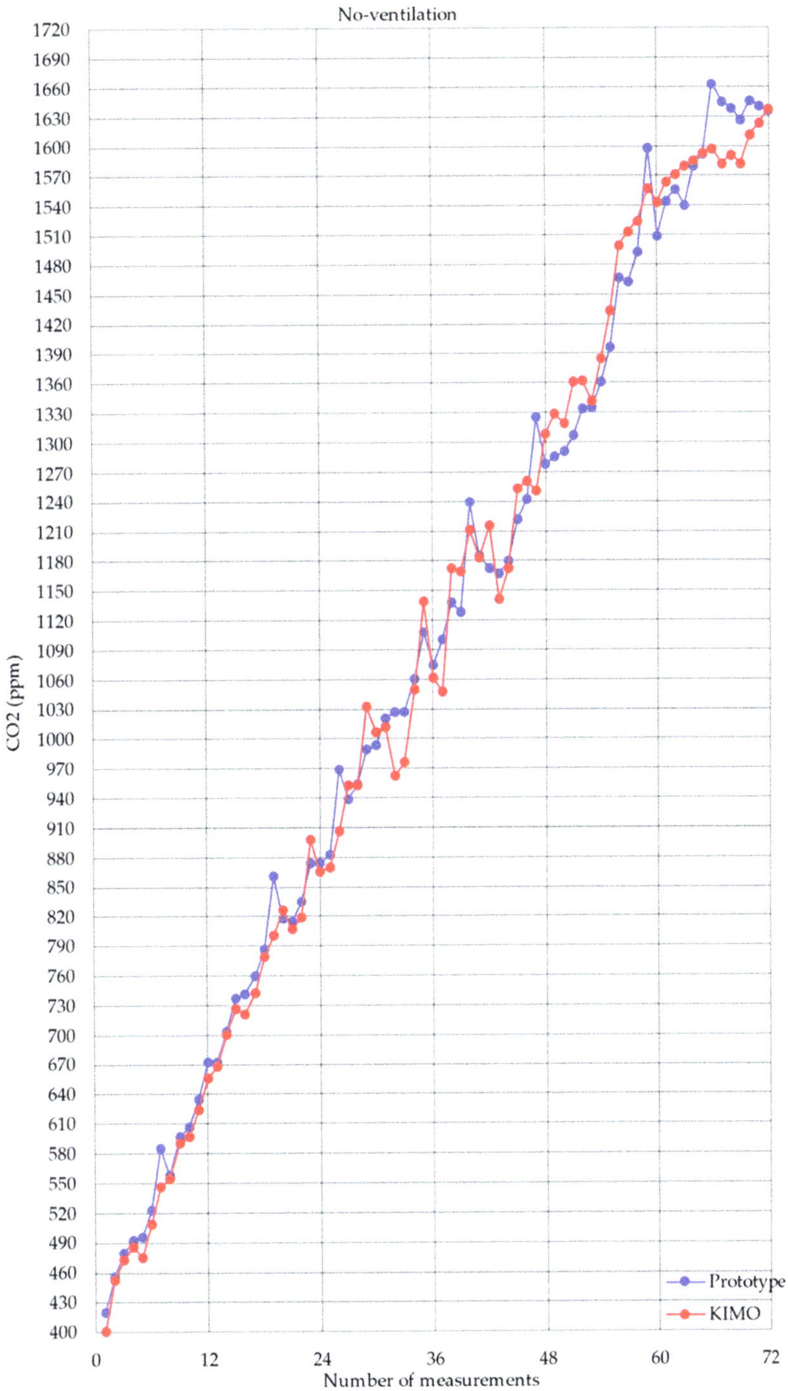

Figure 17. Data for Scenario 4.

As explained before, an average outdoor concentration of 400 ppm was assumed, according to the technical report "CEN/TR 16798-2". Then, normality (with the Kolmogorov–Smirnov test) and homoscedasticity (with the Fisher F-test) was probed. Table 7 shows the values of the statistics for each of the scenarios.

Table 7. Normality and homoscedasticity proofs between KIMO and prototype (p-value).

Scenario	Kolmogorov-Smirnov	Fisher
Cross-ventilation	0.995	0.554
Outdoor-ventilation	0.95	0.883
Indoor-ventilation	0.491	0.341
No-ventilation	0.964	0.905

Then, an equivalence test (the two-one sided test (TOST test)) was used to determine whether the means for both equipment measurements were close enough to be considered equivalent. Finally, the concordance between both samples was checked with the Bland-Altman plot, identifying any systematic difference between the measurements (as fixed bias) or possible outliers. Table 8 shows the values of the statistics for each of the scenarios.

Table 8. Equivalence and concordance proofs between KIMO and prototype (p-value).

Scenario	TOST Top	TOST Lower	Bland & Altman
Cross-ventilation	<0.001	<0.001	0.153
Outdoor-ventilation	<0.001	<0.001	0.251
Indoor-ventilation	0.001	0.005	0.195
No-ventilation	0.031	0.024	0.347

All the statistical indices shown in Tables 7 and 8 as well as in Figure 18, indicate that there was normality and homoscedasticity between the values obtained by the prototype sensor device and the reference standard equipment. In addition, equivalence and concordance were also satisfactorily tested. Finally, the instantaneous error, which is the difference between the measurement of the low-cost prototype sensor device and the KIMO equipment, was calculated. This calculation was made for the 72 measurements taken every 5 min. The mean value was also considered. This study helped to determine the percentage error in each of the scenarios. Data are shown in Table 9.

Table 9. Comparative study of the error with KIMO equipment vs. prototype device.

Scenario	Mean	Min	Max
Cross-ventilation	0.62%	0.00%	1.34%
Outdoor-ventilation	0.44%	0.00%	1.52%
Indoor-ventilation	2.34%	0.00%	7.39%
No-ventilation	2.38%	0.00%	7.49%

As can be observed, the comparative measurement error in all scenarios was less than 8%. If the experiment was performed in a case whose ventilation is maximum (cross-ventilation is 1 m/s), the values obtained were similar to the results of the study by Villanueva et al. [77] for other clonic devices.

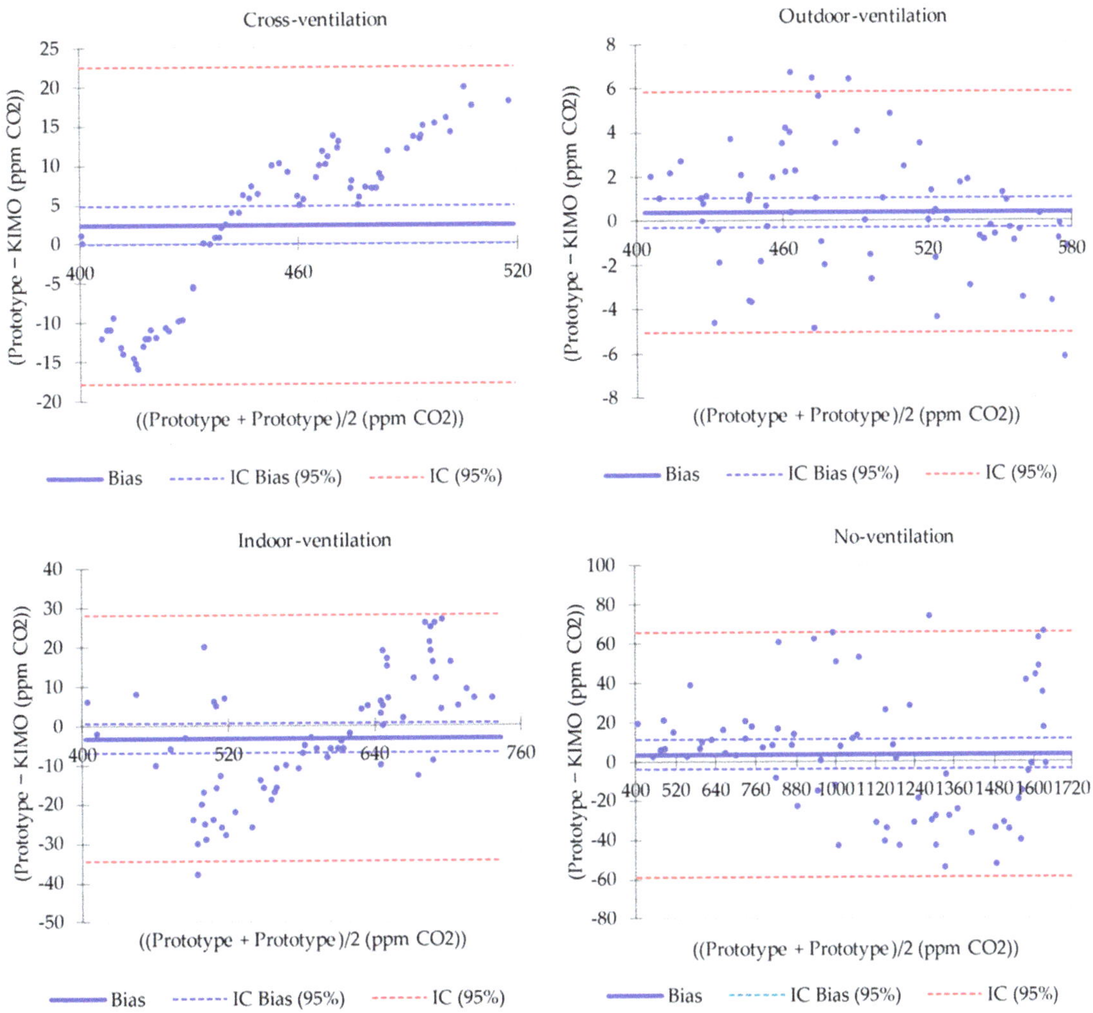

Figure 18. Bland and Altman analysis of different scenarios.

3.5. Phase 5: Integrate into a Smart Building

Smart building infrastructures aim to maximise building efficiency in different fields of application. Specifically, this research focused on the assessment of IAQ to maintain a safe atmosphere for building occupants. However, given that the main issue at present day is the transmission of the COVID-19 disease, the parameter related to CO_2 concentration was used as a reference. For this purpose, a low-cost device was designed to transfer the captured data to the cloud. In this way, it is possible to retrieve the information for decision making or integration with other systems, such as those in charge of ventilation and air conditioning (HVAC) or even other systems for automatic opening and closing of elements for natural air renewal.

The other issue that has been addressed is the definition of a location for this equipment that allows it to be properly measured. In this regard, CFD studies have shown that, although the general recommendation of up to 1.2 m above the floor is correct, there are many more points where reliable data on CO_2 concentration averages can be obtained. In addition, vortices were determined and excluded from the potential location of the device, as its measurements would be distorted from the actual average. A functional scheme of the investigation is presented graphically in Figure 19. Integration with other systems in smart building was not addressed within the scope of this research, although this would allow better monitoring and control of these systems, promoting efficient energy use and helping to control indoor quality for disease prevention.

Figure 19. Communication scheme between devices and systems in a smart building.

4. Conclusions

In the COVID-19 era, CO_2 concentration is the most important indicator related to IAQ to measure the degree of air pollution. This research addressed how to monitor and control CO_2 concentration in order to implement a clonic device to help prevent disease transmission. From commercial components (sensor and motherboard), a low-cost device to measure the concentration of the pollutant CO_2 was designed, developed, assembled, openly programmed, prototyped, and calibrated. The statistical analysis of normality, homoscedasticity, equivalence, and concordance performed determined that the measurements made by the equipment certified (KIMO) used as reference standard and the low-cost clonic device were equivalent, with a maximum error of less than 8% and an average of less than 3%. Therefore, these devices can be used to monitor IAQ by measuring CO_2 concentration.

To the best of our knowledge, they can do so at a lower price than other similar devices proposed by previous research, and much lower than those offered by existing commercial equipment on the market. In addition, it is worth noting that the open-source programming included is easily replicable by other researchers or even end users. This eliminates the economic barriers derived from the high prices of calibrated instruments and equipment that prevent their widespread use, as well as the technical barriers derived from the programming.

On the other hand, a CFD study was carried out to choose the most suitable locations for data collection, identifying and discarding locations that may provide non-representative and unreliable measurements due to the existence of vortices in the air flows, depending on the ventilation scenarios proposed (cross, outdoor, indoor-indirect). The study enabled the selection of a series of points that respond to the needs of CO_2 monitoring and control which go beyond those established in the legislation or those collected in previous research, as far as we are aware. Therefore, a methodology was proposed that helps to select the CO_2 control points according to the existing conditions in the indoor spaces under study.

The data collected wirelessly for interpretation could be evaluated on an Internet of Things (IoT) platform, in real time or deferred. As a result, IAQ could be controlled, interfacing IAQ devices with other systems (such as HVAC). The integration of this type of low-cost clonic devices into existing buildings allows regulation of ventilation, balancing safety through air renewal with comfort and energy efficiency. This ensures adequate sanitary conditions for the occupants while also favoring the optimization of energy resources. In this way, a smarter building was achieved, adding value to its refurbishment and modernization.

Future lines of research should focus on cataloging and parameterizing the location and distribution of the devices according to the behavior of indoor air in other scenarios and measuring the influence of this control on the comfort and energy sustainability of the building, since ventilation, as a means of natural air renewal, can exert a negative influence on energy efficiency when spaces are conditioned. Finally, to facilitate the replicability and comparability of this research, data analysis organised by spreadsheets is included as Supplementary Materials.

Supplementary Materials: The following are available online at https://www.mdpi.com/article/10.3390/app12083927/s1. Process of Calibration, Scenario 1: Cross-ventilation, Scenario 2: Outdoor-ventilation, Scenario 3: Indoor-ventilation, Scenario 4: No-ventilation.

Author Contributions: Conceptualization, A.P.-F. and A.C.-N.; methodology, A.P.-F., P.M.-G. and A.C.-N.; software, P.M.-G.; validation, A.P.-F., P.M.-G. and P.B.-P.; formal analysis, A.P.-F. and A.C.-N.; investigation, A.P.-F., A.C.-N. and M.O.-M.; resources, M.O.-M. and P.M.-G.; data curation, M.O.-M. and P.M.-G.; writing—original draft preparation, A.P.-F. and P.M.-G.; writing—review and editing, A.C.-N. and P.B.-P.; visualization, M.O.-M. and P.M.-G.; supervision, A.P.-F., A.C.-N. and P.B.-P.; project administration, A.P.-F. and A.C.-N.; funding acquisition, A.P.-F., A.C.-N. and M.O.-M. All authors have read and agreed to the published version of the manuscript.

Funding: This research received no external funding.

Institutional Review Board Statement: Not applicable.

Informed Consent Statement: Not applicable.

Data Availability Statement: All the data are included in the article (main text or Appendices A–C) or in Supplementary Materials.

Acknowledgments: All authors acknowledge the help received by the research group TEP-955 from the PAIDI, the ERGOMET Project of the Program for the Promotion of Research Activity of the UCA, the Project "Design of a low-cost non-invasive ergonomic capture system for the analysis of musculoskeletal disorders" of the Program for the Promotion of Research and Transfer of the UCA and the National Plan Research Project PID2019-108669RB-100/AEI/10.13039/501100011033.

Conflicts of Interest: The authors declare no conflict of interest.

Appendix A. Criteria Matrices for Components Selection

Appendix A.1. Sensor Selection

The factors considered for the choice of the sensor that suits the needs of the research are (which are weighted according to a scale from 0 (lowest score) to 2 (best score)):

- Measuring range: This is the range of values (in ppm) capable of being detected by the sensor.
- 2: Sensor capable of detecting any concentration of CO_2, even exceeding by far the limit values established by the standards and regulations in force. Upper limit value of the measuring range greater than 5000 ppm.
- 1: Sensor capable of detecting CO_2 concentrations below the limit values established by the standards and regulations in force. Upper limit value of the measuring range greater than 2000 ppm and smaller than 5000 ppm.

- 0: Sensor capable of detecting CO_2 concentration values of no more than 2000 ppm.
- Accuracy: This is the error of the sensor. This value is shown in the sensor datasheet, defining the accuracy with which the sensor performs the measurements. A higher precision will result in higher accuracy in the data collected.
- 2: Accuracy $\leq \pm 0.1$ (percent)
- 1: $\pm 0.1 <$ Accuracy $\leq \pm 1$
- 0: Accuracy $\geq \pm 1$
- Number of parameters: Monitoring more than one parameter is not desirable. In multi-gas or multi-parameter sensors, a laboratory calibration must be performed (although self-calibration is mentioned in the datasheet, this is obtained with a regression line, which does not necessarily will coincide with measurements by conventional devices already calibrated), because of the presence of several gases or different parameters in the same enclosure may influence data acquisition and interpretation.
- 2: Sensor capable of measuring a single gas.
- 1: Sensor capable of measuring between 2 and 4 different gases.
- 0: Sensor capable of measuring more than 5 gases.
- Response speed: This is the capacity to display changes in the values of the sensor output without delays depending on the variations of the input. It is interesting that the sensitivity is high, as it will show greater flexibility to measure sudden changes in CO_2.
- 2: Response speed $\leq$ 1 min.
- 1: 1 min < Response rate $\leq$ 2 min.
- 0: Response speed > 2 min.
- Warm-up time: This is the period of time during which the sensor does not take reliable measurements and adapts to the surrounding atmosphere. This is for ensuring the sensor takes measurements in accordance with its calibration line, waiting until the start of data acquisition. The shorter the warm-up time, the greater the autonomy, reliability and accuracy of the data collected.
- 2: The warm-up time is several minutes. Maximum 1 h.
- 1: The warm-up time is between 1 h and 8 h.
- 0: The warm-up time is longer than 8 h and can last at least 1 day.
- Price: This is one of the main objectives for creating an affordable device with the highest possible number of features.
- 2: Sensor priced between EUR 0–EUR 30.
- 1: Sensor with price between EUR 30–EUR 60.
- 0: Sensor priced over EUR 60.

The results are shown in Table A1:

Table A1. Criteria matrix for sensor selection.

Factor	T6713	SCD30	SEN0219	MiCS-VZ-89TE
Measuring Range	2	2	2	0
Accuracy	1	1	1	0
N° parameters	2	0	2	2
Response speed	1	2	1	2
Warm-up time	2	2	2	2
Price	0	0	1	2
Total	8	7	9	8

Appendix A.2. Motherboard Selection

The compilation of the different characteristics extracted, which are of interest for the elaboration of the corresponding criteria matrix, are shown in Table A2:

Table A2. Criteria matrix for sensor selection.

Characteristics	ELEGOO MEGA 2560 R3	Raspberry Pi 3 Model B+
Dimensions	101.52 × 53.3 mm	85 × 53 mm
Memory capacity	8 kb	1 Gb
Language	C, C++	Python, C, Java, Javascript, ...
Processor speed	16 MHz	1400 MHz
Inputs and outputs	54 digital I/O pins and 16 analog inputs	40 pins (known as GPIO)
Other connections	USB Type B	4 USB ports, HDMI, Internet, Memory card
Price	EUR 13.99	EUR 30–40

The factors considered for the choice of the motherboard encompass its main characteristics (mentioned above). In addition, factors related to the functionality and intended uses of the programmable device have also been added (the importance is weighted with values from 0 to 2, the latter being the best score depending on the factor considered):

- General characteristics of the board: Compact dimensions and low weight. Storage and autonomy.
- Software and hardware: Hardware easy to use and open-sourced programming language.
- Innovation: Although a completely new device is already being realised with both board options, it is desired that the device and programming platform provide functions that add value to the prototype.
- Interconnection: Suitable for direct connection of the SEN0219 sensor and other devices.
- Price: Competitive price/performance ratio.

The results are shown in Table A3:

Table A3. Criteria matrix for sensor selection.

Factor	ELEGOO MEGA 2560 R3	Raspberry Pi 3 Model B+
Characteristics	1	2
Software/Hardware	2	0
Innovation	2	2
Interconnection	2	2
Sensor integration	2	1
Price	2	2
Total	11	9

Appendix B. Sensor Connection

ELEGOO Mega 2560 R3 motherboard is an open source ATMega2560-ATMEGA16U2 microcontroller board with 54 digital input and output pins, 16 analogue inputs, USB connection, RESET button and external power input. When purchasing the ELEGOO MEGA board, the USB cable for connection via serial port with the computer where the sensor programming is to be carried out is included.

As the SEN0219 sensor is analogue, one of the analogue inputs of the board must be used. Specifically, the analogue input A0 is selected. This input could be any other input as long as it is reflected in the programming code (from A1 to A7). In the same vein, the ground (GND) and voltage pins are essential to receive data from the sensor. The SEN0219 sensor works with a voltage between 4.5 and 5.5 V, so the connection will be made to the 5 V (Power) connection pin on the motherboard.

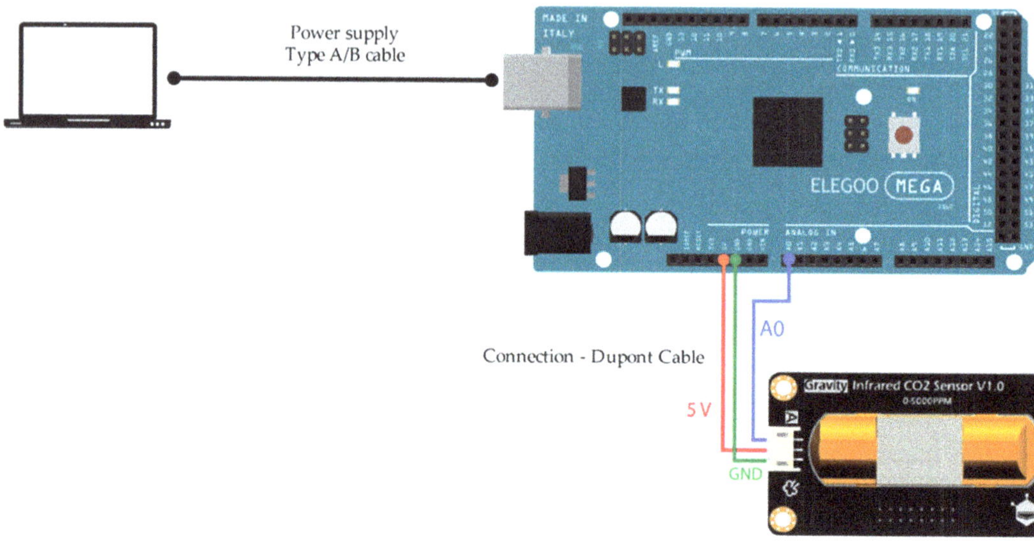

Figure A1. Wiring diagram between SEN0219 sensor and ELEGOO MEGA.

Appendix C. Programming Code in Arduino Software and Visual Studio Code

Appendix C.1. Arduino Base Code

```
int sensorIn = A0;
void setup(){
  Serial.begin(115200);
  analogReference(DEFAULT);
}
void loop(){
  //Lectura de la tensión
  int sensorValue = analogRead(sensorIn);
  // La señal analogica es convertida a tension
  float voltage = sensorValue*(5000/1024.0);
  if(voltage == 0)
  {
    Serial.println("Fault");
  }
  else if(voltage < 400)
  {
    Serial.println("preheating");
  }
  else
  {
    //Recta de regresion lineal utilizada para la calibracion del sensor
    float p_cero = 2.9713;
    float p_uno = −1150.3;
    float concentration=voltage*p_cero+p_uno;
    Serial.print(voltage);
    Serial.print(",");
    Serial.print(concentration);
    Serial.println(";");
  }
```

```
    delay(1000);
}
```

Appendix C.2. Visual Studio Code–Python Programming to Create Data Storage

```
import serial
import xlwt
from datetime import datetime
class SerialToExcel:
    def __init__(self,port,speed):
        self.port = port
        self.speed = speed
        self.wb = xlwt.Workbook()
        self.ws = self.wb.add_sheet("Data from Serial",cell_overwrite_ok = True)
        self.ws.write(0, 0, "Data from Serial")
        self.columns = ["Date Time"]
        self.number = 100
    def setColumns(self,col):
        self.columns.extend(col)
    def setRecordsNumber(self,number):
        self.number = number
    def readPort(self):
        ser = serial.Serial(self.port, self.speed, timeout = 1)
        c = 0
        for col in self.columns:
            self.ws.write(1, c, col)
            c = c + 1
        self.fila = 2
        i = 0
        while(i < self.number):
            line = str(ser.readline())
            if(len(line) > 0):
                now = datetime.now()
                date_time = now.strftime("%m/%d/%Y, %H:%M:%S")
                print(date_time,line)
                if(line.find(",")):
                    c = 1
                    self.ws.write(self.fila, 0, date_time)
                    columnas = line.split(",")
                    for col in columnas:
                        self.ws.write(self.fila, c, col)
                        c = c + 1
                i = i + 1
                self.fila = self.fila + 1
    def writeFile(self,archivo):
        self.wb.save(archivo)
```

Appendix C.3. Visual Studio Code—Data Storage

Define where the data search is performed (Arduino Serial Monitor), number of data to be collected, name of parameters collected according to Arduino programming and time between data storage.

```
from serialToExcel import SerialToExcel
serialToExcel = SerialToExcel("COM3",115200)
columnas = ["voltage","concentration"]
serialToExcel.setColumns(["voltage","concentration"])
```

```
serialToExcel.setRecordsNumber(1800)
serialToExcel.readPort()
serialToExcel.writeFile("archivo2.xls")
```

References

1. Huang, C.; Wang, Y.; Li, X.; Ren, L.; Zhao, J.; Hu, Y.; Zhang, L.; Fan, G.; Xu, J.; Gu, X.; et al. Clinical features of patients infected with 2019 novel coronavirus in Wuhan, China. *Lancet* **2020**, *395*, 497–506. [CrossRef]
2. World Health Organization. Available online: https://www.who.int/home (accessed on 17 February 2021).
3. Razzini, K.; Castrica, M.; Menchetti, L.; Maggi, L.; Negroni, L.; Orfeo, N.V.; Pizzoccheri, A.; Stocco, M.; Muttini, S.; Balzaretti, C.M. SARS-CoV-2 RNA detection in the air and on surfaces in the COVID-19 ward of a hospital in Milan, Italy. *Sci. Total Environ.* **2020**, *742*, 140540. [CrossRef] [PubMed]
4. Lolli, S.; Chen, Y.C.; Wang, S.H.; Vivone, G. Impact of meteorological conditions and air pollution on COVID-19 pandemic transmission in Italy. *Sci. Rep.* **2020**, *10*, 16213. [CrossRef] [PubMed]
5. Pascarella, G.; Strumia, A.; Piliego, C.; Bruno, F.; Del Buono, R.; Costa, F.; Scarlata, S.; Agrò, F.E. COVID-19 diagnosis and management: A comprehensive review. *J. Intern. Med.* **2020**, *288*, 192–206. [CrossRef] [PubMed]
6. *European Commission Council Recommendation (EU) 2021/119 of 1 February 2021 Amending Recommendation (EU) 2020/1475 on a Coordinated Approach to the Restriction of Free Movement in Response to the COVID-19 Pandemic*; Council of the European Union: Brussels, Belgium, 2018; Volume 2016, pp. 48–119.
7. Preparedness, E. *Commission Recommendation (EU) 2021/472 of 17 March 2021 on a Common Approach to Establish a Systematic Surveillance of SARS-CoV-2 and Its Variants in Wastewaters in the EU*; European Commission: Brussels, Belgium, 2021; Volume 1605690513, pp. 3–8.
8. The European Parliament and the Council of the European Union. *Regulation of the European Parliament and of the Council on a Framework for the Issuance, Verification and Acceptance of Interoperable Certificates on Vaccination, Testing and Recovery to Third-Country Nationals Legally Staying or Legally Residing in the territories of Member States during the COVID-19 pandemic (Digital Green Certificate)*; The European Parliament and the Council of the European Union: Brussels, Belgium, 2021.
9. Chu, D.K.; Akl, E.A.; Duda, S.; Solo, K.; Yaacoub, S.; Schünemann, H.J.; El-harakeh, A.; Bognanni, A.; Lotfi, T.; Loeb, M.; et al. Physical distancing, face masks, and eye protection to prevent person-to-person transmission of SARS-CoV-2 and COVID-19: A systematic review and meta-analysis. *Lancet* **2020**, *395*, 1973–1987. [CrossRef]
10. Adhikari, S.P.; Meng, S.; Wu, Y.J.; Mao, Y.P.; Ye, R.X.; Wang, Q.Z.; Sun, C.; Sylvia, S.; Rozelle, S.; Raat, H.; et al. Epidemiology, causes, clinical manifestation and diagnosis, prevention and control of coronavirus disease (COVID-19) during the early outbreak period: A scoping review. *Infect. Dis. Poverty* **2020**, *9*, 29. [CrossRef]
11. Setti, L.; Passarini, F.; De Gennaro, G.; Barbieri, P.; Perrone, M.G.; Borelli, M.; Palmisani, J.; Di Gilio, A.; Piscitelli, P.; Miani, A. Airborne transmission route of COVID-19: Why 2 meters/6 feet of inter-personal distance could not be enough. *Int. J. Environ. Res. Public Health* **2020**, *17*, 2932. [CrossRef]
12. Morawska, L.; Milton, D.K. Displacement ventilation: A viable ventilation strategy for makeshift hospitals and public buildings to contain COVID-19 and other airborne diseases. *Clin. Infect. Dis.* **2020**, *71*, 2311–2313. [CrossRef]
13. Morawska, L.; Tang, J.W.; Bahnfleth, W.; Bluyssen, P.M.; Boerstra, A.; Buonanno, G.; Cao, J.; Dancer, S.; Floto, A.; Franchimon, F.; et al. How can airborne transmission of COVID-19 indoors be minimised? *Environ. Int.* **2020**, *142*, 10583. [CrossRef]
14. Capolongo, S.; Rebecchi, A.; Buffoli, M.; Appolloni, L.; Signorelli, C.; Fara, G.M.; D'Alessandro, D. COVID-19 and cities: From urban health strategies to the pandemic challenge: A decalogue of public health opportunities. *Acta Biomed.* **2020**, *91*, 13–22. [CrossRef]
15. Kampf, G.; Todt, D.; Pfaender, S.; Steinmann, E. Persistence of coronaviruses on inanimate surfaces and their inactivation with biocidal agents. *J. Hosp. Infect.* **2020**, *104*, 246–251. [CrossRef] [PubMed]
16. Przekwas, A.; Chen, Z. Washing hands and the face may reduce COVID-19 infection. *Med. Hypotheses* **2020**, *144*, 110261. [CrossRef] [PubMed]
17. Jayaweera, M.; Perera, H.; Gunawardana, B.; Manatunge, J. Transmission of COVID-19 virus by droplets and aerosols: A critical review on the unresolved dichotomy. *Environ. Res.* **2020**, *188*, 109819. [CrossRef]
18. Chaudhuri, S.; Basu, S.; Kabi, P.; Unni, V.R.; Saha, A. Modeling the role of respiratory droplets in COVID-19 type pandemics. *Phys. Fluids* **2020**, *32*, 063309. [CrossRef]
19. Kohanski, M.A.; Lo, L.J.; Waring, M.S. Review of indoor aerosol generation, transport, and control in the context of COVID-19. *Int. Forum Allergy Rhinol.* **2020**, *10*, 1173–1179. [CrossRef] [PubMed]
20. Azuma, K.; Yanagi, U.; Kagi, N.; Kim, H.; Ogata, M.; Hayashi, M. Environmental factors involved in SARS-CoV-2 transmission: Effect and role of indoor environmental quality in the strategy for COVID-19 infection control. *Environ. Health Prev. Med.* **2020**, *25*, 1–16. [CrossRef]
21. Tan, Z.P.; Silwal, L.; Bhatt, S.P.; Raghav, V. Experimental characterization of speech aerosol dispersion dynamics. *Sci. Rep.* **2021**, *11*, 66. [CrossRef] [PubMed]
22. Kampa, M.; Castanas, E. Human health effects of air pollution. *Environ. Pollut.* **2008**, *151*, 362–367. [CrossRef]
23. *ANSI/ASHRAE Standard 62.1-2019*; Ventilation for Acceptable Indoor Air Quality. American Society of Heating, Refrigerating and Air-Conditioning Engineers: Peachtree Corners, GA, USA, 2019; ISBN 9783642253874.

24. ASHRAE. *Indoor Air Quality: Best Practices for Design, Construction, and Commisionning*; American Society of Heating, Refrigerating and Air-Conditioning Engineers: Peachtree Corners, GA, USA, 2009; ISBN 978-1-933742-59-5.
25. *ISO 16813:2006*; Building Environment Design Indoor Environment. General Principles. International Organization for Standardization: Geneva, Switzerland, 2006.
26. *BS ISO 16814:2008*; Building Environment Design—Indoor Air Quality—Methods of Expressing the Quality of Indoor Air for Human Occupancy. British Standards Institution: London, UK, 2008; p. 66.
27. *EN ISO 20988:2007*; Air Quality—Guidelines for Estimating Measurement\Nuncertainty. International Organization for Standardization: Geneva, Switzerland, 2007.
28. *ISO 16000-40:2019*; Indoor Air—Part 40: Indoor Air Quality Management System. International Organization for Standardization: Geneva, Switzerland, 2019.
29. American Society of Heating, Refrigerating and Air-Conditioning Engineers. *ASHRAE Handbook-HVAC Applications*; American Society of Heating, Refrigerating and Air-Conditioning Engineers: Peachtree Corners, GA, USA, 2019; ISBN 978-1947192133.
30. The European Environment Agency. Air Index EEA Europa. Available online: https://airindex.eea.europa.eu/Map/AQI/Viewer/ (accessed on 26 November 2021).
31. The European Space Agency. *Eur. Air Qual. Astron. Geophys.* **2018**, *59*, 1.5. [CrossRef]
32. Peral, J. The composition of air: The first scientific data. *100cias@uned* **2008**, *12*, 133–139.
33. Stazi, F.; Naspi, F.; Ulpiani, G.; Di Perna, C. Indoor air quality and thermal comfort optimization in classrooms developing an automatic system for windows opening and closing. *Energy Build.* **2017**, *139*, 732–746. [CrossRef]
34. Xie, A.; Skatrud, J.B.; Dempsey, J.A. Effect of hypoxia on the hypopnoeic and apnoeic threshold for CO_2 in sleeping humans. *J. Physiol.* **2001**, *535*, 269–278. [CrossRef] [PubMed]
35. Poulin, M.J.; Cunningham, D.A.; Paterson, D.H.; Kowalchuk, J.M.; Smith, W.D.F. Ventilatory sensitivity to CO_2 in hyperoxia and hypoxia in older aged humans. *J. Appl. Physiol.* **1993**, *75*, 2209–2216. [CrossRef] [PubMed]
36. Novakova, P.; Kraus, M. Carbon Dioxide Concentration in the Bedroom for Various Natural Ventilation Modes. *IOP Conf. Ser. Mater. Sci. Eng.* **2019**, *603*, 052100. [CrossRef]
37. Saad, S.M.; Shakaff, A.Y.M.; Saad, A.R.M.; Yusof, A.M.; Andrew, A.M.; Zakaria, A.; Adom, A.H. Development of indoor environmental index: Air quality index and thermal comfort index. *AIP Conf. Proc.* **2017**, *1808*, 0200. [CrossRef]
38. Kang, J.; Hwang, K.-I. A Comprehensive Real-Time Indoor Air-Quality Level Indicator. *Sustainability* **2016**, *8*, 881. [CrossRef]
39. Domínguez-Amarillo, S.; Fernández-Agüera, J.; Cesteros-García, S.; González-Lezcano, R.A. Bad air can also kill: Residential indoor air quality and pollutant exposure risk during the covid-19 crisis. *Int. J. Environ. Res. Public Health* **2020**, *17*, 7183. [CrossRef]
40. Greenhalgh, T.; Jimenez, J.L.; Prather, K.A.; Tufekci, Z.; Fisman, D.; Schooley, R. Ten scientific reasons in support of airborne transmission of SARS-CoV-2. *Lancet* **2021**, *6736*, 2–4. [CrossRef]
41. Liu, Z.; Ciais, P.; Deng, Z.; Lei, R.; Davis, S.J.; Feng, S.; Zheng, B.; Cui, D.; Dou, X.; Zhu, B.; et al. Near-real-time monitoring of global CO_2 emissions reveals the effects of the COVID-19 pandemic. *Nat. Commun.* **2020**, *11*, 5172. [CrossRef]
42. Pantelic, J.; Liu, S.; Pistore, L.; Licina, D.; Vannucci, M.; Sadrizadeh, S.; Ghahramani, A.; Gilligan, B.; Sternberg, E.; Kampschroer, K.; et al. Personal CO_2 cloud: Laboratory measurements of metabolic CO_2 inhalation zone concentration and dispersion in a typical office desk setting. *J. Expo. Sci. Environ. Epidemiol.* **2020**, *30*, 328–337. [CrossRef]
43. AENOR. *Enery Performance of Buildings-Ventilation for Buldings—Part 2: Interpretation of the Requiirements in EN 16798-1. Indoor Enviromental Input Parameters for Desing and Assessment of Energy Performance of Buildings Addressing Indoor Air Quality, Thermal Environment, Lighting and Acoustics (Module M1-6)*; AENOR: Madrid, Spain, 2019.
44. Kapalo, P.; Mečiarová, L'.; Vilčeková, S.; Krídlová Burdová, E.; Domnita, F.; Bacotiu, C.; Péterfi, K.E. Investigation of CO_2 production depending on physical activity of students. *Int. J. Environ. Health Res.* **2019**, *29*, 31–44. [CrossRef] [PubMed]
45. Yang, L.; Wang, X.; Li, M.; Zhou, X.; Liu, S.; Zhang, H.; Arens, E.; Zhai, Y. Carbon dioxide generation rates of different age and gender under various activity levels. *Build. Environ.* **2020**, *186*, 107317. [CrossRef]
46. Soares, A.; Catita, C.; Silva, C. Exploratory research of CO_2, noise and metabolic energy expenditure in Lisbon commuting. *Energies* **2020**, *13*, 861. [CrossRef]
47. Peng, Z.; Jimenez, J.L. Exhaled CO_2 as a COVID-19 infection risk proxy for different indoor environments and activities. *Environ. Sci. Technol. Lett.* **2021**, *8*, 392–397. [CrossRef]
48. Kissler, S.M.; Tedijanto, C.; Goldstein, E.M.; Grad, Y.H.; Lipsitch, M. Projecting the transmission dynamics of SARS-CoV-2 through the post-pandemic period. *medRxiv* **2020**, *868*, 860–868. [CrossRef]
49. Schieweck, A.; Uhde, E.; Salthammer, T.; Salthammer, L.C.; Morawska, L.; Mazaheri, M.; Kumar, P. Smart homes and the control of indoor air quality. *Renew. Sustain. Energy Rev.* **2018**, *94*, 705–718. [CrossRef]
50. United Nations. Transforming Our World: The 2030 Agenda for Sustainable Development. Available online: https://sdgs.un.org/2030agenda (accessed on 10 January 2022).
51. Stockwell, R.E.; Ballard, E.L.; O'Rourke, P.; Knibbs, L.D.; Morawska, L.; Bell, S.C. Indoor hospital air and the impact of ventilation on bioaerosols: A systematic review. *J. Hosp. Infect.* **2019**, *103*, 175–184. [CrossRef]
52. Stabile, L.; Buonanno, G.; Frattolillo, A.; Dell'Isola, M. The effect of the ventilation retrofit in a school on CO_2, airborne particles, and energy consumptions. *Build. Environ.* **2019**, *156*, 1–11. [CrossRef]

53. Colton, M.D.; Macnaughton, P.; Vallarino, J.; Kane, J.; Bennett-Fripp, M.; Spengler, J.D.; Adamkiewicz, G. Indoor air quality in green vs conventional multifamily low-income housing. *Environ. Sci. Technol.* **2014**, *48*, 7833–7841. [CrossRef]
54. Rickenbacker, H.J.; Vaden, J.M.; Bilec, M.M. Engaging Citizens in Air Pollution Research: Investigating the Built Environment and Indoor Air Quality and Its Impact on Quality of Life. *J. Archit. Eng.* **2020**, *26*, 04020041. [CrossRef]
55. Wang, Y.; Qiao, F.; Zhou, F.; Yuan, Y. Surface Distribution of Severe Acute Respiratory Syndrome Coronavirus 2 in Leishenshan Hospital. *Indoor Built Environ.* **2020**, 1–9. [CrossRef]
56. Fears, A.C.; Klimstra, W.B.; Duprex, P.; Hartman, A.; Weaver, S.C.; Plante, K.S.; Mirchandani, D.; Plante, J.A.; Aguilar, P.V.; Fernández, D.; et al. Persistence of Severe Acute Respiratory Syndrome Coronavirus 2 in Aerosol Suspensions. *Emerg. Infect. Dis.* **2020**, *26*, 2168–2171. [CrossRef] [PubMed]
57. Xu, Y.; Raja, S.; Ferro, A.R.; Jaques, P.A.; Hopke, P.K.; Gressani, C.; Wetzel, L.E. Effectiveness of heating, ventilation and air conditioning system with HEPA filter unit on indoor air quality and asthmatic children's health. *Build. Environ.* **2010**, *45*, 330–337. [CrossRef]
58. Muller, C. Beyond ozone: Cleaning outdoor air for improved IAQ 201. In Proceedings of the Air and Waste Management Association's Annual Conference and Exhibition, Orlando, FL, USA, 24–26 June 2011; AWMA: Durham, NC, USA, 2011; Volume 3, pp. 2653–2662.
59. Lee, S.C.; Guo, H.; Li, W.M.; Chan, L.Y. Inter-comparison of air pollutant concentrations in different indoor environments in Hong Kong. *Atmos. Environ.* **2002**, *36*, 1929–1940. [CrossRef]
60. Vardoulakis, S.; Giagloglou, E.; Steinle, S.; Davis, A.; Sleeuwenhoek, A.; Galea, K.S.; Dixon, K.; Crawford, J.O. Indoor exposure to selected air pollutants in the home environment: A systematic review. *Int. J. Environ. Res. Public Health* **2020**, *17*, 8972. [CrossRef]
61. Huynh, C.K. Building energy saving techniques and indoor air quality—A dilemma. *Int. J. Vent.* **2010**, *9*, 93–98. [CrossRef]
62. Wang, B.; Malkawi, A. Design-based natural ventilation evaluation in early stage for high performance buildings. *Sustain. Cities Soc.* **2019**, *45*, 25–37. [CrossRef]
63. Stabile, L.; Dell'Isola, M.; Frattolillo, A.; Massimo, A.; Russi, A. Effect of natural ventilation and manual airing on indoor air quality in naturally ventilated Italian classrooms. *Build. Environ.* **2016**, *98*, 180–189. [CrossRef]
64. Elhadary, M.I.; Alzahrani, A.M.Y.; Aly, R.M.H.; Elboshy, B. A comparative study for forced ventilation systems in industrial buildings to improve the workers' thermal comfort. *Sustainability* **2021**, *13*, 10267. [CrossRef]
65. Ji, W.; Chen, C.; Zhao, B. A comparative study of the effects of ventilation-purification strategies on air quality and energy consumption in Beijing, China. *Build. Simul.* **2021**, *14*, 813–825. [CrossRef]
66. Kim, J.T.; Yu, C.W.F. Sustainable development and requirements for energy efficiency in buildings—The Korean perspectives. *Indoor Built Environ.* **2018**, *27*, 734–751. [CrossRef]
67. Guyot, G.; Sherman, M.H.; Walker, I.S. Smart ventilation energy and indoor air quality performance in residential buildings: A review. *Energy Build.* **2018**, *165*, 416–430. [CrossRef]
68. Sudhakar, K.; Winderla, M.; Priya, S.S. Net-zero building designs in hot and humid climates: A state-of-art. *Case Stud. Therm. Eng.* **2019**, *13*, 100400. [CrossRef]
69. Feng, W.; Zhang, Q.; Ji, H.; Wang, R.; Zhou, N.; Ye, Q.; Hao, B.; Li, Y.; Luo, D.; Lau, S.S.Y. A review of net zero energy buildings in hot and humid climates: Experience learned from 34 case study buildings. *Renew. Sustain. Energy Rev.* **2019**, *114*, 109303. [CrossRef]
70. Conceição, E.Z.E.; Lúcio, M.M.J.R. Air quality inside a school building: Air exchange monitoring, evolution of carbon dioxide and assessment of ventilation strategies. *Int. J. Vent.* **2006**, *5*, 259–270. [CrossRef]
71. Conceição, E.Z.E.; Farinho, J.P.; Lúcio, M.M.J.R. Evaluation of indoor air quality in classrooms equipped with cross-flow ventilation. *Int. J. Vent.* **2012**, *11*, 53–67. [CrossRef]
72. Sun, Y.; Hou, J.; Cheng, R.; Sheng, Y.; Zhang, X.; Sundell, J. Indoor air quality, ventilation and their associations with sick building syndrome in Chinese homes. *Energy Build.* **2019**, *197*, 112–119. [CrossRef]
73. Hou, J.; Zhang, Y.; Sun, Y.; Wang, P.; Zhang, Q.; Kong, X.; Sundell, J. Air change rates at night in northeast Chinese homes. *Build. Environ.* **2018**, *132*, 273–281. [CrossRef]
74. *ANSI/ASHRAE Standard 62.1*; The Standards for Ventilation and Indoor Air Quality. ASHRAE: Peachtree Corners, GA, USA, 2019.
75. *ANSI/ASHRAE Standard 62.2*; Ventilation and Acceptable Indoor Air Quality in Residential Buildings. ASHRAE: Peachtree Corners, GA, USA, 2019.
76. Ministerio de la Presidencia de España. *Real Decreto 1027/2007: Regulation of Thermal Installations Thermal Installations in Buildings*; BOE: Madrid, Spain, 2007; pp. 35931–35984.
77. Villanueva, F.; Jiménez, E.; Felisi, J.-M.; Garrido, T.; Fiménez, J.-L.; Ródenas, M.; Muñoz, A. Guide about Affordable CO2 detectors for COVID-19 Prevention. 2021. Available online: https://bit.ly/medidoresCO2Objective (accessed on 2 November 2021).
78. Saini, J.; Dutta, M.; Marques, G. Indoor Air Quality Monitoring Systems Based on Internet of Things: A Systematic review. *Int. J. Environ. Res. Public Health* **2020**, *14*, 4942. [CrossRef]
79. Marques, G.; Saini, J.; Dutta, M.; Kumar Singh, P.; Hong, W.-C. Indoor Air Quality Monitoring Systems for Enhanced Living Environments: A Review toward Sustainable Smart Cities. *Sustainability* **2020**, *12*, 4024. [CrossRef]
80. Zhang, H.; Srinivasan, R. A Systematic Review of Air Quality Sensors, Guidelines, and Measurement Studies for Indoor Air Quality Management. *Sustainability* **2020**, *12*, 9045. [CrossRef]

81. Chiesa, G.; Cesari, S.; Garcia, M.; Issa, M.; Li, S. Multisensor IoT platform for optimising IAQ levels in buildings through a smart ventilation system. *Sustainability* **2019**, *11*, 5777. [CrossRef]
82. Sun, S.; Zheng, X.; Villalba-Díez, J.; Ordieres-Meré, J. Indoor air-quality data-monitoring system: Long-term monitoring benefits. *Sensors* **2019**, *19*, 4157. [CrossRef] [PubMed]
83. Benammar, M.; Abdaoui, A.; Ahmad, S.H.M.; Touati, F.; Kadri, A. A modular IoT platform for real-time indoor air quality monitoring. *Sensors* **2018**, *18*, 581. [CrossRef] [PubMed]
84. Marques, G.; Pitarma, R. A cost-effective air quality supervision solution for enhanced living environments through the internet of things. *Electronics* **2019**, *8*, 170. [CrossRef]
85. Marques, G.; Ferreira, C.R.; Pitarma, R. Indoor Air Quality Assessment Using a CO_2 Monitoring System Based. *J. Med. Syst.* **2019**, *43*, 67. [CrossRef]
86. Hsu, C.-N.; Tsai, Y.-L. Experimental Measurement and Computational Simulation Analysis of Indoor Air Quality in Office—Integration of Voltage Adsorption Dust Collection Device and Energy Recovery Ventilator. *Sens. Mater.* **2020**, *32*, 4299–4321. [CrossRef]
87. Omidvarborna, H.; Kumar, P.; Hayward, J.; Gupta, M.; Nascimento, E.G.S. Low-cost air quality sensing towards smart homes. *Atmosphere* **2021**, *12*, 453. [CrossRef]
88. Dong, B.; Prakash, V.; Feng, F.; O'Neill, Z. A review of smart building sensing system for better indoor environment control. *Energy Build.* **2019**, *199*, 29–46. [CrossRef]
89. Cao, S.J.; Ding, J.; Ren, C. Sensor deployment strategy using cluster analysis of Fuzzy C-Means Algorithm: Towards online control of indoor environment's safety and health. *Sustain. Cities Soc.* **2020**, *59*, 102190. [CrossRef]
90. Yang, L.; Ye, M.; He, B.-J. CFD simulation research on residential indoor air quality. *Sci. Total Environ.* **2014**, *472*, 1137–1144. [CrossRef] [PubMed]
91. Bulińska, A.; Buliński, Z. A CFD analysis of different human breathing models and its influence on spatial distribution of indoor air parameters. *Comput. Assist. Methods Eng. Sci.* **2015**, *22*, 213–227.
92. Hyndman, R.J.; Koehler, A.B. Another look at measures of forecast accuracy. *Int. J. Forecast.* **2006**, *22*, 679–688. [CrossRef]
93. Karamirad, M.; Omid, M.; Alimardani, R.; Mousazadeh, H. ANN based simulation and experimental verification of analytical four- and five-parameters models of PV modules. *Simul. Model. Pract. Theory* **2013**, *34*, 86–98. [CrossRef]
94. Gaiser, T.; Barros, I.D.e.; Sereke, F.; Lange, F. Validation and reliability of the EPIC model to simulate maize production in small-holder farming systems in tropical sub-humid West Africa and semi-arid Brazil. *Agric. Ecosyst. Environ.* **2010**, *135*, 318–327. [CrossRef]
95. Wu, X.; Zhu, X.; Cao, G.; Tu, H. Dynamic modeling of SOFC based on a T–S fuzzy model. *Simul. Model. Pract. Theory* **2008**, *16*, 494–504. [CrossRef]
96. Bergmeir, C.; Benítez, J.M. On the use of cross-validation for time series predictor evaluation. *Inf. Sci.* **2012**, *191*, 192–213. [CrossRef]
97. Liu, H.; Tian, H.; Chen, C.; Li, Y. A hybrid statistical method to predict wind speed and wind power. *Renew. Energy* **2010**, *35*, 1857–1861. [CrossRef]
98. Wibowo, T.C.S.; Saad, N. MIMO model of an interacting series process for Robust MPC via System Identification. *ISA Trans.* **2010**, *49*, 335–347. [CrossRef]
99. Edwards, L.J.; Muller, K.E.; Wolfinger, R.D.; Qaqish, B.F.; Schabenberger, O. An R2 statistic for fixed effects in the linear mixed model. *Stat. Med.* **2008**, *27*, 6137–6157. [CrossRef]
100. Chicco, D.; Warrens, M.J.; Jurman, G. The coefficient of determination R-squared is more informative than SMAPE, MAE, MAPE, MSE and RMSE in regression analysis evaluation. *PeerJ Comput. Sci.* **2021**, *7*, e623. [CrossRef] [PubMed]
101. De Myttenaere, A.; Golden, B.; Le Grand, B.; Rossi, F. Mean Absolute Percentage Error for regression models. *Neurocomputing* **2016**, *192*, 38–48. [CrossRef]
102. Sagheer, A.; Kotb, M. Unsupervised Pre-training of a Deep LSTM-based Stacked Autoencoder for Multivariate Time Series Forecasting Problems. *Sci. Rep.* **2019**, *9*, 19038. [CrossRef] [PubMed]
103. Government of Spain. RD 102/2011 of 28th January 2011 on the improvement of air quality. *Bol. Estado* **2011**, *25*, 9574–9626.
104. *UNE 171330-2*; Indoor Environmental Quality. Part 2: Indoor Environmental Quality Inspection Procedures. AENOR: Madrid, Spain, 2014.

applied sciences

MDPI

Article

A New Approach to Assess the Built Environment Risk under the Conjunct Effect of Critical Slow Onset Disasters: A Case Study in Milan, Italy

Juan Diego Blanco Cadena [1], Nicola Moretti [1,†], Graziano Salvalai [1], Enrico Quagliarini [2,*], Fulvio Re Cecconi [1] and Tiziana Poli [1]

1 ABC Department, Politecnico di Milano, 20133 Milan, Italy; juandiego.blanco@polimi.it (J.D.B.C.); nicola.moretti@polimi.it (N.M.); graziano.salvalai@polimi.it (G.S.); fulvio.rececconi@polimi.it (F.R.); tiziana.poli@polimi.it (T.P.)

2 DICEA Department, Università Politecnica Delle Marche, 60131 Ancona, Italy

* Correspondence: e.quagliarini@staff.univpm.it

† Current address: Institute for Manufacturing, Department of Engineering, University of Cambridge, 17 Charles Babbage Road, Cambridge CB3 0FS, UK.

Abstract: Citizens in dense built environments are susceptible to the simultaneous occurrence of Slow Onset Disaster (SLOD) events, being particularly prone to increasing temperatures and air pollution. Previous research works have assessed these events' arousal separately and have identified when their intensity is critical. However, few have integrated their analysis, possibly limited by the quality and granularity of available data, the accessibility and distribution of sensors, and measurements not emulating the surroundings of a pedestrian. Thus, this work performed an outdoor meso-scale multi-hazard-based risk analysis to study the aggregated effects of the SLODs mentioned above. The study was carried out to narrow down the time-frames within 2019 in which these two events could have affected citizens' health the most. A weighted fuzzy logic was applied to superimpose climatic (temperature, humidity, wind speed, and solar irradiance) and air quality (particulate matter, ozone, and ammonium) distress (true risk) on an hourly basis, allocated using set healthy and comfortable ranges for a specific dense urban climate context within Milan (Italy), processing data from Milano via Juvara station. The findings show that sensitive groups were at risk of high temperature and pollution separately during 26% and 29% of summer and mid-season hours, respectively; while multi-hazard risk would arise during 10.93% of summer and mid-season hours, concentrated mainly between 14:00 and 20:00.

Keywords: risk assessment; climate change; health; heat stress; AQI

Citation: Blanco Cadena, J.D.; Moretti, N.; Salvalai, G.; Quagliarini, E.; Re Cecconi, F.; Poli, T. A New Approach to Assess the Built Environment Risk under the Conjunct Effect of Critical Slow Onset Disasters: A Case Study in Milan, Italy. *Appl. Sci.* **2021**, *11*, 1186. https://doi.org/10.3390/app11031186

Academic Editor: Asterios Bakolas
Received: 11 January 2021
Accepted: 25 January 2021
Published: 28 January 2021

1. Introduction

The United Nations [1] reported in 2018 that 55% of the world's population was already living in highly urbanized areas and that this ratio is projected to grow to 68% by 2050. These projections for the urban population, coupled with the projected health threats identified by the World Health Organization (WHO) [2,3] due to climate change, foresee that around 6.7 billion people are at great risk. Hence, identifying the frequency, intensity, and extent of the hazard risk of these threats has become a pillar in the world-wide effort to hold back the consequences of climate change, reflected by the Sustainable Development Goals and in particular by the following three topics: good health and wellbeing, sustainable cities and communities, and climate action [4].

In 2014, the WHO analyzed and projected the effects of climate change on health, identifying heat-related mortality as among the foreseen death risks of the largest impact [2]. Later on, in 2016, the WHO also reported the extent of air pollutants' concentration and their severe potential effect on the decay of health condition and the increase in mortality. For instance, the WHO [2] estimated that by 2030, more than 92,000 additional heat-related deaths are expected; and more than 255,000 for 2050 if no measures are taken. Furthermore, Lee and Kim [5] studied the case of South Korea, comparing the 1992 to 2010 period with the 2090s projections; resulting in a 4× or 6× increase in temperature-related mortality. On the other hand, for air pollutants' concentration, Jackson et al. [6] presented a case study in Washington, in which it was reported that for people over 45 years old, there was an increase of 0.46–1.50% in the risk of mortality for every 10 parts per billion (ppb) increase in eight hour average ozone concentrations.

Climate change itself is an active and slow process, and within this context, increasing temperatures and higher air pollution have been classified in the same way. Both have been allocated into the definition of Slow-Onset Disasters (SLODs) [7], which are easy to perceive, and predict, but harder to mitigate. The arousal frequency and intensity of SLODs' evidence is larger in the urbanized context [8], and this evidence can be worsened given the inherent properties of the Built Environment (BE). A recent report by the Urban Climate Change Research Network [9] shows that mean annual temperatures in 39 cities around the world have increased at a rate of 0.12 to 0.45 °C per decade over the 1961 to 2010 time period. Furthermore, a recently audited air pollution report [10], which surveyed more than 4300 cities worldwide in 2015, concluded that only 20% of the urban population was reported to live in areas that comply with the WHO's established unhealthy concentrations of $PM_{2.5}$. The average particulate air pollution levels in these cities were found to be 4–15-times higher than the WHO air quality guideline limit levels. Dense urban areas increase the risk of the two above-mentioned SLODs for the following main reasons: (1) reduced evaporation, transpiration, and shading, due to limited green areas and disadvantageous geometry; (2) increased surface temperatures with high thermal capacity and/or low albedo; and (3) increased air stagnation due to diminished wind speed [11–15], thus making every portion of the city (e.g., neighborhood) react/provide a different meso-climate to the people within it.

Extensive research has been dedicated to studying these variations with different approaches, at different scales, and with different tools. Udristioiu et al. [16] and Sassi and Fourri [17] are recent examples on how to study the satellite remote sensing data, gathering information on Land Surface Temperature (LST) from infrared radiation measures. Other case studies, such as the one presented by Khamchiangta et al. [18], have used large geographical data to study the land use land coverage and correlated these conditions to estimate the possible response to undisturbed climatic readings (e.g., weather data); also, these databases have been combined with geographical tools (e.g., Geographic information Systems (GISs)) and small-scale on-site measurements to communicate actual point-in-time conditions [19–21].

However, these analyses are limited by the availability, quality, granularity, and applicability of data describing what is the real perception of a pedestrian, or a building occupant, as his/her surroundings can drastically modify the perception of climatic and air quality distress [22,23]. Moreover, integrated and multi-hazard analysis at the meso-scale is hampered by the distribution of sensors and the fact that, mostly, the data collection location does not always gather more than one parameter category (e.g., weather and pollutants), which makes the superposition of the effects hardly applicable.

Therefore, this work intends to propose and perform a preliminary multi-hazard risk analysis of northern Italy (i.e., Milan), using available open-source climate and air quality databases supported by the Lombardy Regional Environmental Protection Agency (ARPA Lombardia) [24]. Milan was considered a relevant location where a large population density is found (large risk exposure) and both considerable increasing temperatures and the decay of air quality and the increase of pollution converge (large risk hazard), based on the preliminary assessment analysis carried out by Salvalai et al. [25] using demographic data from the Italian National Institute of Statistics (ISTAT) [26] and historical temperature and particulate matter measurements from EEA [27,28].

In particular, the study presented in this paper is concentrated on assessing the overall increasing temperatures and air pollution SLODs' risk frequency for the least resilient population: demographic groups that are more prone to suffer severe health consequences when exposed to adverse environmental conditions. Moreover, critical hazard risk timeframes (including daily profiles) are established to acknowledge and communicate those moments of the year in which people are and probably will continue to be critically at risk of their health being affected if exposed to these conditions.

2. Materials and Methods

This work intends to propose a replicable procedure at the city and neighborhood scale, based on available and accessible data repositories (i.e., open data).

2.1. Temperature-Related Risk RiskT

There are few indexes or metrics that integrate the weather parameters with ease to understand the proper sensation of people outdoors. Most of them require rather extensive on-site surveys to feed the model with several measured quantities, or to feed the virtual model for lengthy computer simulations. Fortunately, indexes such as the Universal Thermal Climate Index ($UTCI$) [29] can be used in a simplified manner, which requires rather commonly surveyed parameters to estimate the pedestrian's sensation of his/her surroundings; that is:

- Dry bulb air temperature (t_{db-air})
- Mean Radiant Temperature (MRT)
- Relative Humidity (RH)
- Air/wind velocity (V_a)

However, for a preliminary and large-/meso-scale outdoor human-centered analysis, there is limited information on where the pedestrian is or will be located and how he/she is or will be exposed to solar and emitted surface-heat radiation. Thus, computing spatial-location variables, such as MRT and spot-specific V_a, goes beyond the scope of this study. Moreover, established simplifications for MRT estimations (as the ones listed in ASHRAE 55-2017 [30] for an indoor assessment) cannot be applied given the exposure to direct solar radiation, high registered V_a, high variance of the type of pedestrian activity, and the clothing worn.

Therefore, for surveyed data obtained from open-source data such as national, regional, or local weather station network repositories, another approach is needed to anticipate and identify potential moments during the year in which pedestrians can be more susceptible to risk. A susceptible temperature-related risk is then interpreted as the conjoint frequency of undesirable conditions of t_{db-air}, solar radiation (I_{tot}), RH, and V_a.

To do so, a simple Boolean and weighting analysis was performed following fuzzy logic; that is, for every hour of the year, the condition of each parameter was verified; if so, this was allocated a certain risk weight according to the parameter's influence on thermal stress, and all the weights were summed for every single Hour Of the Year (HOY) (see Equation (1)).

$$RiskT_{HOY} = \left(T_{bool} * T_{weight} + I_{bool} * I_{weight} + RH_{bool} * RH_{weight} + V_{bool} * V_{weight}\right)_{HOY} \quad (1)$$

2.1.1. Undesired Boolean Conditions

The Boolean is considered true (i.e., $i_{bool} = 1$) when certain weather conditions fall outside the comfort ranges. Based on indoor and outdoor comfort assessment metrics or indexes, these true ranges were established as:

- t_{db-air} is recorded above 26 °C (only heat stress (T) is considered);
- I_{tot} is measured over 300 W/m^2;
- RH is registered outside a 30–70% range; and
- V_a speed is below 2 m/s.

An additional condition was included to avoid false positive results, reporting heat-related distress when temperatures are below 18 °C. Hence, the Boolean shall be considered negative (i.e., $T_{bool} = -1$).

2.1.2. Weighting Factors

Based on the effect of the parameter's significance on outdoor thermal stress perception, factors summing to 1 were allocated, giving larger values to those directly related to sky heat exchange (i.e., T_{db-air} and I_{tot}) [22]. Further development of the research work includes a fine-tuning of these weights:

- for temperature, $T_{weight} = 0.4$ is used, given that the temperature is perhaps the most relevant parameter;
- for radiation, $I_{weight} = 0.3$ is used, given that direct solar radiation falling on a body can significantly alter its heat exchange; and
- for both air humidity and wind speed, $RH_{weight} = V_{weight} = 0.15$ is given, as they can intensify or alleviate one's perception of the previous conditions.

2.1.3. Comparison with the Simplified *UTCI* Calculation

To validate the suitability of the assessment process and the weight allocation, the values estimated were compared to the results obtained with the simplified outdoor thermal comfort *UTCI* calculation.

UTCI values were computed using the UTCI Python code produced by Ladybug Tools [22] based on the Fortran code provided by Brode et al. [29], but considering that $t_{db-air} \approx MRT$. The validation was considered successful based on the difference in the number of hours classified as at risk of high thermal stress, and the accuracy was computed using a classification-accuracy assessment (see Equation (2)).

$$Accuracy = \frac{TP + TN}{TP + TN + FP + FN} \tag{2}$$

where each of the following variables is:

- TP the number of True Positive values;
- TN the number of True Negative values;
- FP the number of False Positive values; and
- FN the number of False Negative values.

Heat distress was set to be expected, for sensitive groups, when:

- $RiskT > 0.5$
- $UTCI > 26$ °C

These weights, ranges, and thresholds were set based on the regulations enforced in the case study analyzed, the available databases, and the assumed pedestrian condition within a narrow urban canyon (i.e., shaded and wind protected).

2.2. *Pollution-Related Risk*

The Air Quality Index (AQI) was used as an established and widely used air pollution metric [31]; also, it can be estimated with ease based on the data gathered at air quality weather stations. The metric is based on the concentration of the main health affect-related pollutants; and also, they can be compared in a unique scale. When more than one pollutant is studied, the communicated AQI is the largest value.

The AQI calculation is based on Equation (3), which depends on the current pollutant concentration (C_p) and the data collected in Table 1; which contains the pollutant concentration Break Points (BP_{Hi} and BP_{Lo}) and the AQI range limits (I_{Hi} and I_{Lo}). The pollutant concentrations are normally evaluated from an 8 hour (for O_3 and NH_3) or 24-h exposure (for $PM_{2.5}$ and PM_{10}); thus, depending on the case, a moving mean of the previous 8 h is computed, or 24 h values are assumed for the 24 h period, allowing comparing the results on an hourly basis.

The selected weather and air quality station monitors only some substances; hence, O_3, NH_3, $PM_{2.5}$, and PM_{10} were analyzed.

$$AQI_{p_i} = \frac{I_{Hi} - I_{Lo}}{BP_{Hi} - BP_{Lo}} * (C_p - BP_{Lo}) + I_{Lo} \quad (3)$$

Table 1. Parameters needed for AQI calculation for an 8 hour or 24 hour exposure period (extracted and edited from the Environmental Protection Agency (EPA) guidelines [31]). BP, Break Point.

BPs					
O3 (ppm)	**PM_{10} (μg/m^3)**	**$PM_{2.5}$ (μg/m^3)**	**NH_3 (μg/m^3)**	**AQI**	**Category**
0.000–0.064	0–54	0.0–15.4	0–200	0–50	Good
0.065–0.084	55–154	15.5–40.4	201–400	51–100	Moderate
0.085–0.104	155–254	40.5–65.4	401–600	101–150	Unhealthy for sensitive groups
0.105–0.124	255–354	65.5–150.4	601–800	151–200	Unhealthy
0.125–0.374	355–424	150.5–250.4	801–1200	201–300	Very unhealthy
	425–504	250.5–350.4	1201–1800	301–400	Hazardous
	505–604	350.5–500.4	>1800	401–500	Hazardous

As performed in Section 2.1, the number of hours in which the residents of the nearby area are exposed to hazard risk were computed, assuming the risk threshold for low air quality sensitive groups as $AQI > 100$ (Table 2).

Table 2. Sensitive groups at risk by pollutant type when the AQI goes above 100 for an 8 hour (O_3) or 24 hour (PM_{10}) average exposure period (extracted and edited from the EPA guidelines [31]).

Pollutant	Demographic Groups at Higher Risk
Ozone (O_3)	People with lung disease, children, elderly, and people who are active outdoors
Particulate matter ($PM_{2.5}$)	People with heart or lung disease, elderly, and children
Particulate matter (PM_{10})	People with heart or lung disease, elderly, and children

2.3. Air Temperature and Pollution Confluence Risk

Given that the results for both $RiskT$ and AQI were allocated into 2 macro categories (i.e., at risk and no risk) and that they were reported hourly, the confluence risk was estimated as the product of the Boolean of both risks' existence (i.e., $RiskT > 0.5$ and $AQI > 100$).

Then, the recurrence of these conditions was estimated as the ratio between the hours of risk presented, during:

- the whole year (8760 h);
- the summer period only (15 June–15 September, 2208 h);
- both the mid-season and summer period (15 April–15 October, 4392 h); and,
- mid-season (15 April–15 June and 15 September–15 October, 2184 h).

2.4. Case Study Description

Profiting from the database made available by the regional environmental monitoring institution (ARPA Lombardy Agency [24]), it was possible to retrieve air quality and weather data from a single, but relevant, data source (i.e., Milano via Juvara) near a university district in the city of Milan, Italy (Figure 1). These data were collected and elaborated only for 2019, given the yearly data's completeness and recentness, as described in the following sections.

The weather and air quality station Milano via Juvara is located within the Buenos Aires-Venezia Local Identity Unit (NIL), a neighborhood characterized by low green area coverage and mid-to-high built density, as documented by the Municipality of Milan [32] (see Figure 1). It was expected that the data collected would provide evidence to assume that the area has a low heat and pollution management capacity, which puts its inhabitants at a relevant increasing temperatures and air pollution SLODs risk.

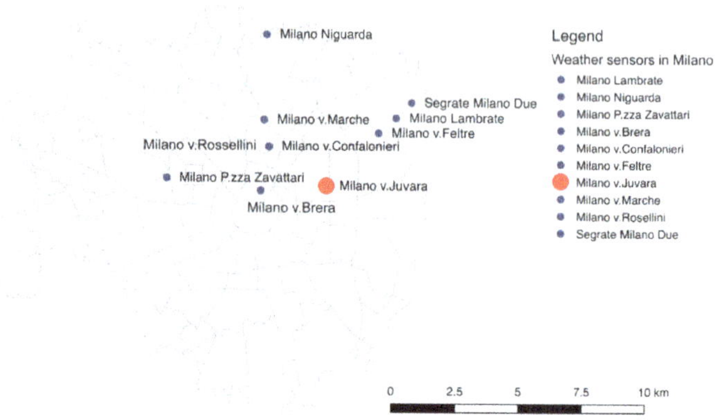

Figure 1. Location of the Milano via Juvara station within the city of Milan divided by NILs.

3. Results

To understand the frequency distribution on a yearly basis, the results are presented on 24 × 365 heat maps for every risk type and daily frequency distribution charts for their confluence (e.g., Figure 2). It was considered that summer is between Day Of the Year (DOY) 166 and 258; and winter ends and starts on DOYs 105 and 288, respectively.

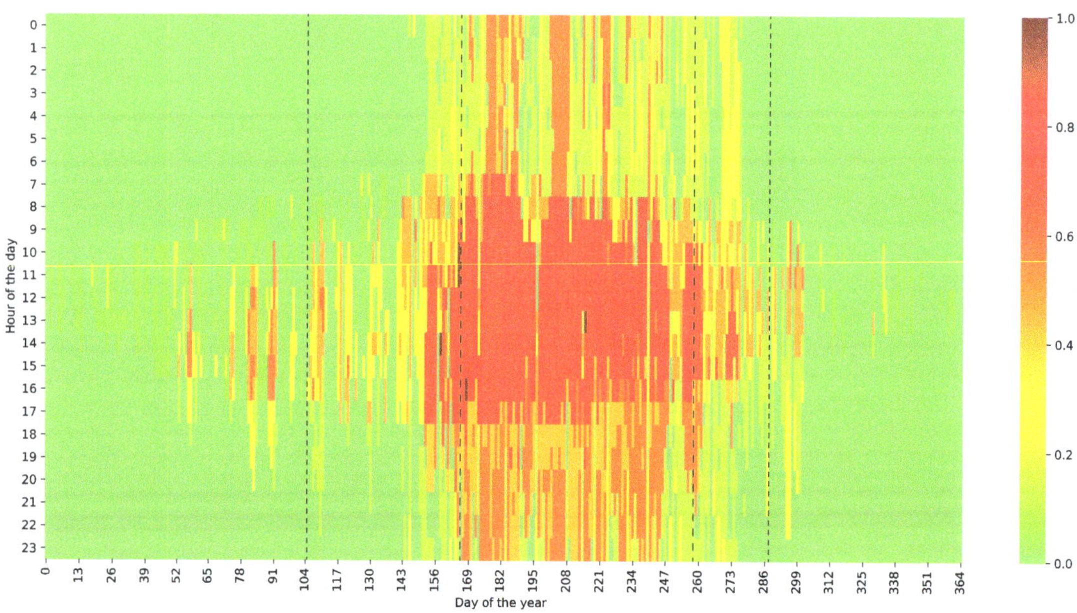

Figure 2. Degree of *RiskT* from values monitored in 2019 at the station Milano v. Juvara, based on the confluence of disadvantageous conditions of t_{db-air}, I_{tot}, RH, and V_a; which are closely related to the SLOD of increasing temperatures.

3.1. Meaningful Thermal Stress Risk Frequency

Following the methodology described in Section 2, $RiskT$ and $UTCI$ values were computed for each hour of the data collected from M. v. Juvara station during 2019. These are plotted in Figures 2 and 3 to qualitatively analyze how this thermal-stress-related risk was manifesting, and quantitative reported frequency data are summarized in Table 3.

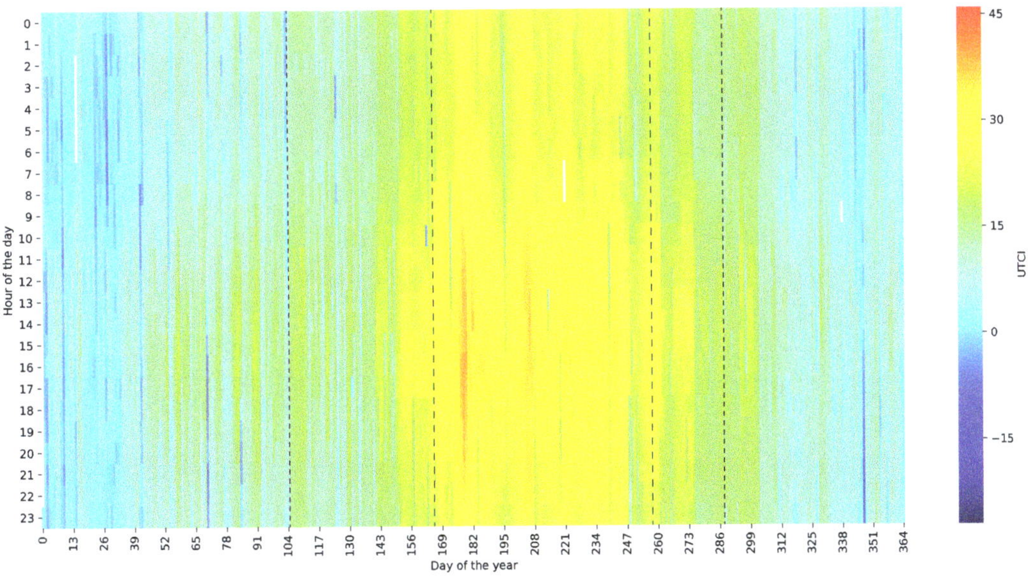

Figure 3. Degree of thermal stress expressed in $UTCI$, based only on the values for 2019 from Milano v. Juvara of t_{db-air}, RH, and V_a.

Table 3. Summary results of the allocated thermal stress, when using $RiskT > 0.5$ or $UTCI > 26$ as the criterion, differentiated by season.

		RiskT		UTCI	
Season	**# of Hours**	**Hours at Risk**	**Ratio**	**Hours at Risk**	**Ratio**
Whole year	8760	1324	15.11%	1343	15.33%
Summer and mid	4392	1295	29.49%	1343	30.58%
Summer	2208	1145	26.21%	1253	28.69%
Mid	2184	150	3.42%	90	2.05%
Winter	4368	29	0.66%	0	0.00%

The risk for both $RiskT$ and $UTCI$ is notably concentrated in summer (29.49% and 30.58%). However, Figure 3 displays that for $UTCI$, rare and low intensity risk arose during the mid-season (2.05%), and no risk was registered during winter. On the other hand, Figure 2 presents that $RiskT$ would communicate a higher risk intensity in summer, a larger frequency and intense risk for the mid-season (3.42%), and a rare risk during winter (0.66%).

The above could result from the assumption made on MRT for computing $UTCI$ and the allocated weights used for estimating $RiskT$. The classification made with the $RiskT$ criterion was used, given the significant effect that direct solar radiation falling on a person has on heat exchange and thermal perception [33].

3.2. Unhealthy Polluted Air Recurrence

Following Section 2, pollutant concentration was screened for every available dataset, and the AQI was computed accordingly. Information on the air concentration behavior, during 2019, of O_3, NH_3, $PM_{2.5}$, and PM_{10} is reported on Figure 4.

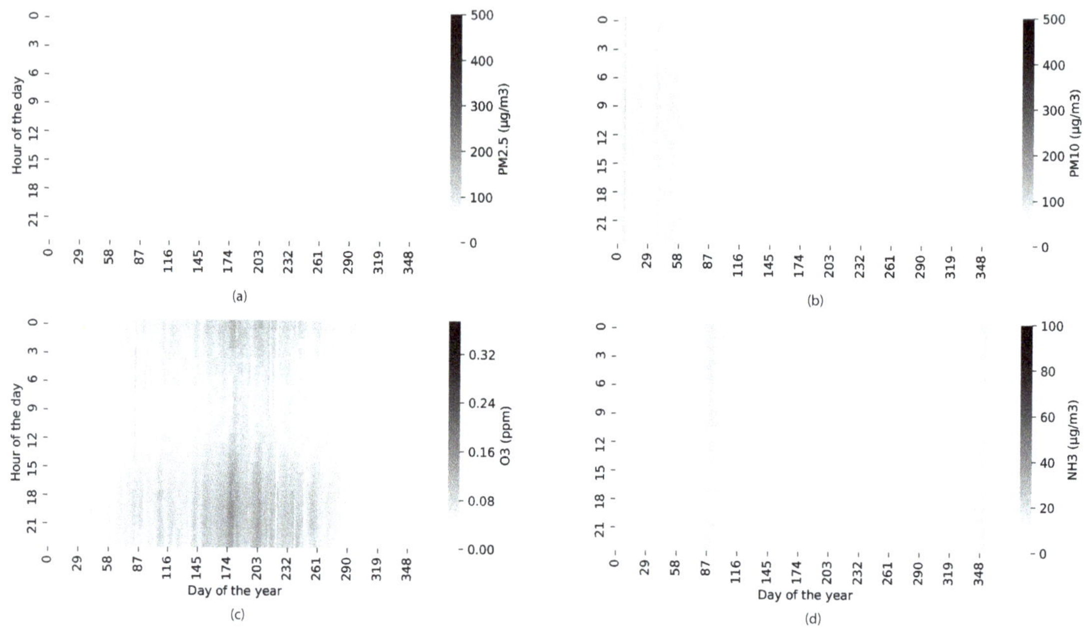

Figure 4. Degree of air pollution concentration in the AQI, based only on the 2019 values from Milano v. Juvara for (**a**) $PM_{2.5}$, (**b**) PM_{10}, (**c**) O_3, and (**d**) NH_3.

These pollutants' concentrations are reported in way that they display as well the gravity of their accumulation, by setting the gradient ranges from zero to the maximum value that would fall under the greatest hazardous category of AQI classification. An exception was made for NH_3 as the reported values are considerably lower, and the maximum value for display was set to be the limit value before generating a risk to sensitive groups.

From these, it is possible to see how PM_{10} and O_3 are those representing a greater risk to the citizens' health in the area. In particular, they are at greater risk of O_3 accumulation during the afternoons and early mornings by the middle of the year (i.e., summer season).

Looking instead to the complete AQI analysis, there is a large hour ratio of low air quality during winter time classified as risky (23.76%), comparable to the one reported for summer and mid-season (26.3%); even the yearly risk/total hours (23.03%); see Table 4. Thus, the risk frequency can be assumed to be rather similar throughout the seasons. Anyway, Figure 5 clearly shows how the intensity, or gravity, of this air pollution SLODs risk measured with the AQI is greater during the summer season.

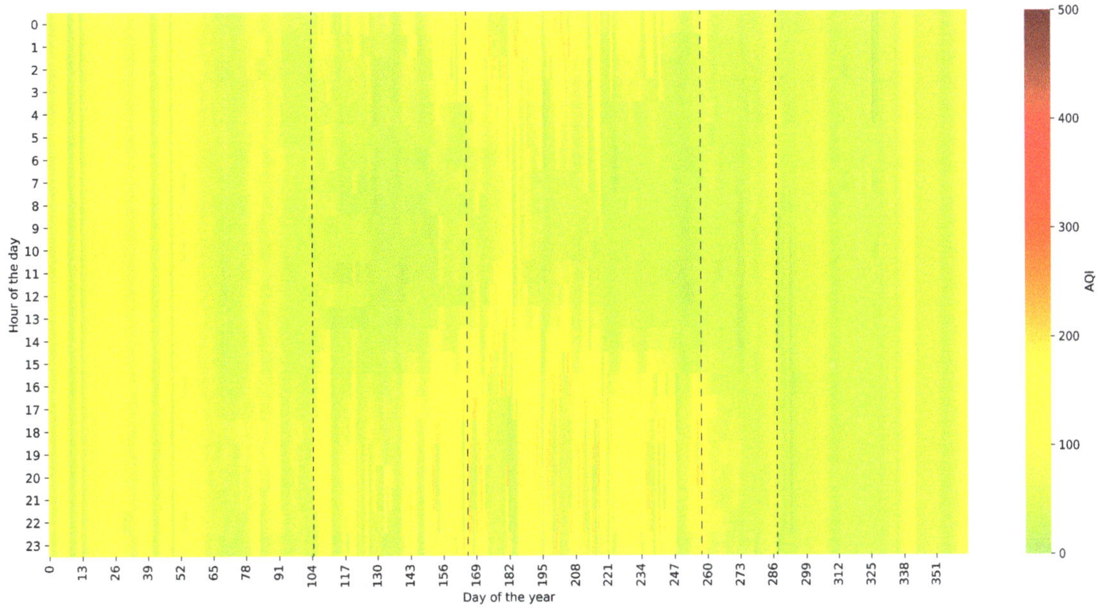

Figure 5. Final degree of air pollution risk in the *AQI* based on 2019's values from Milano v. Juvara.

Table 4. Summary results of the allocated air pollution distress, when using $AQI > 100$ as the criterion, differentiated by seasons.

	AQI		
Season	**# of Hours**	**Hours at risk**	**Ratio**
Whole year	8760	2193	25.03%
Summer and mid	4392	1155	26.30%
Summer	2208	808	18.50%
Mid	2184	347	7.90%
Winter	4368	1038	23.76%

3.3. Boosted Health Risk from the Multiple Hazards Effect

Both SLODs' risks are present and of considerable severity for sensitive groups during summer. However, to clearly state how frequent the potential danger for these demographic groups is, a superimposition assessment was applied as described in Section 2.3 and displayed in Figure 6.

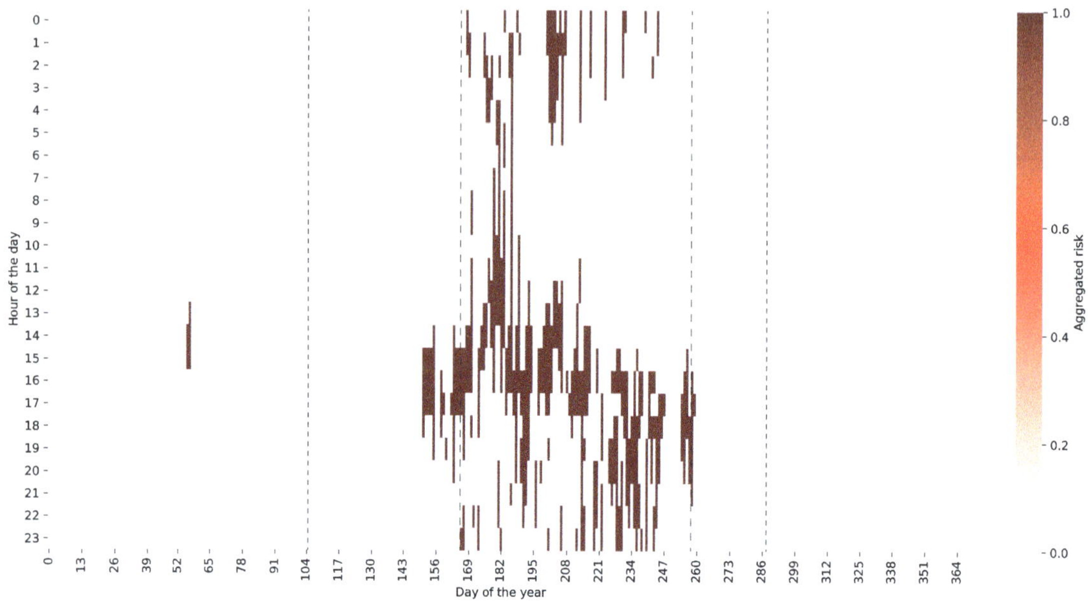

Figure 6. Conjoint risk frequency distribution on a yearly basis for 2019's data extracted from weather and air quality station Milano v. Juvara.

Figure 6 clearly shows how their confluence frequency is concentrated during the summer season (9.96% hour ratio), although some rare conditions were encountered during winter (i.e., last week of February, around DOY ≈ 58), in which there was a severe *RiskT* (>0.5) and *AQI* (>100) representing only 0.11% of the hours of said period. The frequency by seasons has been compared and condensed in Table 5.

In addition, one can note that the hazard risk confluence is concentrated in the early mornings and the afternoons.

Table 5. Summary results of the allocated multi-hazard risk, when using $AQI > 100$ as the criterion, differentiated by seasons.

Season	# of Hours	RiskT and AQI Hours at Risk	Ratio
Whole year	8760	485	5.54%
Summer and mid	4392	480	10.93%
summer	2208	435	9.96%
Mid	2184	45	1.02%
Winter	4368	5	0.11%

4. Discussion

4.1. Validity of the Preliminary Assessment

A direct comparison was done between the values obtained for *RiskT* and *UTCI* condensed in Table 3, to understand how different the collected samples of the proposed risk criteria and the established, but simplified *UTCI* were.

UTCI presented a higher risk frequency overall with an absolute algorithmic difference of 0.22% when comparing yearly data, 1.09% during summer and mid-season, or 2.48% during summer. Yet, *RiskT* presented an absolute 1.37% more hours of potential thermal stress risk during mid-seasons and 0.66% during winter when compared to *UTCI*.

The absolute algorithmic difference between risk and no risk hours ranged between 19 and 108 h, being least for the 8760 yearly period (i.e., 19) and highest for the 2208 h in summer. This was considered non-negligible; thus, the classification accuracy was computed, assuming as true values those computed for $UTCI$, using Equation (2), as 91.56%; from: 360 False Positives (FPs), 379 False Negatives (FNs), 964 True Positives (TPs), and 7057 True Negatives (TNs).

These results were expected as the assumption of $MRT \approx t_{db-air}$ did not truly express the condition of a person outdoors, and $RiskT$ was configured in a way in which it gave a large significance to the presence of intense I_{tot} (i.e., $I_{weight} = 0.3$). However, with an accuracy over 90% on a large sample (>8000 measurements), it could be assumed that the approach is feasible for preliminary analyses.

4.2. Considerable Prevailing Pedestrian Risk

Certainly, during the summer period in Milan (approximately 18–29% of the season hours), sensitive groups among citizens or visitors in the areas adjacent to the weather and air quality station Milano v. Juvara were subjected to critical hazards (see Table 2). In addition, in this same period, people outdoors were vulnerable to the conjunct effects of SLODs risk hazards (increasing temperatures and air pollution) for a considerable amount of hours (9.96%).

On the other hand, multi-hazard risk arousal during the winter period was unlikely. The seeming outliers were further studied by filtering all collected data; thus, this hazard risk reported for winter resulted from high I_{tot} (>300 W/m^2), dry air RH (<30%), low wind speed (<2 m/s^2), and a high concentration of particulate matter ($PM_{2.5} > 40.5$ $\mu g/m^3$). Although, in these hours, high t_{db-air} were not reported (maximum values were registered around 22.3 °C), this does not exclude the possibility of significant thermal stress given the exposure to such solar irradiation [30,33,34].

In addition, winter shall not be assumed as a low risk season as air pollution hazard was present during 23.76% of the winter time (see Table 4). Figures 5 and 4a show how most of the hazardous AQI was correlated to the particulate matter (i.e., $PM_{2.5}$ and PM_{10}) concentration trend; given that these values are normally reported as the daily average, larger concentrations could have been experienced by pedestrians in the station's surroundings.

In brief, pedestrians within the built environment close to Milano v. Juvara were subjected to either one of the SLODs hazards for 3032 h during 2019 (34.61% of the time) and contemporaneously 485 h only (5.54% of the time). However, this gives only an overview of the risk; thus, a frequency profile was drawn to better understand the occurrence trend of these hazards on a daily basis.

Figure 7 displays the frequency density distribution of each and the conjunct SLODs hazard during the day for that location during 2019, allowing identifying the peak moments of the day in which pedestrians were more prone to health deterioration. It is worth noting how the peaks for increasing temperatures and air pollution have a similar growing trend, but a distinct dissipation behavior; this could be attributed to the low capacity of the zone to deposit, absorb, dilute, or dissipate air pollutants; or simply, a pollution concentration increase due to higher pollutant source loads during these time-frames related to anthropogenic activities.

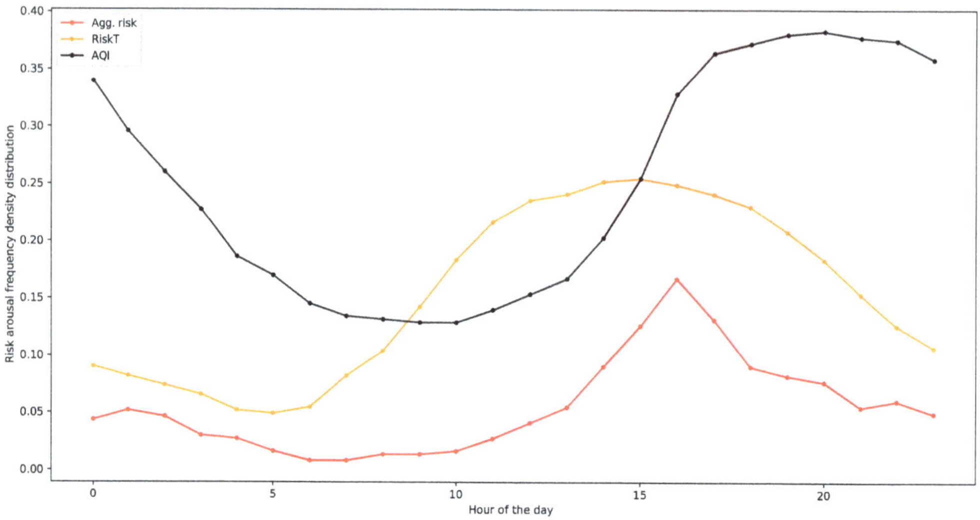

Figure 7. Conjoint risk (Agg. risk) daily profile of the hourly frequency distribution on a yearly basis for 2019's data extracted from Milano v. Juvara.

The constructed profile also allowed identifying the great risk that a person can be exposed during the night-time. People outdoors would be frequently subjected to poor air quality and likewise for those indoors who rely on night cooling or natural ventilation.

Instead, when coupled, the profile shows that pedestrians were most vulnerable at 16:00 (maximum frequency distribution value). Moreover, the most unhealthy period can be established during 14:00 and 18:00, having more than 9% of those specific i-th hours classified as a conjunct hazard moment; and the second most hazardous time-frame was found between 18:00 and 20:00 with a $\approx$8% ratio.

4.3. Limitations and Further Work

The 91.56% classification accuracy of the proposed criterion is not yet fully satisfactory, and adopting a simplified $UTCI$ result as the true values is only applicable under a few surrounding outdoor conditions within the built environment. The accuracy could be improved by further calibrating the weighting factors used by performing in situ measurements and adjusting MRT for computing $UTCI$ accordingly. These tests and data elaboration will reduce the number of both FP and FN, thus increasing the accuracy.

In addition, a normed value could allow capturing and communicating more information on the intensity of the thermal stress and air pollution hazard; as well as providing a rather continuous function, instead of such a discrete trend when applying Booleans.

Further work is planned to include also the physiological characteristics of sensitive groups to further understand the thermal stress variability among those sensitive groups at risk. Furthermore, gathering larger yearly databases and handling time-series missing data would allow making more reliable assumptions and conclusions from the results.

In addition, generating hazard risk frequency daily profiles from environmental data eases future work on the combination of other risk-related analysis, which have a narrower arousal temporal scale (e.g., earthquakes, terrorism attacks), for the superimposing of multiple risk assessment.

5. Conclusions

People frequenting dense built environments are susceptible to SLODs, and some of these could arise simultaneously; in particular, people are prone to increasing temperatures and air pollution. Coupling their analysis to an integrated approach is limited by the data accessibility, applicability, quality, and granularity. This work proposes a framework that enables and facilitates performing, for a specific site (i.e., Milano v. Juvara's surroundings, Milan, Italy), an outdoor meso-scale multi-hazard-based risk analysis; with which it is possible to obtain a narrower time-frame in which these two events could have affected citizens' health the most.

A weighted fuzzy logic analysis permitted superimposing climatic and air quality distress on an hourly basis, allocated using set healthy and comfortable ranges. Then, it is possible to delineate an average daily multi-hazard arousal profile that can be later used as the input in other risk-related assessments.

Specifically, for the selected case study, findings show that sensitive groups were frequently and greatly at risk during 2019; in fact:

- Summer was found to be the most stressful period for both increasing temperatures and air pollution using a simplified thermal-related index (i.e., *RiskT*) and the *AQI* as separate criteria and/or combined criteria.
- Winter prevalently had low air quality; along with late afternoon and early morning hours on a daily average.
- Thermal stress was reported more frequently instead during midday and afternoon hours.
- During the day, the conjoint effect of climatic and air quality distress was found to be more recurrent between 14:00 and 18:00, with a slight decay during 18:00 and 20:00 (as highlighted in Figure 7).

Author Contributions: Conceptualization, J.D.B.C., N.M., G.S., T.P., E.Q., and F.R.; methodology, J.D.B.C., N.M., and G.S.; validation, J.D.B.C., G.S., and E.Q.; formal analysis, J.D.B.C., N.M., and F.R.; investigation, J.D.B.C. and N.M.; resources, G.S.; data curation, J.D.B.C.; writing, original draft preparation, J.D.B.C.; writing, review and editing, J.D.B.C., N.M., G.S., and E.Q.; visualization, J.D.B.C. and N.M.; supervision, G.S. and E.Q.; project administration, G.S. and E.Q.; funding acquisition, G.S., E.Q., and T.P. All authors read and agreed to the published version of the manuscript.

Funding: This research was funded by MIUR (the Italian Ministry of Education, University and Research) under the project BE S2ECURe - (make) Built Environment Safer in Slow and Emergency Conditions through behavioral assessed/designed Resilient solutions, Grant Number 2017LR75XK.

Institutional Review Board Statement: Not applicable.

Informed Consent Statement: Not applicable.

Data Availability Statement: Publicly available datasets were analyzed in this study. This data can be found here: https://www.arpalombardia.it/Pages/Temi-Ambientali.aspx.

Conflicts of Interest: The authors declare no conflict of interest.

References

1. United Nations, Department of Economic and Social Affairs, Population Division. *World Urbanization Prospects: The 2018 Revision*; Technical report; United Nations: New York, NY, USA, 2019.
2. World Health Organization. *Quantitative Risk Assessment of the Effects of Climate Change on Selected Causes of Death, 2030s and 2050s*; WHO Press: Geneva, Switzerland, 2014.
3. World Health Organization. *Ambient Air Pollution: A Global Assessment of Exposure and Burden of Disease*; World Health Organization: Geneva, Switzerland, 2016; p. 121.
4. United Nations. *Transforming Our World: the 2030 Agenda for Sustainable Development*; United Nations: New York, NY, USA, 2015.
5. Lee, J.Y.; Kim, H. Projection of future temperature-related mortality due to climate and demographic changes. *Environ. Int.* **2016**, *94*, 489–494. [CrossRef] [PubMed]
6. Jackson, J.E.; Yost, M.G.; Karr, C.; Fitzpatrick, C.; Lamb, B.K.; Chung, S.H.; Chen, J.; Avise, J.; Rosenblatt, R.A.; Fenske, R.A. Public health impacts of climate change in Washington State: Projected mortality risks due to heat events and air pollution. *Clim. Chang.* **2010**, *102*, 159–186. [CrossRef]
7. UNFCCC. *Slow Onset Events*; Technical report; United Nations: New York, NY, USA, 2012.

8. U.S. Environmental Protection Agency. Urban heat island basics. In *Reducing Urban Heat Islands: Compendium of Strategies | Heat Island Effect | US EPA*; U.S. Environmental Protection Agency: Washington, DC, USA, 2008.
9. Rosenzweig, C.; Solecki, W.D.; Romero-Lankao, P.; Mehrotra, S.; Dhakal, S.; Ibrahim, S.A. *Climate Change and Cities: Second Assessment Report of the Urban Climate Change Research Network*; Cambridge University Press: Cambridge, UK, 2018.
10. Niemenmaa, V.; Happach, B.; Kubat, J.; Otto, J.; Pirelli, L.; Simeonova, R.; Zalega, A.; Wisniewska-Danek, K.; Radecka-Moroz, K.; Wojciechowski, J.; et al. *Air pollution: Our Health Still Insufficiently Protected*; Technical report; ECA: Luxembourg, 2018. [CrossRef]
11. Colaninno, N.; Morello, E. Modelling the impact of green solutions upon the urban heat island phenomenon by means of satellite data. *J. Physics: Conf. Ser.* **2019**, *1343*, 012010. [CrossRef]
12. Fuladlu, K.; Riza, M.; İlkan, M. The Effect of Rapid Urbanization on the Physical Modification of Urban Area. In Proceedings of the International Conference on Architecture and Built Environment with AWARDs, Venice, Italy, 22–24 May 2018; pp. 1–9.
13. Paolini, R.; Mainini, A.G.; Poli, T.; Vercesi, L. Assessment of thermal stress in a street canyon in pedestrian area with or without canopy shading. *Energy Procedia* **2014**, *48*, 1570–1575. [CrossRef]
14. Jamei, E.; Rajagopalan, P.; Seyedmahmoudian, M.; Jamei, Y. Review on the impact of urban geometry and pedestrian level greening on outdoor thermal comfort. *Renew. Sustain. Energy Rev.* **2016**, *54*, 1002–1017. [CrossRef]
15. Stewart, I.D.; Oke, T.R. Local climate zones for urban temperature studies. *Bull. Am. Meteorol. Soc.* **2012**, *93*, 1879–1900. [CrossRef]
16. Udristioiu, M.T.; Velea, L.; Bojariu, R.; Sararu, S.C. Assessment of urban heat Island for Craiova from satellite-based LST. In *AIP Conference Proceedings*; AIP Publishing LLC: Melville, NY, USA, 2017; Volume 1916, p. 040004.
17. Sassi, M.Z.; Fourrié, N.; Guidard, V.; Birman, C. Use of Infrared Satellite Observations for the Surface Temperature Retrieval over Land in a NWP Context. *Remote. Sens.* **2019**, *11*, 2371. [CrossRef]
18. Khamchiangta, D.; Dhakal, S. Time series analysis of land use and land cover changes related to urban heat island intensity: Case of Bangkok Metropolitan Area in Thailand. *J. Urban Manag.* **2020**, *9*, 383–395. [CrossRef]
19. Ketterer, C.; Matzarakis, A. Mapping the Physiologically Equivalent Temperature in urban areas using artificial neural network. *Landsc. Urban Plan.* **2016**, *150*, 1–9. [CrossRef]
20. Sharifi, E.; Lehmann, S. Correlation analysis of surface temperature of rooftops, streetscapes and urban heat island effect: Case study of central Sydney. *J. Urban Environ. Eng.* **2015**, *9*, 3–11. [CrossRef]
21. Vienneau, D.; De Hoogh, K.; Briggs, D. A GIS-based method for modelling air pollution exposures across Europe. *Sci. Total. Environ.* **2009**, *408*, 255–266. [CrossRef] [PubMed]
22. Mackey, C.; Galanos, T.; Norford, L.; Roudsari, M.S.; Bhd, N.S. Wind, Sun, Surface Temperature, and Heat Island: Critical Variables for High-Resolution Outdoor Thermal Comfort Payette Architects, United States of America Massachusetts Institute of Technology, United States of America University of Pennsylvania. In Proceedings of the 15th IBPSA Conference, San Francisco, CA, USA, 7–9 August 2017; pp. 985–993.
23. Salmond, J.A.; Williams, D.E.; Laing, G.; Kingham, S.; Dirks, K.; Longley, I.; Henshaw, G.S. The influence of vegetation on the horizontal and vertical distribution of pollutants in a street canyon. *Sci. Total. Environ.* **2013**, *443*, 287–298. [CrossRef] [PubMed]
24. ARPA Lombardia. Dati e Indicatori, ARPA Lombardia: Rosellini, Milano, Italy. Available online: https://www.arpalombardia.it (accessed on 25 March 2020).
25. Salvalai, G.; Enrico, Q.; Blanco Cadena, J.D. Built environment and human behavior boosting Slow-onset disaster risk. In *Heritage 2020*; Green Lines Institute for Sustainable Development: Coimbra, Portugal, 2020; pp. 199–209.
26. ISTAT. Statistiche Istat, ISTAT: Rome, Italy. Available online: http://dati.istat.it/ (accessed on 16 January 2020).
27. (EEA), Observed Annual Mean Temperature Change from 1960 to 2019 and Projected 21st Century Change Under Different Emissions Scenarios in Europe. Available online: https://www.eea.europa.eu/data-and-maps/figures/trends-in-annual-temperature-across-1 (accessed on 31 March 2020).
28. European Environment Agency. *Air Quality in Europe—2019 Report. EEA Report No 10/2019*; Technical report; European Environment Agency: Copenhagen, Denmark, 2019. [CrossRef]
29. Bröde, P.; Fiala, D.; Blazejczyk, K.; Epstein, Y.; Holmér, I.; Jendritzky, G.; Kampmann, B.; Richards, M.; Rintamäki, H.; Shitzer, A.; et al. Calculating UTCI equivalent temperature. In *Environmental Ergonomics XIII*; University of Wollongong: Wollongong, Australia, 2009; pp. 49–53.
30. ANSI/ASHRAE. *ANSI/ASHRAE Standard 55-2017 : Thermal Environmental Conditions for Human Occupancy*; ASHRAE Inc.: Atlanta, GA, USA, 2017.
31. U.S. Environmental Protection Agency. *Guidelines for the Reporting of Daily Air Quality—The Air Quality Index (AQI)*; Technical report; EPA: Washington, DC, USA, 2006.
32. Comune di Milano. Portale Open Data | Comune di Milano, Comune di Milano: Milano Italia. Available online: http://dati.comune.milano.it/ (accessed on 16 January 2020).
33. Zani, A.; Mainini, A.G.; Blanco Cadena, J.D.; Schiavon, S.; Arens, E. A New Modeling Approach for the Assessment of the Effect of Solar Radiation on Indoor Thermal Comfort. In Procceeding of the Building Performance Analysis Conference and SimBuild, Chicago, IL, USA, 26–28 September 2018.
34. Arens, E.; Hoyt, T.; Zhou, X.; Huang, L.; Zhang, H.; Schiavon, S. Modeling the comfort effects of short-wave solar radiation indoors. *Build. Environ.* **2015**, *88*, 3–9. [CrossRef]

applied sciences

MDPI

Article

Electrical Energy Prediction in Residential Buildings for Short-Term Horizons Using Hybrid Deep Learning Strategy

Zulfiqar Ahmad Khan, Amin Ullah, Waseem Ullah, Seungmin Rho, Miyoung Lee and Sung Wook Baik *

Sejong University, Seoul 143-747, Korea; mzulfiqar3797@gmail.com (Z.A.K.); qamin3797@gmail.com (A.U.); kwaseem3797@gmail.com (W.U.); smrho@sejong.ac.kr (S.R.); miylee@sejong.ac.kr (M.L.)

* Correspondence: sbaik@sejong.ac.kr

Received: 9 November 2020; Accepted: 29 November 2020; Published: 2 December 2020

Abstract: Smart grid technology based on renewable energy and energy storage systems are attracting considerable attention towards energy crises. Accurate and reliable model for electricity prediction is considered a key factor for a suitable energy management policy. Currently, electricity consumption is rapidly increasing due to the rise in human population and technology development. Therefore, in this study, we established a two-step methodology for residential building load prediction, which comprises two stages: in the first stage, the raw data of electricity consumption are refined for effective training; and the second step includes a hybrid model with the integration of convolutional neural network (CNN) and multilayer bidirectional gated recurrent unit (MB-GRU). The CNN layers are incorporated into the model as a feature extractor, while MB-GRU learns the sequences between electricity consumption data. The proposed model is evaluated using the root mean square error (RMSE), mean square error (MSE), and mean absolute error (MAE) metrics. Finally, our model is assessed over benchmark datasets that exhibited an extensive drop in the error rate in comparison to other techniques. The results indicated that the proposed model reduced errors over the individual household electricity consumption prediction (IHEPC) dataset (i.e., RMSE (5%), MSE (4%), and MAE (4%)), and for the appliances load prediction (AEP) dataset (i.e., RMSE (2%), and MAE (1%)).

Keywords: bidirectional gated recurrent unit; convolutional neural networks; electricity consumption prediction; hybrid deep learning model; residential load prediction

1. Introduction

The electric power industry plays an important role in the economic development of a country, and its decisive operation provides significant societal wellbeing. As reported in [1] the global energy consumption is increasing for sustainable advancement in society; therefore, the effectiveness of electricity consumption prediction needs to be improved [2]. As reported by the World Energy Outlook in 2017, the compound annual growth rate (CAGR) of electricity demand will have a global incremental rise of 1.0% in the period of 2016–2040 [3]. Another report presented in [4] described that residential buildings generally consume 27% of the total energy consumption, whereas buildings in the United States (US) consume 40% of their national energy [5]. Owing to high energy consumption levels in residential buildings, efficient management of their consumption is essential. Therefore, proper planning of energy is vital for energy saving, which is possible through effective energy consumption prediction models.

The electricity prediction strategies for short-term horizons are categorized into four types: very-short, short, medium, and long-term [6,7]. Short and very short-term predictions refer to minutely ahead predictions up to several days. Medium-term load prediction means one week ahead prediction or up to a year, whereas annual or several years' ahead prediction is called very long-term prediction. Each electricity prediction horizon has its own applications; however, in this study, we focus on short-term and very-short-term predictions [6,8–10]. The major applications of short-term electric load prediction are power plant's reliable and secure operation, reliability and economic dispatch, and power system generation scheduling. Short-term load prediction ensures power system security, and it is an essential tool for the determination of the optimal operational state. Reliability and economic dispatch are also important applications of power systems in short-term horizons, wherein the abrupt variation of load demand fluctuates its reliability. This causes a power supply shortage if the load demand is underestimated, which makes it difficult to manage overload conditions and the quality of the overall power supply system. Another application of short-term load prediction is generation scheduling, which can be achieved through accurate load prediction to verify the allocation of operational limitations, generation resources, equipment usage, and environmental constraints. In the literature, several studies have been conducted for short-term load prediction, electricity load forecasting, electricity demand forecasting, electricity storage, and occupant behavior [11–15]. The mainstream electricity load prediction models are mainly grouped into two categories: statistical models and artificial intelligence models [16,17]. Statistical models such as Auto Regressive Integrated Moving Average (ARIMA) [18], linear regression [19], Kalman filtering [20], and clustering [21,22] etc. were used for load prediction in the early days. These models are effective in learning linear data but inadequate to learn the nonlinear complex electricity load. Besides this, the artificial intelligence models can learn nonlinear complex and linear electricity loads, which are further divided into shallow and deep structure methods. Shallow based methods include random forest [23], wavelet neural networks [24], support vector machines (SVMs) [25], artificial neural networks (ANN) [26], and extreme learning machines [27]. Shallow based methods perform well compared to statistical methods but perform poorly in feature mining. Hence, these methods require more features and selects strong features to enhance the prediction accuracy. Obtaining optimal feature extraction is a challenging task for these methods. Further, these methods have insufficient generalization ability over different datasets due to small hypothesis owing to less number of parameters. Deep structure methods have the ability to address the aforementioned concerns of shallow-based methods using multi-layer processing and hierarchical feature learning from electricity historical data. Recently, recurrent neural networks (RNNs) and convolutional neural networks (CNNs) are two powerful architecture proposed in the literature for the analysis of time series data. For instance, Amarasinghe et al. [28] developed a CNN-based methodology for electricity load forecasting and compared their results with a factored restricted Boltzmann machine, sequence-to-sequence long short term memory (LSTM), support vector regressor (SVR), and ANN. Another study presented in [29] proposed a deep CNN network for day-ahead load forecasting and compared the results with an extreme learning machine, ARIMA, CNN, and RNN. Several studies also used RNN models for electricity load prediction, whereas Tokgoz et al. [30] used RNN, gated recurrent unit (GRU), and LSTM models for electricity load prediction in Turkey and extensively decreased the error. Furthermore, the authors of [31] developed an LSTM-based model for periodic energy prediction and compared their results with other models. Another study presented in [32] developed an RNN-based model for medium- and long-term electricity load prediction.

The electricity consumption data is time-series data, which comprises spatial and temporal information. The CNN models perform well for spatial information extraction, but insufficient for temporal information, whereas the RNN models are insufficient for spatial information and can learn temporal information. Therefore, to develop an optimal model for electricity load prediction, hybrid models are introduced in the recent literature. For instance, Kim et al. [33] developed a hybrid model combining CNN with LSTM for short-term load prediction and compared their results with GRU, attention LSTM, LSTM, and bidirectional LSTM. Ullah et al. [34] also developed a hybrid model with a combination of CNN and multi-layer bidirectional LSTM and compared their results with bidirectional LSTM, LSTM, and CNN-LSTM. Similarly, another study presented in [17] integrated a CNN with an

LSTM auto-encoder and compared the final results with LSTM, LSTM autoencoder, and CNN-LSTM. Moreover, Sajjad et al. [35] and Afrasiabi et al. [36] presented the performance of a CNN-GRU based model for electricity forecasting. The performance of hybrid models is quite promising and has achieved state-of-the-art accuracy; however, further improvements are needed for optimal electricity load prediction. Therefore, in the current study, we established a two-step framework for predicting the electricity load that includes data preprocessing and proposed a hybrid model. In the first step, the historical data of electricity is refined to remove abnormalities that are then passed to the next step CNN along with MB-GRU model for learning. To extract the spatial information, we used CNN layers where MB-GRU is used to learn the temporal information. The contributions of the proposed research are summarized below:

- The electricity consumption data are gathered from smart meter sensors, which include missing values, redundant values, outlier values, etc., due to several reasons, such as faults in meter sensors, variable weather conditions, abnormal customer consumption patterns, etc., which need to be refined before training the model. Therefore, in this work, the input raw datasets are refined before training to fill the missing values and remove the outlier values from the dataset. Similarly, the electricity consumption patterns are of very diverse nature where the neural networks are sensitive to it, so a data normalization technique is applied to bring the dataset into a standard range.
- The mainstream methods use solo models for electricity consumption prediction, which are unable to precisely extract spatiotemporal patterns and have high error rates. Therefore, in this study, we proposed a hybrid model with a combination of CNN and MB-GRU that helps to improve the accuracy of electricity consumption prediction.
- The performance of the model was evaluated using the root mean square error (RMSE), mean square error (MSE), and mean absolute error (MAE). The experimental results show that the proposed model extensively decreases the error rate when compared to baseline models.

The main goal of this work is to improve the prediction accuracy for short-term electrical load prediction in residential buildings, which reduces customer consumption and provides economic benefits. The experimental section shows the effectiveness of the proposed method, which ensures the best performance of the proposed method as compared to other baseline models.

The remainder of the paper is arranged as follows: Section 2 provides a detailed explanation of the proposed method; Section 3 includes the experimental results of the proposed method and comparison with other state-of-the-art models. Finally, the manuscript is concluded in Section 4.

2. Proposed Framework

Accurate electricity load prediction is very important for electricity saving and vital economic implications [37]. As reported by [38] a 1% decrease in the error rate of the electricity prediction model can profit 1.6M dollars and can save 10K MW of electricity annually. For accurate load prediction, an appropriate learning methodology is required. The electricity load prediction models are learned from historical data generated from smart meter sensors. However, some times, due to weather conditions, meter faults, etc., they generate some abnormal data that should be refined before training. Therefore, this work represents a two-step framework that includes data preprocessing and the proposed hybrid model. The preprocessing step refines the input raw data of electricity consumption and then passes it to the proposed model to learn it, as demonstrated in Figure 1, where the details of each step are further discussed in the following sections.

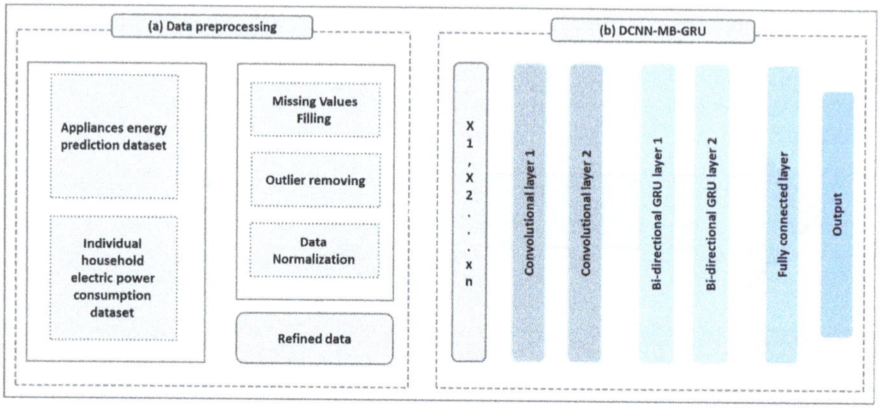

Figure 1. Two steps framework for electricity load prediction. (**a**) Applying different preprocessing techniques to refine the input raw dataset. (**b**) The proposed convolutional neural network (CNN)-multilayer bidirectional (MB)-gated recurrent unit (GRU) architecture to learn the patterns of electricity consumption.

2.1. Data Preprocessing

For better performance of the electricity load prediction models, the training data should be analyzed before training. As previously mentioned, the historical data of electricity consumption include abnormalities that affect the model performance. In this study, we used individual household electricity consumption prediction (IHEPC) and appliances load prediction (AEP) datasets, which consist of missing and outlier values. These abnormalities are removed from the data in the preprocessing step of the proposed framework. For missing values filling, NAN interpolation techniques are used, whereas for outlier values, three sigma rules of thumb [39] are applied. After the outlier reduction and filling missing values, the datasets are normalized using the min–max normalization technique to transform the dataset into a particular range which the neural network can learn easily.

2.2. Proposed CNN and MBGRU Architecture

This work combines a CNN with a multilayered bidirectional GRU (MB-GRU) for short-term electricity load prediction, where the CNN layers are incorporated to extract features from the preprocessed input data, and the MB-GRU model is used to learn the sequences between them. CNN is a neural network architecture that learns in a hierarchical manner whereby each layer learns more and more abstract features. The first layers learn atomic/primitive representations, while the intermediate-level layers learn intermediate abstract representations, and finally, fully connected layers learn high-level patterns. Therefore, the depth of the network is defined by the number of such layers. The higher the layer count, the deeper the network and can learn tiny representations. A CNN is a particular type of deep neural network that employs alternating layers of convolutions and pooling. It contains trainable filter banks per layer. Each individual filter in a filter bank, which is called a kernel and has a fixed receptive field (window) that is scanned over a layer below it, to compute an output feature map. The kernel performs a simple dot product and bias computation as it scans the layer below it and then feeds the result through an activation function, a rectifier, for example, to compute the output map. The output map is then subsampled using sum or max pooling, the latter being more common in order to reduce sensitivity to distortions in the upper layers. This process is alternated up to some point when the features become specific to the problem at hand. Thus, the CNN is a deep neural network for learning increase and more compact features that can later be used for recognition problems. The last few layers in a typical CNN comprise a typical fully connected neural network or support vector machines in order to recognize different combinations of features from the convolutional layers. The CNN architecture is used in different domains, such as image and video recognition [17,35,40,41], language processing [42,43], electricity load forecasting [44,45], crowed counting [46], etc. In the time series domain, the CNN layers are used to extract spatial information and then pass the output into sequential learning algorithms such as RNN LSTM and GRU.

RNN [47] is a sequence learning architecture with backward connections among hidden layers that include some kind of memory and is extensively used in several domains such as natural language processing [48], time series analysis [49], and speech recognition [50], visual data processing [51–53], etc. The RNN models generate output at each time stamp from the input data, which leads to the vanishing gradient problem. The RNN model forgets the long sequence of electricity data, such as 60-min resolution, which leads to loss of important information.

$$f_T = \odot(\acute{\omega}_f \cdot [h_{T-1}, a_T] + b_j) \tag{1}$$

$$i_T = \odot(\acute{\omega}_i \cdot [h_{T-1}, a_T] + b_i) \tag{2}$$

$$c_T = \tan h\,(\acute{\omega}_c \cdot [h_{T-1}, a_T] + b_c) \tag{3}$$

$$C_T = f_T \times C_{T-1} + i_T \times c_T \tag{4}$$

$$O_T = \odot(\acute{\omega}_O \cdot [h_{T-1}, a_T] + b_O) \tag{5}$$

$$h_T = O_T \times \tan h(\odot(C_T)). \tag{6}$$

The problem of losing long sequence information is addressed by LSTM using the three-gate mechanism input, output, and forget. The mathematical representation of each gate is shown in Equations (1)–(6). In these equations the output of each gates is represented through "i", "f" and "O" where "$\odot$" represents the activation function. The weights of the gates are represented through "$\acute{\omega}$", whereas" h_{T-1}" refers to the output of previous LSTM block and "b" represents the bias of the gates. The LSTM structure is complex and computationally expensive due to these gates' units and memory cells. To overcome the concern of LSTM, another lightweight architecture is developed called GRU [54] which comprises the reset and update gates. The mathematical representation of GRU gates are shown in Equations (7)–(10) where the update gate examines the earlier cell memory to remain active,

$$z_T = \odot(\acute{\omega}_z \,.\, [h_{T-1}, a_T] + b_z) \tag{7}$$

$$r_T = \odot(\acute{\omega}_r \,.\, [h_{T-1}, a_T] + b_r) \tag{8}$$

$$\tilde{h}_t = \tan h(\acute{\omega}_h \,.\, [r_T \cdot h_{T-1}, a_T] + b_h) \tag{9}$$

$$h_T = (1 - z_T) \cdot h_{T-1} + z_T \cdot \tilde{h}_T) \tag{10}$$

and the reset gate merges the next cell input sequence with previous cell memory. In this study, we used MB-GRU, which processes the sequence of input data in both backward and forward directions [55]. The bidirectional RNN models perform better in several domains such as classification, summarization [56], and load forecasting [57]. Therefore, in this study, we incorporate bidirectional GRU layers that contain both backward and forward layers, where the output sequence of the foreword layer is iteratively calculated through input in the positive sequence. The output of the backward layer is calculated through the reverse of the input.

The electricity consumption patterns include spatial and temporal features. Some researchers deployed a solo model that is insufficient to extract both types of features at a time. Therefore, in this work, we established a hybrid model that combines CNN with MB-GRU, as shown in Figure 1b. The proposed hybrid model includes an input layer, CNN layers, and bidirectional GRU layers. Two CNN layers are incorporated after several experiments over different layers and different parameters. Finally, we select filters of 8 and 4 for the first and second CNN layers with a kernel size of 3.1, respectively and used ReLU as an activation function in these layers. After convolutional layers, two bidirectional GRU layers are incorporated to learn the temporal information of the electricity historical data. Finally, the fully connected layers are integrated for the final output prediction.

3. Results and Discussion

In this section, we provide a detailed description of the dataset, evaluation metrics, and experimentation over the IHEPC and AEP datasets, and compare them with other baseline models. The model was trained over a GeFore GTX 2060 GPU with 64 GB RAM using the Keras framework with backend TensorFlow.

3.1. Datasets

The model's performance is assessed on two benchmark datasets, AEP and IHEPC [58,59]. The AEP dataset was recorded in 4.5 months in a residential house in 10 min resolution. This dataset comprises 29 various parameters of weather information (wind speed, humidity, dew point, temperature, and pressure), light, and appliance energy consumption, as presented in Table 1. The data samples were collected from both indoor and outdoor environments through a wireless sensor network. The outdoor data are collected from nearby airport. The building includes 9 indoor and 1 outdoor temperature sensors, 9 humidity sensors in which 7 are integrated with indoor environment and one is in outdoor environment. The outdoor pressure, visibility, temperature, humidity, and dew point are recorded nearby airport region. The IHEPC dataset includes 9 parameters which are; date, time, voltage, global-active-power (GAP), intensity, global-reactive-power (GRP) and three sub-metering as shown in Table 2. The dataset was recorded in a residential house in France during 2006 and 2010 for one-minute resolution.

Table 1. AEP dataset variables, a short description and its units.

S.no	Data Features	Units
1	Appliances: total energy consumption by appliances.	Wh
2	Light: total energy consumption by lights.	Wh
3	T1: demonstrate the temperature of kitchen.	C
4	RH1: demonstrate the humidity in kitchen.	%
5	T2: demonstrate the temperature of living room.	C
6	RH2: demonstrate the humidity of living room.	%
7	T3: demonstrate the temperature of laundry room.	C
8	RH3: demonstrate the humidity of laundry room.	%
9	T4: demonstrate the temperature of office room.	C
10	RH4: demonstrate the humidity of office room.	%
11	T5: demonstrate the temperature of bathroom.	C
12	RH5: demonstrate the humidity of bathroom.	%
13	RH6: demonstrate the outside temperature of building.	C
14	RH6: demonstrate the outside humidity of building.	%
15	T7: demonstrate the temperature of ironing room.	C
16	RH7: demonstrate the humidity in ironing room.	%
17	T8: demonstrate the temperature of teenager room.	C
18	RH8: demonstrate the humidity of teenager room.	%
19	T9: demonstrate the temperature of parent room.	C
20	RH9: demonstrate the humidity of parent room.	%
21	To: demonstrate the outside temperature which are collected from Chievres Weather Station (CWS).	C
22	Pressure: outside pressure which are collected from CWS.	Mm Hg
23	Rho: demonstrate the outside humidity from CWS.	%
24	Wind speed: outside wind speed which are collected from CWS.	m/s
25	Visibility: outside visibility from CWS.	Km
26	Tdewpoint: outside Tdewpoint from CWS.	C

3.2. Metrics of Evaluation

To evaluate the performance of the model, we used RMSE, MSE, and MAE metrics. The mathematical representation of these metrics is depicted in Equations (11)–(13). RMSE calculates the difference between all predicted data points and the actual data point, then compute the mean of these square errors and finally calculate the square root of the mean values. The MSE calculates the mean disparity between the actual and model output values. The MAE calculates the mean absolute difference between the actual and predicted values.

$$RMSE = \sqrt{\frac{1}{n}\sum_{1}^{n}(y-\hat{y})^2} \tag{11}$$

$$MSE = \frac{1}{n}\sum_{1}^{n}(y-\hat{y})^2 \tag{12}$$

$$\mathrm{MAE} = \frac{1}{n}\sum_{1}^{n}|y-\hat{y}| \tag{13}$$

3.3. Experimentations over IHEPC, AEP Dataset and Comparison with other Models

In this section, we evaluate the performance comparison of the proposed model with existing models for short-term load prediction (one hour ahead) over the IHEPC and AEP datasets. For the IHEPC dataset, the proposed model achieved 0.42, 0.18, and 0.29 RMSE, MSE, and MAE, respectively. The performance prediction over the test data is displayed in Figure 2a. A comparison of the proposed model over IHEPC dataset for short-term load prediction with other baseline models is shown in Figure 3. For more detail the performance of the proposed model is compared with [1,17,33–35,60,61] in the short-term horizon. In [60] the authors used deep learning methodology for residential load prediction and obtained 0.79 RMSE and 0.59 MAE, whereas [33] used a CNN-LSTM hybrid network for short-term residential load prediction and achieved 0.59, 0.35, and 0.33 RMSE, MSE, and MAE, respectively. 0.47, 0.19 and 0.31 RMSE, MSE and MAE was reported in [17] whereas [34] reports 0.56, 0.31 and 0.34 values for these metrics. In [61] the authors achieved 0.38 MSE and 0.39 MAE, whereas [1] reported 0.66 RMSE. Another strategy presented in [35] attained 0.47, 0.22, and 0.33 RMSE, MSE, and MAE, respectively. Among these results, the proposed model achieved the lowest error rate for short-term electric load prediction.

Table 2. Individual household electricity consumption prediction (IHEPC) dataset variables description and units.

S.no	Data Features	Units
1	Date	dd/mm/yyyy
2	Time	hh: mm: ss
3	GAP	Kw
4	GRP	Kw
5	Voltage	V
6	Global intensity	Amp
7	Submeetring-1	W
8	Submeetring-2	W
9	Submeetring-3	W

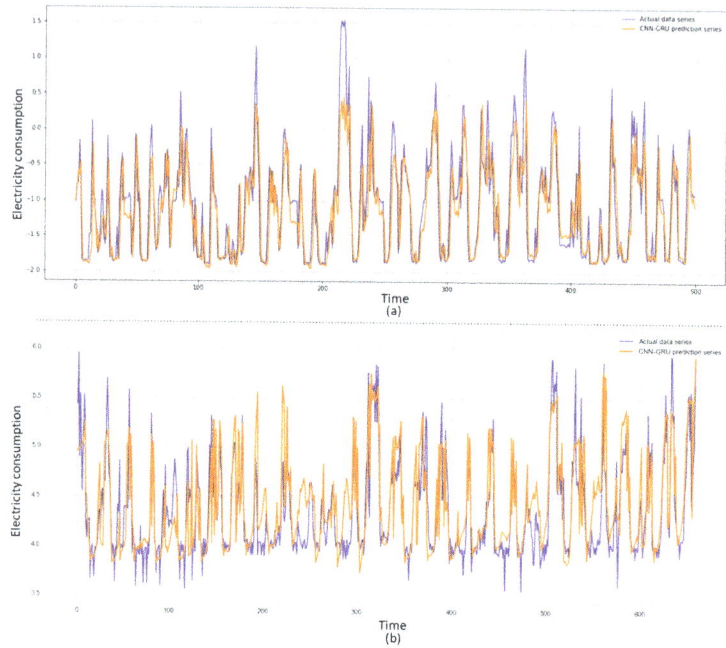

Figure 2. Predictions results of our framework over test data (**a**) actual and predicted values for IHEPC dataset (**b**) actual and predicted values for AEP dataset.

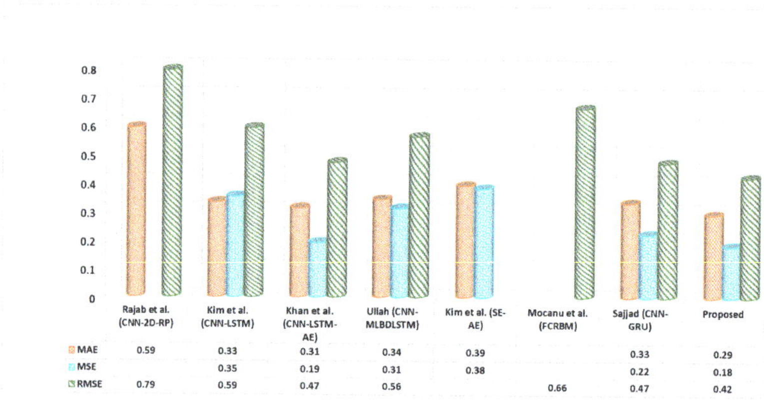

Figure 3. Performance comparison of the proposed model with the other state-of-the-art models over IHEPC dataset.

Furthermore, the effectiveness of the proposed model is evaluated over the AEP dataset for a short-term horizon. For the AEP dataset, the proposed model attained 0.31, 0.10, 0.33 RMSE, MSE, and MAE, respectively, whereas the prediction results are shown in Figure 2b. Similarly, the effectiveness of the proposed model over the AEP dataset is also compared with other baseline models, as shown in Figure 4. For instance, the results are compared with [34,62–64]. For further details, Ref. [62] achieved 0.59 RMSE and 0.26 MSE, Ref. [63] achieved 0.35 RMSE and 0.66 MSE, and [64] achieved a 0.59 RMSE. In [35], authors achieved 0.31, 0.09, and 0.24 scores for RMSE, MSE, and MAE, respectively. Compared to these models, the proposed model performed better in reducing the RMSE and MAE error rate, while only the results of Sajjad et al. [35] in terms of MSE is better than the proposed model in the short-term horizon.

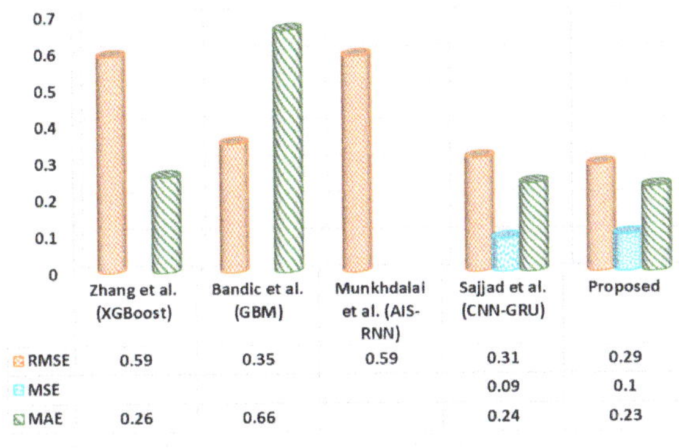

Figure 4. Performance contrast of the proposed model with the other baseline models over AEP dataset.

4. Conclusions

In this study, we established a two-step methodology for short-term load prediction. In the first step, we performed data preprocessing over raw data to refine it for training. The refinement of the data is important because the historical data of energy consumption is generated from smart meter sensors, which include abnormalities such as outliers, missing values etc. These abnormalities from the raw data are extracted in this step, and finally, the normalization technique is applied to transform the data into a specific range. The second step is the hybrid model, which is a combination of CNN and multilayered bi-directional GRU (MB-GRU). The CNN layers are incorporated to extract important features from the refined data, while the MB-GRU layers are used to learn the temporal information of electricity consumption data. The proposed methodology is tested over two challenging datasets and achieves better performance when compared to other methods, as demonstrated in the results section.

Author Contributions: Conceptualization, Z.A.K.; methodology, Z.A.K.; software, Z.A.K.; validation, W.U. and A.U.; formal analysis, M.Y.L.; investigation, A.U.; resources, S.W.B.; data curation, Z.A.K.; writing—original draft preparation, Z.A.K.; writing—review and editing, Z.A.K., W.U. and A.U.; visualization, W.U.; supervision, S.W.B.; project administration, M.L.; funding acquisition, S.W.B. and S.R. All authors have read and agreed to the published version of the manuscript.

Funding: This work was supported by the National Research Foundation of Korea (NRF) grant funded by the Korean government (MSIT) (No. 2019M3F2A1073179).

Conflicts of Interest: The authors declare no conflict of interest.

References

1. Mocanu, E.; Nguyen, P.H.; Gibescu, M.; Kling, W.W. Deep Learning for Estimating Building Energy Consumption. *Sustain. Energy Grids Netw.* **2016**, *6*, 91–99. [CrossRef]
2. Guo, Z.; Zhou, K.; Zhang, C.; Lu, X.; Chen, W.; Yang, S. Residential Electricity Consumption Behavior: Influencing Factors, Related Theories and Intervention Strategies. *Renew. Sustain. Energy Rev.* **2018**, *81*, 399–412. [CrossRef]
3. IEA. *International Energy Outlook*; IEA: Paris, France, 2014; Volume 18.
4. Nejat, P.; Jomehzadeh, F.; Taheri, M.M.; Gohari, M.; Majid, M.Z.A. A Global Review of Energy Consumption, CO 2 Emissions and Policy in the Residential Sector (with an Overview of the Top Ten CO 2 Emitting Countries). *Renew. Sustain. Energy Rev.* **2015**, *43*, 843–862. [CrossRef]

5. Amarasinghe, K.; Wijayasekara, D.; Carey, H.; Manic, M.; He, D.; Chen, W.-P. Artificial Neural Networks Based Thermal Energy Storage Control for Buildings. In Proceedings of the 41st Annual Conference of the IEEE Industrial Electronics Society, Jokohama, Japan, 9–12 November 2015; pp. 005421–005426.
6. Yang, A.; Li, W.; Yang, X. Short-Term Electricity Load Forecasting Based on Feature Selection and Least Squares Support Vector Machines. *Knowl. Based Syst.* **2019**, *163*, 159–173. [CrossRef]
7. Koprinska, I.; Rana, M.; Agelidis, V.G. Correlation and Instance Based Feature Selection for Electricity Load Forecasting. *Knowl. Based Syst.* **2015**, *82*, 29–40. [CrossRef]
8. Khwaja, A.S.; Anpalagan, A.; Naeem, M.; Venkatesh, B. Joint Bagged-Boosted Artificial Neural Networks: Using Ensemble Machine Learning to Improve Short-Term Electricity Load Forecasting. *Electr. Power Syst. Res.* **2020**, *179*, 106080. [CrossRef]
9. Heydari, A.; Nezhad, M.M.; Pirshayan, E.; Garcia, D.A.; Keynia, F.; De Santoli, L. Short-Term Electricity Price and Load Forecasting in Isolated Power Grids Based on Composite Neural Network and Gravitational Search Optimization Algorithm. *Appl. Energy* **2020**, *277*, 115503. [CrossRef]
10. Zhang, C.; Li, J.; Zhao, Y.; Li, T.; Chen, Q.; Zhang, X. A Hybrid Deep Learning-Based Method for Short-Term Building Energy Load Prediction Combined with an Interpretation Process. *Energy Build.* **2020**, *225*, 110301. [CrossRef]
11. Naspi, F.; Arnesano, M.; Zampetti, L.; Stazi, F.; Revel, G.M.; D'Orazio, M. Experimental Study on occupants' Interaction with Windows and Lights in Mediterranean Offices during the Non-Heating Season. *Build. Environ.* **2018**, *127*, 221–238. [CrossRef]
12. Stazi, F. *Thermal Inertia in Energy Efficient Building Envelopes*; Elsevier: Amsterdam, The Netherlands, 2017.
13. Rupp, R.F.; Ghisi, E. Assessing Window Area and Potential for Electricity Savings by Using Daylighting and Hybrid Ventilation in Office Buildings in Southern Brazil. *Simulation* **2017**, *93*, 935–949. [CrossRef]
14. Pereira, P.F.; Ramos, N.M.M. Influence of Occupant Behaviour on the State of Charge of a Storage Battery in a Nearly-Zero Energy Building. In *E3S Web of Conferences*; EDP Sciences: Paris, France, 2020; Volume 172, p. 16010.
15. Bot, K.; Ramos, N.M.; Almeida, R.M.; Pereira, P.F.; Monteiro, C. Energy Performance of Buildings With on-site Energy Generation and Storage—An Integrated Assessment Using Dynamic Simulation. *J. Build. Eng.* **2019**, *24*, 100769. [CrossRef]
16. Han, T.; Muhammad, K.; Hussain, T.; Lloret, J.; Baik, S.W. An Efficient Deep Learning Framework for Intelligent Energy Management in IoT Networks. *IEEE Internet Things J.* **2020**, *99*, 1. [CrossRef]
17. Khan, Z.A.; Hussain, T.; Ullah, A.; Rho, S.; Lee, M.; Baik, S.W. Towards Efficient Electricity Forecasting in Residential and Commercial Buildings: A Novel Hybrid CNN with a LSTM-AE Based Framework. *Sensors* **2020**, *20*, 1399. [CrossRef]
18. Wei, L.; Zhen-Gang, Z. Based on Time Sequence of ARIMA Model in the Application of Short-Term Electricity Load Forecasting. In Proceedings of the International Conference on Research Challenges in Computer Science, Shanghai, China, 28–29 December 2009; pp. 11–14.
19. Hong, T.; Gui, M.; Baran, M.E.; Willis, H.L. Modeling and Forecasting Hourly Electric Load by Multiple Linear Regression with Interactions. In Proceedings of the IEEE PES General Meeting, Minneapolis, MN, USA, 25–29 July 2010; pp. 1–8.
20. Al-Hamadi, H.; Soliman, S. Fuzzy Short-Term Electric Load Forecasting Using Kalman Filter. *IEE Proc. Gener. Transm. Distrib.* **2006**, *153*, 217–227. [CrossRef]
21. Ullah, A.; Haydarov, K.; Haq, I.U.; Muhammad, S.; Rho, S.; Lee, M.Y.; Baik, S.W. Deep Learning Assisted Buildings Energy Consumption Profiling Using Smart Meter Data. *Sensors* **2020**, *20*, 873. [CrossRef]
22. Lu, Y.; Tian, Z.; Peng, P.; Niu, J.; Li, W.; Zhang, H. GMM Clustering for Heating Load Patterns in-depth Identification and Prediction Model Accuracy Improvement of District Heating System. *Energy Build.* **2019**, *190*, 49–60. [CrossRef]
23. Lahouar, A.; Slama, J.B.H. Day-Ahead Load Forecast Using Random Forest and Expert Input Selection. *Energy Convers. Manag.* **2015**, *103*, 1040–1051. [CrossRef]
24. Chen, Y.; Luh, P.B.; Guan, C.; Zhao, Y.; Michel, L.D.; Coolbeth, M.A.; Friedland, S.; Rourke, S.J. Short-Term Load Forecasting: Similar Day-Based Wavelet Neural networks. *IEEE Trans. Power Syst.* **2009**, *25*, 322–330. [CrossRef]
25. Wang, Y.; Xia, Q.; Kang, C. Secondary Forecasting Based on Deviation Analysis for Short-Term Load Forecasting. *IEEE Trans. Power Syst.* **2011**, *26*, 500–507. [CrossRef]

26. Tsekouras, G.; Hatziargyriou, N.; Dialynas, E. An Optimized Adaptive Neural Network for Annual Midterm Energy Forecasting. *IEEE Trans. Power Syst.* **2006**, *21*, 385–391. [CrossRef]
27. Li, S.; Wang, P.; Goel, L. A Novel Wavelet-Based Ensemble Method for Short-Term Load Forecasting With Hybrid Neural Networks and Feature Selection. *IEEE Trans. Power Syst.* **2015**, *31*, 1788–1798. [CrossRef]
28. Amarasinghe, K.; Marino, D.L.; Manic, M. Deep Neural Networks for Energy Load Forecasting. In Proceedings of the 26th International Symposium on Industrial Electronics (ISIE), Edinburgh, UK, 19–21 June 2017; pp. 1483–1488.
29. Khan, S.; Javaid, N.; Chand, A.; Khan, A.B.M.; Rashid, F.; Afridi, I.U. Electricity Load Forecasting for Each Day of Week Using Deep CNN. In *Advances in Intelligent Systems and Computing*; Springer: Berlin/Heidelberg, Germany, 2019; pp. 1107–1119.
30. Tokgoz, A.; Unal, G. A RNN Based Time Series Approach for Forecasting Turkish Electricity Load. In Proceedings of the 26th Signal Processing and Communications Applications Conference (SIU), Izmir, Turkey, 2–5 May 2018; pp. 1–4.
31. Wang, J.Q.; Du, Y.; Wang, J. LSTM Based Long-Term Energy Consumption Prediction with Periodicity. *Energy* **2020**, *197*, 117197. [CrossRef]
32. Rahman, A.; Srikumar, V.; Smith, A.D. Predicting Electricity Consumption for Commercial and Residential Buildings Using Deep Recurrent Neural Networks. *Appl. Energy* **2018**, *212*, 372–385. [CrossRef]
33. Kim, T.-Y.; Cho, S.-B. Predicting Residential Energy Consumption Using CNN-LSTM Neural Networks. *Energy* **2019**, *182*, 72–81. [CrossRef]
34. Ullah, F.U.M.; Ullah, A.; Haq, I.U.; Rho, S.; Baik, S.W. Short-Term Prediction of Residential Power Energy Consumption via CNN and Multi-Layer Bi-Directional LSTM Networks. *IEEE Access* **2020**, *8*, 123369–123380. [CrossRef]
35. Sajjad, M.; Khan, Z.A.; Ullah, A.; Hussain, T.; Ullah, W.; Lee, M.Y.; Baik, S.W. A Novel CNN-GRU-Based Hybrid Approach for Short-Term Residential Load Forecasting. *IEEE Access* **2020**, *8*, 143759–143768. [CrossRef]
36. Afrasiabi, M.; Mohammadi, M.; Rastegar, M.; Stankovic, L.; Afrasiabi, S.; Khazaei, M. Deep-Based Conditional Probability Density Function Forecasting of Residential Loads. *IEEE Trans. Smart Grid* **2020**, *11*, 3646–3657. [CrossRef]
37. Bunn, D. Forecasting Loads and Prices in Competitive Power Markets. *Proc. IEEE* **2000**, *88*, 163–169. [CrossRef]
38. Hobbs, B.F.; Jitprapaikulsarn, S.; Konda, S.; Chankong, V.; Loparo, K.A.; Maratukulam, D.J. Analysis of the Value for Unit Commitment of Improved Load Forecasts. *IEEE Trans. Power Syst.* **1999**, *14*, 1342–1348. [CrossRef]
39. Chandola, V.; Banerjee, A.; Kumar, V. Anomaly Detection: A survey. *ACM Comput. Surv.* **2009**, *41*, 1–58. [CrossRef]
40. Sajjad, M.; Zahir, S.; Ullah, A.; Akhtar, Z.; Muhammad, K. Human Behavior Understanding in Big Multimedia Data Using CNN Based Facial Expression Recognition. *Mob. Netw. Appl.* **2019**, *25*, 1611–1621. [CrossRef]
41. Haq, I.U.; Ullah, A.; Muhammad, K.; Lee, M.Y.; Baik, S.W. Personalized Movie Summarization Using Deep CNN-Assisted Facial Expression Recognition. *Complexity* **2019**, *2019*, 1–10. [CrossRef]
42. Young, T.; Hazarika, D.; Poria, S.; Cambria, E. Recent Trends in Deep Learning Based Natural Language Processing [Review Article]. *IEEE Comput. Intell. Mag.* **2018**, *13*, 55–75. [CrossRef]
43. Mustaqeem; Kwon, S. MLT-DNet: Speech Emotion Recognition Using 1D Dilated CNN Based on Multi-Learning Trick Approach. *Expert Syst. Appl.* **2020**, 114177. [CrossRef]
44. Chen, K.; Chen, K.; Wang, Q.; He, Z.; Hu, J.; He, J. Short-Term Load Forecasting with Deep Residual Networks. *IEEE Trans. Smart Grid* **2018**, *10*, 3943–3952. [CrossRef]
45. Wang, H.; Yi, H.; Peng, J.; Wang, G.; Liu, Y.; Jiang, H.; Liu, W. Deterministic and Probabilistic Forecasting of Photovoltaic Power Based on Deep Convolutional Neural Network. *Energy Convers. Manag.* **2017**, *153*, 409–422. [CrossRef]
46. Khan, N.; Ullah, A.; Haq, I.U.; Menon, V.G.; Baik, S.W. SD-Net: Understanding Overcrowded Scenes in Real-Time via an Efficient Dilated Convolutional Neural Network. *J. Real Time Image Process.* **2020**, 1–15. [CrossRef]

47. Jozefowicz, R.; Zaremba, W.; Sutskever, I. An Empirical Exploration of Recurrent Network architectures. In Proceedings of the International Conference on Machine Learning, Lille, France, 6–11 July 2015; pp. 2342–2350.
48. Yin, W.; Kann, K.; Yu, M.; Schütze, H.J.A.P.A. Comparative Study of cnn and rnn for Natural Language processing. *arXiv* **2017**, arXiv:1702:01923.
49. Liu, Y.; Gong, C.; Yang, L.; Chen, Y. DSTP-RNN: A Dual-Stage Two-Phase Attention-Based Recurrent Neural Network for Long-Term and Multivariate Time Series Prediction. *Expert Syst. Appl.* **2020**, *143*, 113082. [CrossRef]
50. Guo, J.; Tiwari, G.; Droppo, J.; Van Segbroeck, M.; Huang, C.-W.; Stolcke, A.; Maas, R. Efficient Minimum Word Error Rate Training of RNN-Transducer for End-to-End Speech Recognition. *Interspeech* **2020**. [CrossRef]
51. Ullah, A.; Muhammad, K.; Hussain, T.; Baik, S.W. Conflux LSTMs Network: A Novel Approach for Multi-View Action Recognition. *Neurocomputing* **2020**.
52. Ullah, A.; Muhammad, K.; Hussain, T.; Lee, M.; Baik, S.W. Deep LSTM-Based Sequence Learning Approaches for Action and Activity Recognition. In *Deep Learning in Computer Vision*; Informa UK Limited: Colchester, UK, 2020; pp. 127–150.
53. Ullah, W.; Ullah, A.; Haq, I.U.; Muhammad, K.; Sajjad, M.; Baik, S.W. CNN Features with Bi-Directional LSTM for Real-Time Anomaly Detection in Surveillance Networks. *Multimed. Tools Appl.* **2020**, 1–17. [CrossRef]
54. Chung, J.; Gulcehre, C.; Cho, K.; Bengio, Y. Empirical evaluation of gated recurrent neural networks on sequence modeling. *arXiv* **2014**, arXiv:1412:3555.
55. Schuster, M.; Paliwal, K. Bidirectional Recurrent Neural Networks. *IEEE Trans. Signal Process.* **1997**, *45*, 2673–2681. [CrossRef]
56. Hussain, T.; Muhammad, K.; Ullah, A.; Cao, Z.; Baik, S.W.; De Albuquerque, V.H.C. Cloud-Assisted Multi-View Video Summarization Using CNN and Bi-Directional LSTM. *IEEE Trans. Ind. Inform.* **2019**. [CrossRef]
57. Tang, X.-L.; Dai, Y.; Wang, T.; Chen, Y. Short-Term Power Load Forecasting Based on Multi-Layer Bidirectional Recurrent Neural Network. *IET Gener. Transm. Distrib.* **2019**, *13*, 3847–3854. [CrossRef]
58. Repository Appliances Energy Prediction Data Set. Available online: https://archive.ics.uci.edu/ml/Datasets/Appliances+energy+prediction (accessed on 15 October 2020).
59. UCI. Individual Household Electric Power Consumption Data Set. Available online: https://archive.ics.uci.edu/ml/datasets/Individual+household+electric+power+consumption (accessed on 15 October 2020).
60. Rajabi, R.; Estebsari, A. Deep Learning Based Forecasting of Individual Residential Loads Using Recurrence Plots. In *IEEE Milan PowerTech*; IEEE: New York, NY, USA, 2019; pp. 1–5.
61. Kim, J.-Y.; Cho, S.-B. Electric Energy Consumption Prediction by Deep Learning with State Explainable Autoencoder. *Energies* **2019**, *12*, 739. [CrossRef]
62. Zhang, T.; Liao, L.; Lai, H.; Liu, J.; Zou, F.; Cai, Q. Electrical Energy Prediction with Regression-Oriented Models. In *Advances in Intelligent Systems and Computing*; Springer: Berlin/Heidelberg, Germany, 2018; pp. 146–154.
63. Bandic, L.; Kevric, J. Near Zero-Energy Home Prediction of Appliances Energy Consumption Using the Reduced Set of Features and Random Decision Tree Algorithms. In *Advances on P2P, Parallel, Grid, Cloud and Internet Computing*; Springer: Berlin/Heidelberg, Germany, 2018; pp. 164–171.
64. MunkhDalai, L.; MunkhDalai, T.; Park, K.H.; Amarbayasgalan, T.; Erdenebaatar, E.; Park, H.W.; Ryu, K.H. An End-to-End Adaptive Input Selection with Dynamic Weights for Forecasting Multivariate Time Series. *IEEE Access* **2019**, *7*, 99099–99114. [CrossRef]

applied sciences

MDPI

Article

Evaluating the Impact of the COVID-19 Pandemic on Residential Energy Use in Los Angeles

Michael J. Klopfer [1], Joy E. Pixley [1], Armen Saiyan [2], Amir Tabakh [2], David Jacot [2] and Guann-Pyng Li [1,*]

[1] California Plug Load Research Center (CalPlug), California Institute of Telecommunications and Information Technology (Calit2), University of California, Irvine, CA 92697, USA; mklopfer@uci.edu (M.J.K.); jpixley@uci.edu (J.E.P.)

[2] Los Angeles Department of Water and Power (LADWP), Los Angeles, CA 90051, USA; Armen.Saiyan@ladwp.com (A.S.); Amir.Tabakh@ladwp.com (A.T.); david.jacot@ladwp.com (D.J.)

* Correspondence: gpli@uci.edu; Tel.: +1-(949)-824-9073

Featured Application: This work is potentially useful for updating residential energy usage models, considering presented energy impacts of the COVID-19 pandemic.

Abstract: The 2020 COVID-19 pandemic provided an opportunity to assess energy use during times of emergency that disrupt daily and seasonal patterns. The authors present findings from a regional evaluation in the city of Los Angeles (California, USA) with broad application to other areas and demonstrate an approach for isolating and analyzing residential loads from community-level electric utility feeder data. The study addresses effects on residential energy use and the implications for future energy use models, energy planning, and device energy standards and utility program development. In this study we review changes in residential energy use during the progression of the COVID-19 pandemic from four residential communities across Los Angeles covering approximately 6603 households within two microclimate sub regional areas (Los Angeles Basin and San Fernando Valley). Analyses address both absolute and seasonal temperature-corrected energy use changes while assessing estimated changes on energy usage from both temperature-sensitive loads (e.g., air conditioning and electric heating) and non-temperature-sensitive loads (e.g., consumer electronics and major appliance use). An average 5.1% increase in total residential energy use was observed for non-temperature sensitive loads during the pandemic period compared to a 2018–2019 baseline. During mid-spring when shelter in place activity was highest a peak monthly energy use of 20.9% increase was seen compared to a 2018–2019 composite baseline. Considering an average of the top five warmest summer days, a 9.5% increase in energy use was observed for events during summer 2020 compared to summer 2018 (a year with similar magnitude summer high heat events). Based on these results, a potential trend is identified for increased residential load during pandemics and other shelter-in-place disruptions, net of any temperature-sensitive load shifts with greater impacts expected for lower-income communities.

Keywords: residential energy modeling; COVID-19; coronavirus pandemic; temperature sensitivity; energy security

Citation: Klopfer, M.J.; Pixley, J.E.; Saiyan, A.; Tabakh, A.; Jacot, D.; Li, G.-P. Evaluating the Impact of the COVID-19 Pandemic on Residential Energy Use in Los Angeles. *Appl. Sci.* **2021**, *11*, 4476. https://doi.org/10.3390/app11104476

Academic Editor: Matti Lehtonen

Received: 12 April 2021
Accepted: 7 May 2021
Published: 14 May 2021

1. Introduction

In 2020, changes in energy use and emissions were seen worldwide as a direct effect of the COVID-19 pandemic [1–6]. Mandatory stay-at-home periods globally reduced jet and aviation fuel by 50%, gasoline by 30%, and electricity (on average) about 10 percent during the early pandemic where shelter-in-place (SIP) orders were widespread across many regions. This reduction was followed by partial rebounds for all mentioned energy types later in 2020 [2,7–11]. While commercial transport and mobility to support commercial activities (e.g., commuting for work) were greatly reduced by a curtailment in overall business activities, the impact on residential energy use is harder to directly assess from publicly available electrical grid regional operator data. Preliminary results from studies early in the pandemic suggest increased residential energy use, but results vary [12–14]. Further, little attention has been paid to understanding the mechanisms leading to this change in energy use during both the early pandemic SIP periods and periods following, in addition to regressive periods due to regional re-closures due to increased COVID-19 cases.

Analysis of total energy use for a given region provides conclusions for macro trends. However, analyzing data comprised of heavily mixed sectors (residential and commercial loads) and as a combined set across all day types (weekends and weekdays) provides limited utility for sector-based analysis, and complicates actionable model adjustments for energy planning and conservation efforts. Approximately 21% of energy use nationwide is from residential customers [15]. Residential energy efficiency is a substantial focus for utility programs, but sector changes can be obscured within direct regional load figures. While a general decrease in energy use was broadly observed across most regions worldwide during the 2020 COVID-19 pandemic, modeling and planning difficulties when predicting future demand led to service disruptions. Most notably, poor forecasting models for pandemic-related changes in energy use directly led to widespread rolling blackouts in California in mid-August of 2020 during a substantial heatwave [16,17]. The 2020 pandemic period exhibited increased reliance on non-dispatchable, low carbon energy sources, with increases of 22.3% solar production and 13.5% wind production in the US compared to 2019 [18]. Understanding sector-focused changes in energy use helps improve demand predictions for future widespread lockdown events in an era of increasing effects of climate change and increased reliance on non-dispatchable and distributed generation.

Residential electric load is primarily comprised of the following major load categories: electricity-driven space conditioning (air conditioning, ventilation/forced air circulation, and electric heaters), lighting, major appliances, miscellaneous (plug) loads, constant building loads, and electric transportation. Of these categories, only space conditioning is directly temperature sensitive. Demand from three other categories—lighting, major appliances, and plug loads—is largely driven by occupancy without substantial regard to ambient temperature. With 42% of US residential use due to space conditioning, ambient temperature is a primary driver of residential electricity use, especially with high air conditioning penetration [15,18,19]. Despite the mild climate in Southern California, Chen et al. assessed a substantial (69% estimated) regional household penetration for air conditioning [19,20]. This includes residential air conditioning systems in different form factors and cooling capacities. For temperature-sensitive loads, both increased occupancy and the reaction of occupants to change in the ambient temperature affect energy use. For the remaining categories, changes in daily occupancy rates (occupied by none versus one or more individuals) and resulting changes in device use behavior (i.e., which loads or devices are used and how they are used) are the main considerations.

Residential occupancy shifted substantially for much of the population during the pandemic, particularly early in the pandemic timeline. While exact assessments of stay-at-home rates are difficult, general trends show higher rates of SIP compliance early-on following the first COVID-19 case wave with proportional compliance (SIP compliance compared to present active COVID-19 cases) generally dropping during the following COVID-19 case waves throughout 2020. In Los Angeles County, mobility data indicates estimated stay-at-home rates of 50.6% of individuals on April 11 and dropping to 35.5% of individuals by September 1 (compared to approximately 25% during mid-February) [21]. Similarly, in a national Gallup study, 49% of respondents reported being likely to shelter in place if asked to during a third surge in late 2020, compared to 67% in early April 2020 during the first surge [22]. SIP restrictions reduced leisure activities in evenings and especially on weekends, but primarily impacted weekday occupancy through three mechanisms: a shift toward working from home, reduced access to educational facilities for students and educators, and increased unemployment [23]. During 2020, Los Angeles experienced a maximum unemployment rate of 18.8% in May 2020 with a recovery to 12.3% by December 2020 compared to a pre-pandemic level of 4.9% in February 2020 (nonseasonally adjusted) [24]. The majority of jobs lost across the USA (as in other countries) were in leisure, hospitality, entertainment, manufacturing, and food services sectors, with pandemic-related job loss disproportionally impacting women, younger workers, and workers with less education [25]. Minor shifts in population impacting household size also occurred during the early pandemic: in a June 2020 Pew Research Center study 6% of respondents reported gaining a household member and 3% reported moving because of the pandemic. Of those who moved, 61% of respondents reporting moving into a family member's home. The shutdown of college campuses (25%), the desire to be with family (20%), and financial related reasons (18%) were major relocation catalysts, and relocations were highest among young adults (ages 18–29) [26].

The current study analyzed energy use data from distribution station feeder loads, specific to defined geographic areas in the city of Los Angeles, accessed using generalized utility supervisor control data acquisition (SCADA). Such grouped load data is often the only measure available. Prior investigations have identified limitations in using it in standard linear regression-based energy prediction models due to autocorrelation and homoscedasticity. There are also limits when relying on temperature data at high time scale resolutions (e.g., per day), given the shifts in energy use corresponding to behavior variation over the course of the day. However, for certain use cases comparing daily average energy use to daily temperature data has been demonstrated to provide satisfactory estimation figures [27,28]. Here, the authors demonstrate an approach for analyzing grouped load data and daily temperature values to provide insight into how energy use changes due to widespread emergency conditions such as the COVID-19 pandemic.

With a diverse population and a warm, dry climate and typically temperate spring, Los Angeles provides a near-ideal environment to assess the impact of the pandemic on residential utility customers, especially assessing non-temperature sensitive load contribution to total residential energy use. In addition, the city of Los Angeles provides a useful case study because it was substantially impacted by the COVID-19 pandemic in both number of COVID-19 cases in addition to state and local restrictions on business, services, and travel. California's aggressive stay-at-home order was initiated on 19 March and was followed by a relaxation in June, a partial reinstatement in July (following the start of a second wave of COVID-19 cases), a relaxation in September and an amended limited stay at home order issued in late November following through the end of the year (in response to a third wave of COVID-19 cases). Los Angeles County (the major regional health reporting resource covering the city of Los Angeles) suffered three successively increasing waves of COVID-19 case peaks in 2020, occurring on 8 April, 22 July, and 27 December with this one county representing 32% of all cases statewide (note that approximately 25.5% of California's population lives in Los Angeles County) [29,30]. As of 31 December, 7.7% of the LA county population had been infected with COVID-19. The city of Los Angeles regularly maintained stricter controls on business activities to reduce population movement compared to both state and county COVID-19 guidelines [30,31]. A follow-up SIP order to the one issued in spring focused on Los Angeles, beginning 30 November and continuing through 31 December, this was the strictest order in the state of California, effectively banning most outdoor gatherings, restricting employment travel, and reducing retail capacity. Accordingly, the city of Los Angeles provides a rich opportunity to draw transferrable lessons on energy responses to major behavioral shifts.

2. Materials and Methods

2.1. Feeder and Data Selection

This study used Los Angeles Department of Water and Power (LADWP) municipal electrical substation 4.7 kV customer distribution feeder net loads (reported in hourly average kW load values) servicing a designated geographic territory within the city of Los Angeles to provide serviced population load data. Comparison baselines were created from composites of 2018 and 2019 load data from individual feeders, while the evaluation period was initiated with the California SIP order on 19 March 2020 and continued through the end of reporting on 31 December 2020 [31,32]. For comparison as needed an evaluation period beginning on 1 January 2020 is used to report energy use change with respect to calendar year. To compare heatwave-related events, period to period comparisons between single years were used. From a pool of all available feeders the authors performed a two-tier screening process. The first tier selected for feeders that serviced primarily residential customers (greater than 90% residential customers with largely negligible pre-pandemic observable commercial loading patterns). The second tier selected for diversity across the city considering community location, community income, and community building types. Feeders were then excluded for substantial service interruptions or for major changes from 2018 through 2020 rendering those periods non-comparable, including substantial changes in customer base, major new construction, building demolition, or zoning changes. Feeders were also excluded for exhibiting great heterogeneity of income across neighborhoods served by the same feeder, with the exception of Feeder A providing service to Section 8 subsidized low-income housing in Watts. Four feeders were ultimately selected for the current analysis, serving distinct communities across Los Angeles with a total residential customer base of approximately 6603 combined residential customers covering areas with a range of median incomes (see Table 1).

Table 1. Summary of sampled feeders including feeder service area and service demographic information.

Feeder	Location (ZIP Code) [1]	Residential Customers	Region, WS ICAO Code [2]	Median Income (Specific Feeder) [3]
A (primary)	Watts (90059)	2563 (91.1%) [4,5]	LA Basin, KCQT	$51,635 ($15,584)
B (primary)	Southeast LA (90037)	1086 (90.1%) [4]	LA Basin, KCQT	$44,965 ($37,004)
C (primary)	Toluca Lake (91602)	1214 (96.5%) [4]	SF Valley, KVNY	$109,254 ($49,039)
D (primary)	Burbank (91601) [6]	1740 (90.0%) [4,7]	SF Valley, KBUR	$72,868 ($51,003)
E (example)	Central Wilshire (90036)	1320 (94.6%) [8]	LA Basin, KCQT	$117,596 ($103,242)
F (example)	Downtown (90014)	0 (0.0%) [9]	LA Basin, KCQT	N/A
NPL	All Los Angeles City	1.24 M (90.7%)	Entire city (KCQT, reference)	$62,142

Refer to Supplementary Table S1 for additional details. [1] Presented with community name, reference zip/postal code tabulation area (ZCTA) inclusive of served feeder area; note that Los Angeles-Long Beach Census tract codes inclusive of feeder service area are presented in the supplementary extension of this table. [2] Los Angeles (LA) Basin or San Fernando (SF) Valley; NWS weather station (WS), ICAO airport code used for identification; temperature reference and corresponding microclimate region. [3] Median income of service area inclusive ZCTA, and specific feeder service area median income. [4] Feeder service area includes single and multi-family homes, small apartment complexes. [5] Feeder service area includes public housing. [6] This feeder corresponds to a service area bordering North Hollywood (Los Angeles) and Burbank and is served by LADWP. [7] Single and multi-family homes, small apartment complexes near commercial district. [8] Feeder service area includes a large apartment community, low rise with numerous common facilities. [9] Pair of mid-rise buildings mixed retail, mercantile, offices, buildings in LA Jewelry District.

Feeders were evaluated across the period of investigation from 2018 through 2020. Customer construction permit records indicate <6% mid-day solar load contribution total, with slow growth, and 2017 motor vehicle records showed <5% average customer EV penetration average across all primary evaluated feeder service areas (see Table S1). Both factors suggest a low overall impact such that the change between the pandemic and pre-pandemic periods for the evaluated feeders and accordingly the differential impact from solar and EV loads are treated as negligible. The majority of the building types represented were a mixture of single-family homes and small multi-family properties hosting several units, with a smaller proportion of low-rise apartment complexes. The communities assessed represented two microclimates: the Los Angeles Basin and San Gabriel Valley. The California Energy Commission designates two of these feeders (A and

B) in climate zone (CZ) #8 and two (C and D) in CZ #9 [33]. Typically, temperature data would be collected from a weather station in the same CZ. However, the neighborhoods served by Feeders A and B are on the border of CZ #9 and exhibit more similarity with the weather station in CZ #8 than with the closest weather station in CZ #9, which is farther away and on the coast. For this reason, the closest weather station is used for all analyses, regardless of designated CZ.

Along with the residential distribution feeder data (listed as primary feeders) used in these analyses, two example feeders are provided for additional context in the discussion section, representing a large apartment complex and a commercial zone.

System-wide net power load (NPL) was sourced directly from LADWP. Reported NPL summarizes full system net load (not including customer onsite co-generation) on an hourly basis for 2018 through 2020. All power data was analyzed with ambient temperature data, which was sourced from local National Weather Service (NWS) weather stations via a third party sourcing utility, MesoWest/SynopticLabs [34]. Load data was temporally correlated with weather data interpolated to the nearest hour using Universal Translator 3 (UT Online, Pacific Energy Center, Pacific Gas and Electric Corp., San Francisco, CA, USA) [35] and Easy Data Transform (Oryx Digital Ltd., Swindon, Wiltshire, England, UK) [36] software packages. When city data is not available with respect to COVID-19 caseloads and stay-at-home rates, data scoped at the inclusive Los Angeles County or California state level is used.

2.2. Load Evaluation

First, individual feeders' average loads were compared on a monthly or weekly basis (using an ISO 8601 defined week—see Table S4) across the period of study without temperature normalization or restriction. Major holidays were excluded from categorization. Analysis of input data and calculations were performed in kW and kWh.

For temperature analyses, hourly average temperature values were used along with monthly degree day values, which were assessed from local NWS observation weather station monthly reports (see Table 1 for data source information). Interpolative re-sampling was used to correlate temperature data to load data. Temperature data and derivative units were converted from °F to °C for final reporting and rounded to the nearest 0.1 °C for reported values.

Two effects of temperature on load must be distinguished in these analyses. First, the expected effect of temperature on electricity use (particularly cooling on hot days) must be considered when comparing across periods with different temperature patterns. Second, higher residential occupancy rates can increase households' response to temperature, making the effect of hot days stronger during the stay-at-home periods than otherwise.

Temperature models to assess sensitivity to load change due to temperature change were created using 2018 and 2019 daily average load data (for counterfactual models) discretely processed with ambient temperatures corresponding to feeder weather station source. As individual household loads are not available, reporting is performed in percent change compared to the counterfactual model used as a baseline for 2020 observed data. Depending on the specific application, temperature data was used as average period temperature or relative to heating or cooling degree days with a customary balance point of 18.3 °C (65 °F). In average-temperature regression models, the mean static temperature (MST) temperature was used rather than the customary degree day balance point value in calculation. Processing was performed as an average daily load considering daily average temperature (computed average of all periods as opposed to average of minimum and maximum daily observed temperature approach, which is used with degree-day calculation). Hourly models were used for direct comparison of specific, short-term periods. Being more stable, daily models were used in the linear modeling methods used in this report, consistent with similar observations in previous method comparisons [19,27].

For the relationship of energy usage to ambient temperature, piecewise regression corresponding to ASHRAE RP-1050 type linear change-point regression [37,38] was used to determine the 5-parameter models used (corresponding to three segments marked by two change points (CPs)–representing three regressed periods of load versus temperature), see Equation (1):

$$\begin{gathered} y = m_1x_1 + b_1 \text{ (inf. to low CP bound, } t_1\text{)}, \\ y = b_2 \text{ (low CP bound, } t_1 \text{ to high CP bound, } t_2\text{)}, \\ y = m_2x_2 + b_3 \text{ (high CP bound, } t_2 \text{ to inf)}, \end{gathered} \tag{1}$$

where y is an average feeder load (kW), m_1 is a regressed constant (kW/temp), x_1 is a period average temperature value (below the low CP bound), m_2 is a regressed constant (kW/temp), x_2 is a period average temperature value (above the high CP bound), b_1 and b_3 are a set of regressed intercept values corresponding to load at the CP bounds (kW) and b_2 is average constant load (kW) across temperature range between CP bounds.

This model accounts for temperature effects on energy use for heating and cooling as well as temperature ranges where load is not substantially affected by temperature. Regression analyses were performed using the open-source Energy Charting and Metrics (ECAM) (ECAM v.6.6, Bonneville Power Administration, Portland Oregon, OR, USA) calculation engine for Microsoft Excel (Excel v.14.0 (32-bit), Microsoft Corp., Redmond, WA, USA) [39] with an 80% confidence interval (CI) used for both temperature change point determination and data boundary determination. Testing showed that an 80% CI provided a balance between valid model calculation convergence and data inclusivity for all feeders analyzed. The calculated midpoint temperature between the determined change points corresponds to the MST. As data is analyzed, CI boundaries are similarly passed through calculations to provide error estimation for multi-step calculations. Temperature-based correction was used to normalize the influence of temperature such that all data sets are corrected to a value representing MST on a daily or hourly basis (as previously discussed) and compared. This approach estimates non-temperature-sensitive load. In addition, a reporting-period basis calculation provides estimates of post-period energy difference considering pre-period basis. This calculation used ECAM's internal engine implementing modified ASHRAE Guideline 14, model guidelines [28,40]. Analyses compared 2020 energy use to baseline data in 2018–2019 (for either the 2020 calendar year period or 2020 COVID-19 pandemic subset period) to normalize the impact of temperature between the evaluation and baseline periods in comparison. By removing this substantial factor, this provides a means to assess differences in load due to the changed factors (namely occupancy) during the COVID-19 pandemic period compared to the baseline period.

A separate two-term linear regression (see Equation (2)) was performed to model the impact of temperature on load as a function of heating degree day (HDD) or cooling degree day (CDD) values on a daily basis. Raw calculated values were limited such that values with CDD or HDD values less than 1.1 °C (2 °F) were removed from the model to reduce the bias from non-temperature related load variance. A CDD or HDD value would be mutually exclusive for a given day. Analyses were performed using a multiple linear regression in Origin Pro 9.0 (Origin Lab Corp., Northampton, MA, USA). Regression results were modeled for impact across an inclusive range of HDD and CDD values for both the baseline and evaluation period and presented as a simple percent difference for change comparing the differences between evaluation and comparison period with the same change in simulated CDD and HDD values:

$$y = m_1x_1 + m_2x_2 + b, \tag{2}$$

where y is the total feeder daily energy use (kWh), m_1 is a regressed constant (in kWh per HDD), x_1 is the HDD (single day) value, m_2 is a regressed constant (kWh per CDD), x_2 is the CDD (single day) value and b is regressed energy independent of HDD or CDD change (kWh).

2.3. Temperature Restricted Load Calculation

Temperature restriction is an approach used to filter values outside a pre-defined temperature range where limited correlation exists between temperature and elevated energy usage for each hourly temperature value. This method is appropriate when ambient temperatures largely remain near 18.3 °C (65 °F), which is the conventional degree-day calculation reference value customarily used by the US NWS. In the current analyses, a 4 °C range above and below the balance-point temperature was used for the restriction cut-off. Temperature-restricted 2020 evaluation period load data was compared to the combined 2018 and 2019 composite counterfactual model on a monthly or weekly basis considering day-type scope (all days, weekdays, or weekends/weekend days) or illustratively to a 2018 or 2019 single year baseline. Calculations of energy usage change were performed in the same manner as that used in the previously discussed temperature normalization process.

3. Results

Stay-at-home behavior generally tracks early public directives and provides the framework for interpreting shelter in place (SIP) response and the impact on energy usage. An LA County state of emergency was declared on 3 March while a California-wide state of emergency was declared on 4 March in response to rising regional case numbers. An SIP executive order was initiated in California on 19 March, and modified for provisions for essential workers on 4 May [32]. A follow-up tightening of restrictions followed on 2 July. Estimates of SIP response rates based on smartphone data (reported from early February through early September) show approximate alignment with LA County first wave COVID-19 reported case values (see Figure 1).

Figure 1. Comparison of LA County and all of California for shelter-in-place response and COVID-19 diagnosed cases over time. Data sources: California Department of Public Health [41], SafeGraph, Inc. [21]. The SIP Index represents the change (as a difference) in the % of people staying home compared to pre-pandemic baseline. The index ranges from −100% to 100%, where 0 (zero) is no change from a pre-pandemic baseline.

SIP response for the observed period peaked on 12 April [21,41], and decreased through late June. SIP response, measured as stay-at-home rate, is designated as no commuting or transit observed via mobile phone tracking. A pre-pandemic baseline rate of approximately 25% stay-at-home corresponds to a SIP index of 0. On 13 July commerce was restricted during the second case wave. Compared to the initial SIP response and despite the severity of the second wave (July through August), at nearly an order of magnitude higher than the first wave (mid-March through April), the population reaction was weaker, with less than a 5% increase in SIP response as compared to California and LA County at the pandemic onset, with a nearly a 15% decrease comparing the peak of COVID-19 case count during the second wave to that of the first wave. The magnitude increase of successive COVID-19 case peaks for each wave is so substantial that Figure 1 uses a y-axis logarithmic plot scaling to present this. Comparatively, SIP data is presented with a y-axis linear plot scaling. This smartphone based measure of SIP response over time closely resembles other indicators of stay-at-home behavior, such as keyword search histories for topics related to baking and home improvement, providing anecdotal evidence on activities performed by individuals with more available time and resources during the peak SIP period [42,43].

3.1. Unnormalized Load Comparison

The first set of load analyses use gross energy use data, not normalized for temperature. Energy use for Feeders A, B, C, and D for the pandemic period compared to the comparison period was higher by 10.0% for all days of the week considered together and by 10.4% during weekdays alone (see Table 2).

Table 2. Change in energy use for 2020 compared to a comparison baseline for Feeders A-D showing monthly energy use. Values are not normalized for temperature. Positive values indicate higher 2020 energy use compared to the counterfactual model constructed using the 2018–2019 baseline during the comparable monthly period. See Figure S2 for the yearly summary of individual feeders and Figure S3 for a weekly summary chart for individual feeders.

Month	All Days	Weekdays	Weekend Days
January	0.7%	1.2%	−0.4%
February	−10.3%	−10.1%	−10.6%
March (14–31 March)	2.6% (8.6%)	3.2% (9.5%)	1.1% (6.2%)
April	13.4%	13.1%	15.6%
May	20.9%	22.4%	17.3%
June	6.2%	9.1%	−0.4%
July	−10.5%	−12.7%	−5.3%
August	12.1%	8.0%	22.5%
September	25.4%	27.6%	20.6%
October	18.0%	20.9%	10.8%
November	5.0%	5.6%	3.6%
December	1.4%	1.0%	2.2%
Yearly Average	6.9%	7.2%	6.3%
COVID-19 Period Average	10.0%	10.4%	9.3%

Evaluating temperature differences while considering occupancy differences for the same period helps differentiate the causes of energy use change (see Figure S1 for monthly summarized temperature information for the LA Basin feeders). As shown earlier, stay-at-home rates for LA County rose swiftly in late March, peaked in April, reduced but remained high in May and June, and fell to a lower plateau for the rest of the summer. As shown in Figure 2, average temperatures were fairly similar in the 2020 period as in the 2018–2019 comparison period. Energy use was 2.6% higher for the whole month of March, but 8.6% higher for the second half of the month, after the initial SIP order (see Figure 2). Average temperatures were somewhat higher in April (1.8 °C, not significant) than in the composite 2018–2019 comparison period.

(**a**)

(**b**)

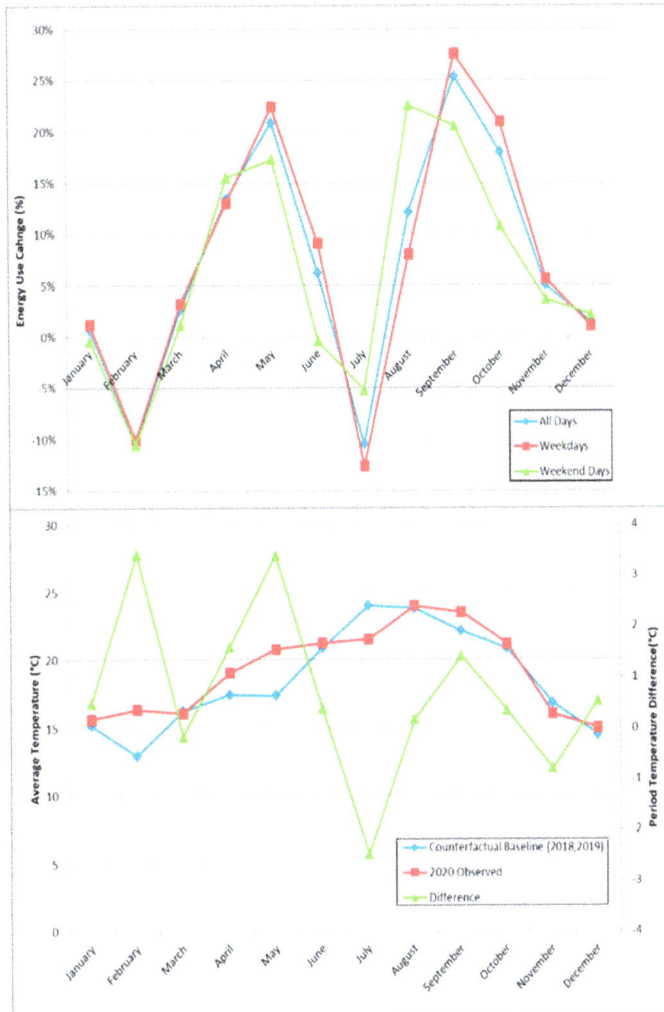

Figure 2. (**a**) Energy usage across all evaluated communities (simple composite average, not temperature normalized) with 2020 observation compared to a 2018–2019 comparison baseline, showing higher energy use. (**b**) Average monthly temperature as measured at weather data source KCQT in downtown Los Angeles observed for 2020 and comparison periods. Energy use strongly follows temperature change in relation to the mean static temperature (MST).

However, during most parts of the day and night temperatures were near the 18.3 °C (65 °F) nominal balance point, where the load is least impacted by temperature. Temperatures were much higher in May: a weighted average of 20.9% warmer (4.2 °C) with an average 2020 temperature above the balance point of 18.3 °C, indicating cooling-related energy use as a driver for the increase of 13.4% in average load that month. June 2020 had an average temperature within 1 °C of the counterfactual (weighted), but an average of 6.2% increase for 2020 against the counterfactual, suggesting increases in non-temperature-sensitive loads. Summer 2020 had generally reduced stay-at-home rates compared to spring with a substantially cooler July compared to the same period in the counterfactual. During August 2020, an extended warm period mid-month increased the average monthly temperature, which would have otherwise been a month cooler than the comparison monthly period. During this month, yearly record-high energy use in California was recorded. Increased occupancy compared to the comparison period with extended periods of high temperature led to increased energy use during these extreme heat events.

In fall and early winter, October and November both had monthly averages for 2020 within 1 °C of the monthly comparison periods but have 18% and 5% respective increases in energy use over the comparison periods for each month. December, with <1 °C of the monthly comparison period, despite the high COVID-19 cases had an energy usage increase within 2% as compared to the comparison period.

In general, monthly average load correlates with temperature change, consistent with expected temperature-driven load increases in hotter periods, particularly if higher occupancy rates lead to stronger response to ambient temperature. However, higher energy use in March provides a tell-tale indicator of increased load in these residential neighborhoods due to SIP activity during a period of relatively consistent temperature. By comparison, the overall LADWP NPL decreased during March and April in large part due to a reduction in commercial activities, which use a higher proportion of total energy load than residential customers (see Figure 3, top portion).

3.2. Temperature Normalization

Temperature normalization compensates for the impact of temperature on energy use, to better estimate the impact of non-temperature sensitive loads. However, as temperatures can vary across larger measured areas that combine residential and commercial loads, use of this technique on highly distributed loads such as NPL can lead to poor correlation (see Figure 3, bottom portion). Correlations between temperature and commercial loads are generally weaker than for residential because commercial buildings tend to have a higher proportion of temperature-insensitive process loads and large scheduled or sensed ventilation loads regardless of ambient temperature.

Residential energy use presented as a total for the evaluated feeders is shown in Figure 4 and Table 3. Total load yearly average difference against the baseline is 3.6% for 2020 for a scope of all days and 5.1% for the pandemic period against the comparison baseline. During the pandemic period, the average increase due to non-temperature sensitive loads is estimated at 5.6% for weekdays and 4.8% for weekend days. During the spring months of March through June, when SIP response was the highest, average total loads for these residential feeders were higher by 5.2% for all days, with a much higher increase for weekdays (6.2%) than for weekends (3.6%). When the 80% CI regression coefficients are evaluated for temperature and normalized for each MST value, a general pattern develops in the 2020 pandemic period of a smaller static temperature range with a higher comparable static load (greater temperature insensitive load proportion) compared to the baseline. Energy use is higher at low temperatures for all 4 feeders for temperatures adjacent to the upper temperature boundary for 2020 weekdays compared to counterfactual model values for weekdays. The nature of the data shows a distribution for 2020 with a large spread and bias to high load shifts in early spring compared to the comparison data considering the same sub-periods of evaluation. With lower temperatures in July 2020 compared to the counterfactual baseline, temperature range under-sampling occurred, resulting in low temperature data biasing the 2020 data. The limited number of days with high average temperatures in July 2020 compared to the baseline period results in variability as low temperature data is substantially influencing average daily the temperature-to-load relationship.

(a)

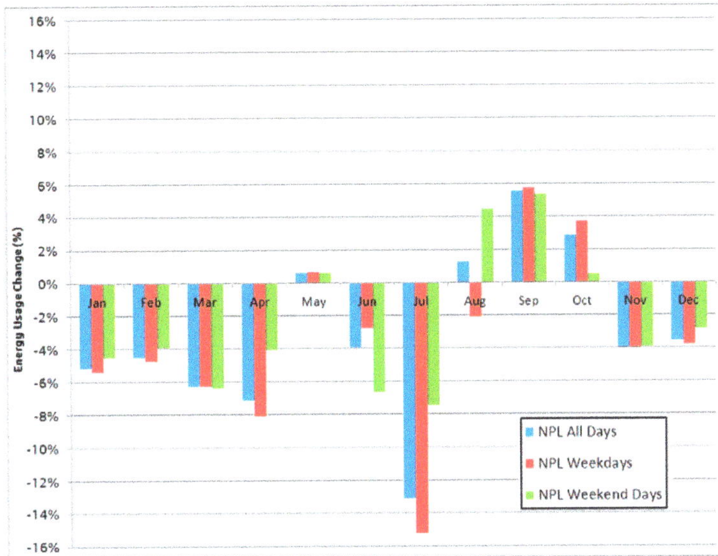

(b)

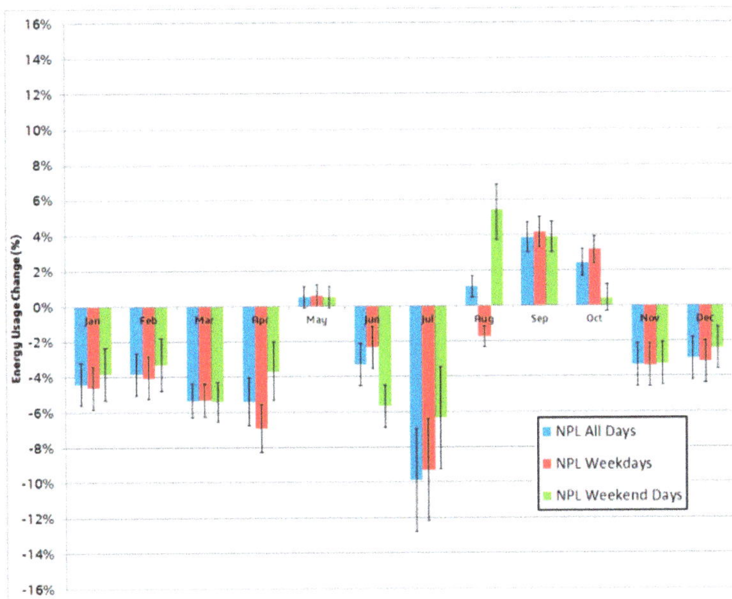

Figure 3. Monthly net load (NPL) including residential and commercial customers for Los Angeles Department of Water and Power for the pandemic compared to a counterfactual model using a 2018–2019 baseline. (**a**) Presented without temperature correction, and (**b**) presented with normalization to MST against a corresponding monthly counterfactual value, showing 80% CI boundaries in error bars as a result of temperature normalization.

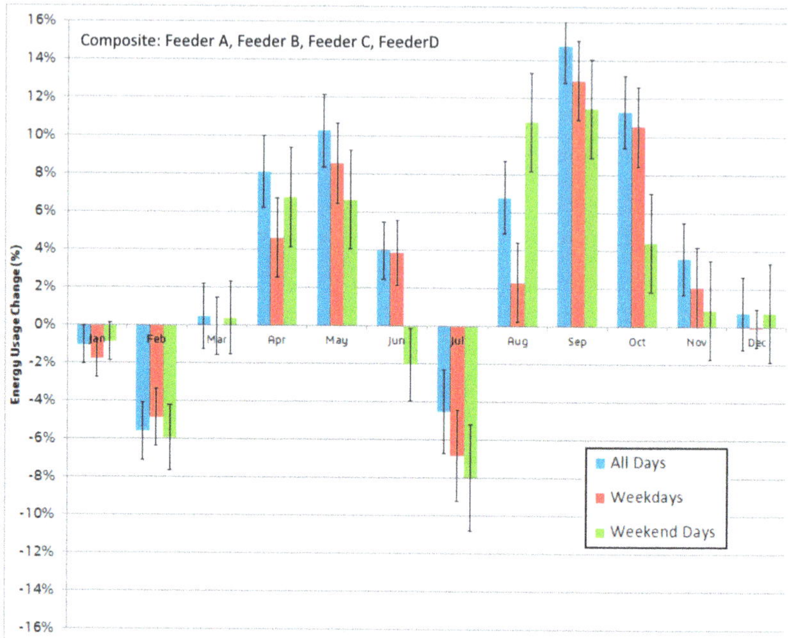

Figure 4. Total residential estimated non-temperature sensitive energy change for 2020 compared to a counterfactual using a 2018–2019 baseline, presented on a monthly basis as a simple composite average of feeders.

Table 3. Change in energy use for 2020 compared to a 2018–2019 baseline for Feeders A–D showing monthly energy use after temperature normalization. Positive values indicate higher 2020 energy use compared to the counterfactual model values.

Month	All Days	Weekdays	Weekend Days
January	−1.1%	−1.0%	−1.6%
February	−4.4%	−5.1%	−6.9%
March (14–31 March)	1.0% (3.5%)	1.3% (3.9%)	0.1% (2.3%)
April	6.9%	7.5%	7.5%
May	9.4%	10.5%	7.4%
June	3.4%	5.4%	−0.4%
July	−4.2%	−6.2%	−3.6%
August	5.0%	3.3%	11.2%
September	13.2%	15.1%	12.5%
October	10.0%	12.4%	6.3%
November	2.9%	3.4%	2.2%
December	0.7%	0.7%	1.9%
Yearly Average	3.6%	3.9%	3.0%
COVID-19 Period Average	5.1%	5.6%	4.8%

ECAM's native engine was used to generate a predictive model of total load change for the entire pandemic period against a counterfactual model of the comparison period (Figure 5). Energy use change reported is consistent with the temperature normalization method and within 2% for all individual feeders across the evaluation period. Results show relatively constant non-temperature load for the COVID-19 pandemic period in 2020 compared to the counterfactual in the 1–5% range considering all days (weekends and weekdays) (see Figure 5).

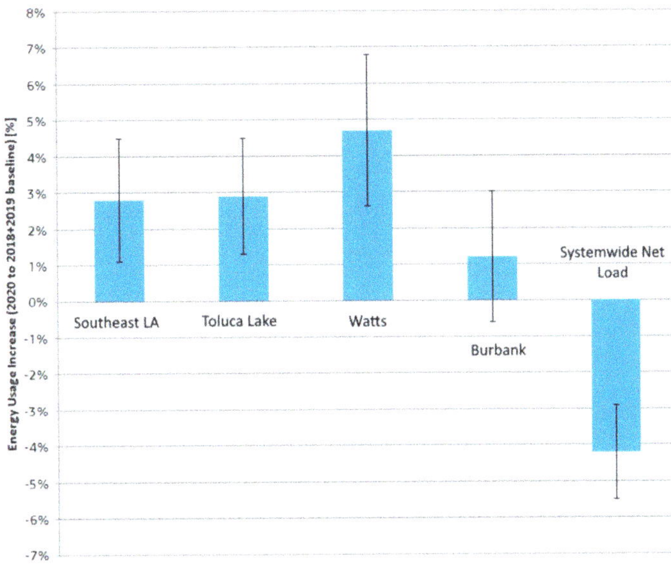

Figure 5. Total estimated non-temperature sensitive energy change during the COVID-19 pandemic period compared to a 2018–2019 comparison baseline for each analyzed feeder in addition to system wide LADWP NPL load. Error bars represent 80% CI bounds propagated.

Comparing change in energy use to median household income for each feeder (Figure 6), a weak trend develops suggesting higher impacts for temperature-insensitive loads for feeders in communities with lower median income. This may be due to disproportionate impact within this population of unemployment or population shift due to the pandemic. The Burbank feeder (Feeder D), while servicing primarily residential buildings, has a business artifact from an auto dealership on the periphery of the feeder territory which caused a small reduction in load early during the early COVID-19 pandemic period in 2020 compared to the counterfactual baseline.

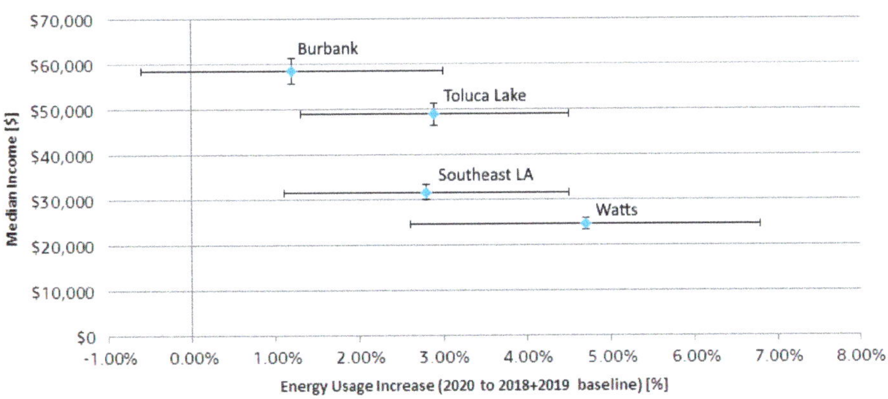

Figure 6. Energy change compared to feeder service community median income with a consistent 5% shown in the vertical error bars and the 80% CI shown in the horizontal error bars. The analysis scope was the COVID-19 pandemic period of Mid-March through December.

Estimation of energy use as a function of heating and cooling use change showed modest changes in the impact of load as a function of average HDD and CDD compared to the counterfactual period considering only the COVID-19 pandemic period as well as all of 2020 considering weekends and weekdays separately or combined (Figure 7).

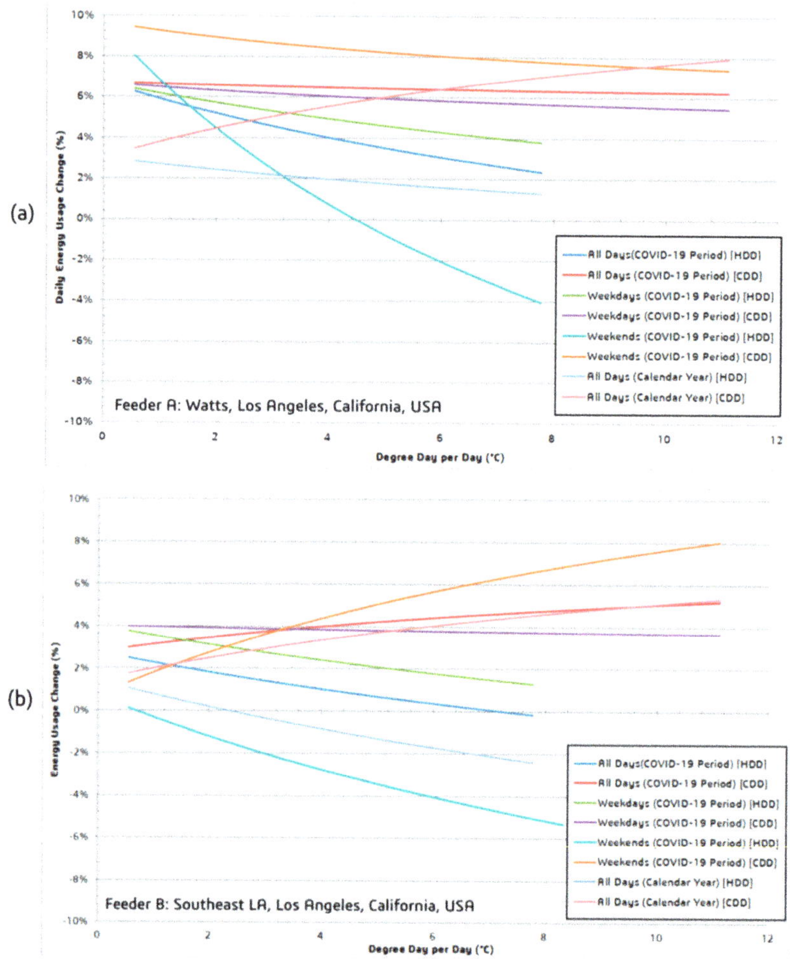

Figure 7. *Cont.*

(c) 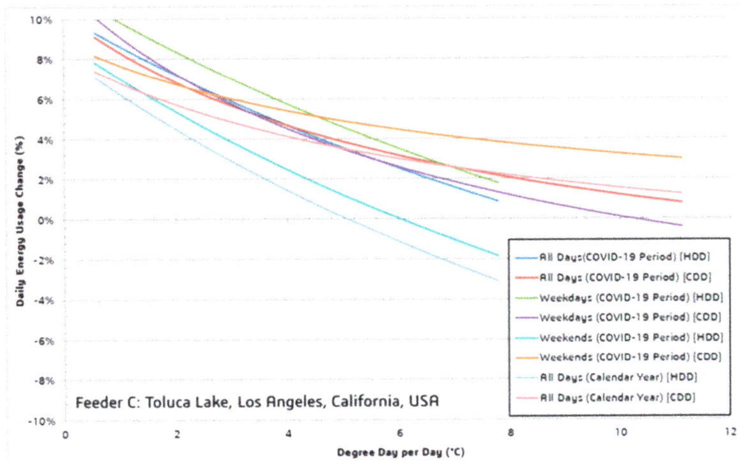

(d) 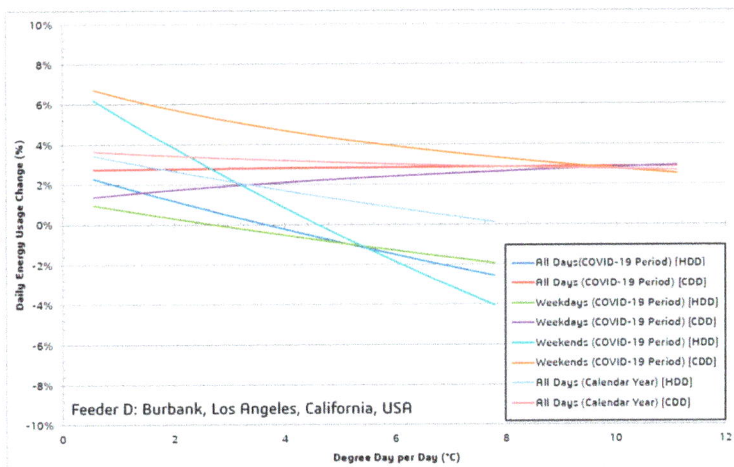

Figure 7. (**a**–**d**) Modeled normalized load change for 2020 compared to the baseline period for both calendar year periods (2020 to a 2018–2019 baseline) and subsets of mid-March through December for all periods comparing change in load relative to the baseline for a range of CDD and HDD values for each of the four feeders (**a**) Feeder A, (**b**) Feeder B, (**c**) Feeder C, (**d**) Feeder D. See Figure S2/Table S3 for a similar presentation of this data using normalized MST values and average daily temperatures as opposed to HDD and CDD values.

The HDD impact from heating loads decreased in all cases as presented. As noted earlier, 2020 was warmer in early spring leading to potential model bias during the period where SIP would have had the greatest impact on energy use. Electric heating (primarily portable space heaters) is a minor heat source in the region, with natural gas heating being predominant. Another region with higher heating requirements may provide better data for impact analysis. As expected, cooling loads for most scopes increase as temperatures rise from moderate to high, but plateau at very high temperatures, after air conditioning use is saturated. With this said, high heat events did distinctly show an increase in load for a given CDD value; this is especially apparent in the feeders in the LA Basin. For the Burbank feeder, a leveling off of increasing load is observed as the result of limited reserve cooling capacity–all available cooling having already been activated and in use (see

Figure 8). Per Chen et al., warmer areas in Southern California, such as the San Fernando Valley, are less temperature sensitive compared to cooler areas. The current results suggest this phenomenon similarly carries over to a more limited change in energy use during extreme heat events during the COVID-19 pandemic period as compared to other more temperature sensitive areas.

(a)

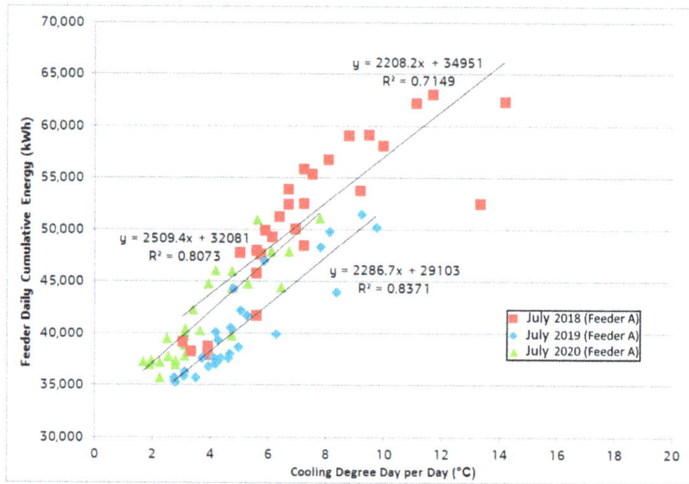

(b)

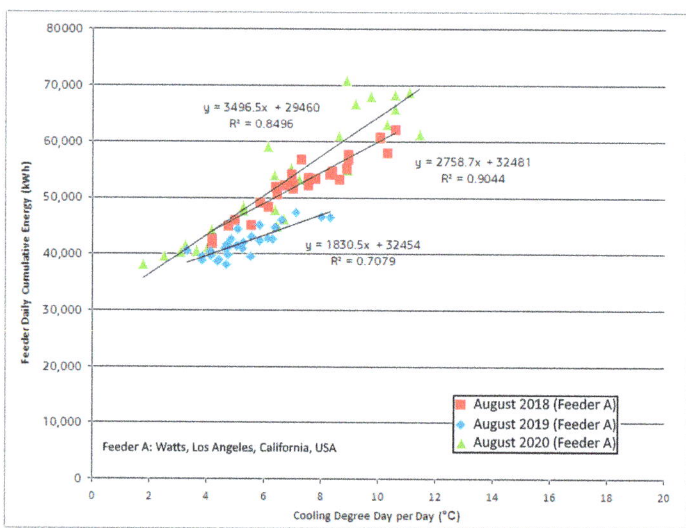

Figure 8. *Cont.*

(c)

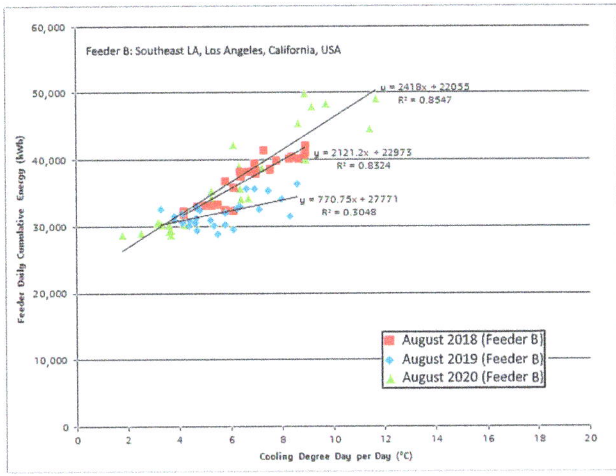

(d)

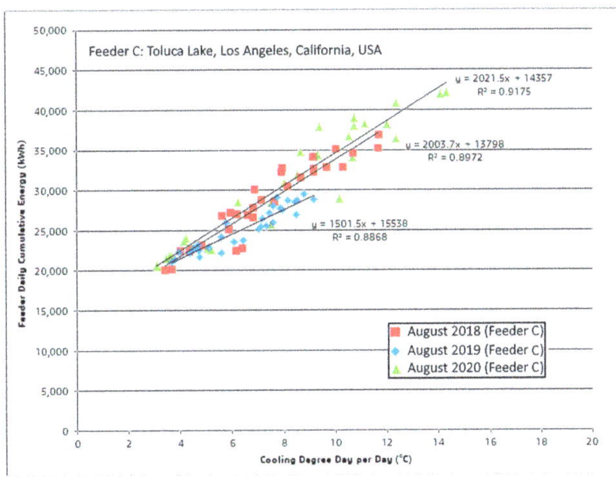

(e)

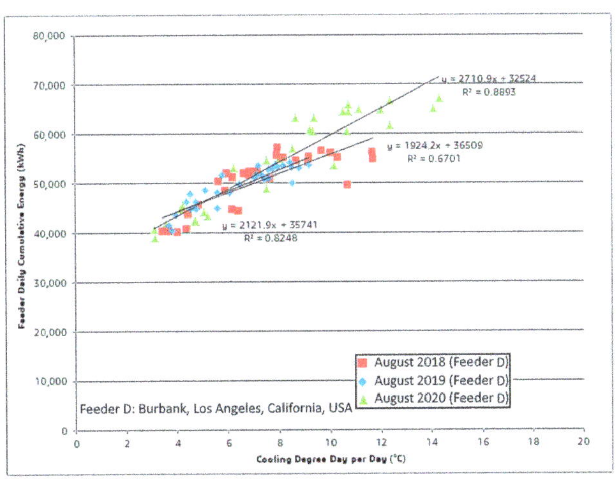

Figure 8. Comparison of feeder daily energy use for 2018, 2019, and 2020 for observed CDDs for (**a**) the month of July for Feeder A; and (**b–e**) presenting the month of August for Feeders A–D.

3.3. Temperature Restriction

Estimation of non-temperature sensitive loads on an hourly basis provides indication for granular energy use change based on changes in behavioral patterns that can only be observed at an hourly (versus a daily) level. Removing heating and cooling loads by restricting points when these loads are likely active reduces the temperature variability and helps present impact due to behavior change during SIP and the impact on non-temperature sensitive loads. Mid-day energy use is increased on weekdays (Figure 9) for most feeders. Weekend data is typically noisier than weekday data given relative under sampling compared to weekdays. Early evening peaks are moderately higher and weekday morning peaks are reduced. The values found via this direct analysis (Table 4) are largely similar to the estimated change due to non-temperature sensitive loads (Table 3).

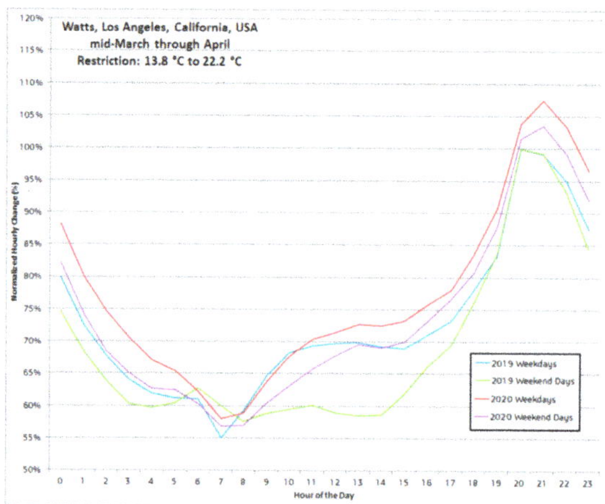

(a)

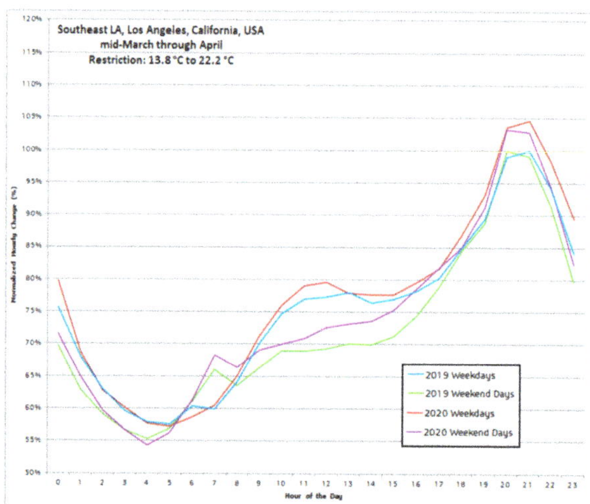

(b)

Figure 9. Baseline peak normalized (separately for weekdays and weekends) energy use change comparing 2020 to 2019 baseline for mid-March through April using a temperature restriction. Examples presented for (**a**) Feeder A and (**b**) Feeder B within the LA Basin microclimatic region.

Table 4. Energy use difference for mid-March through April comparing 2020 against a counterfactual baseline of 2018–2019. The all-day average for Feeders A–D was 5.3%.

Source	Weekday Change	Weekend Day Change
Feeder A	5.0%	3.1%
Feeder B	2.8%	2.0%
Feeder C	9.9%	13.9%
Feeder D	2.8%	2.0%
Average (A–D)	5.1%	6.1%
Feeder E	7.2%	4.8%
Feeder F	−28.0%	−24.8%
NPL	−6.2%	−4.2%

4. Discussion

While major fuel and energy sources were observed to show a net decrease in use early in the pandemic, the opposite was largely observed for residential energy use. These findings were consistent with that of earlier studies such as those performed by Pecan Street [14] in Austin, TX, with 113 panel-instrumented homes: study results showed an approximate 42% (~300 W) mid-day increase in April 2020 for non-temperature sensitive loads such as consumer electronics, appliances, miscellaneous electric loads (plug loads), and lighting, compared to a baseline of the previous year, reflecting increased occupancy with increased load during both weekdays and weekends. Full-day energy use increase is likely closer to ~14%, estimating from Pecan Street provided figures. Similarly, this Pecan Street study identified an increase in temperature sensitivity across March and April identified by average home kWh/cooling degree day (CDD) of the evaluated period with a value of 0.7 in April 2020 compared to a value of 0.56 for the average of April 2017, April 2018, and April 2019, a comparative 25% increase in load for each CDD change [14]. These results match the general trends observed in our study, albeit with higher magnitude changes between 2020 observations and past baselines. Much of this difference is likely related to Pecan Street's use of instrumented single-family, higher-income housing combined with regional climatic variance (e.g., impact of humidity and higher regional temperatures on cooling behaviors). Also, days with potential heating and cooling activity in shoulder periods (often with low CDD or HDD values) can incur bias from the dominant space conditioning energy load used during the period, as previously mentioned. Energy use for this scenario can increase for low HDD or CDD values; our tests showed that using a threshold value of 2 CDD or HDD substantially reduced this impact. The temperature in Los Angeles in April rarely requires air conditioning usage, whereas Austin, Texas experienced a warm and humid spring during the highest SIP period.

Load impact from non-temperature-sensitive loads during the early pandemic were estimated from sampled feeders through both temperature restriction (Table 4) and temperature normalization (Figure 4) resulting in estimated increases of 5.3% and 5.7%, respectively (mid-Mar through April, all days), less than that reported by Pecan Street. With the exception of Feeder C, change in weekday load was more impacted than weekend load compared to the 2018–2019 baseline during the early pandemic (Table 4). Non-temperature loads were a substantial component of energy used which is evidenced by the similarity in total load change (Table 3) to temperature restricted load change (Table 4). Heavy mixtures of both HDD and CDD during this period complicate regression analyses (of the type used in Figure 7). This is because the nature of the degree day metric is not exclusive to heating or cooling, but is the balance point difference computed between the range from daily highs and lows. When temperature fluctuates enough over a 24-h period to require both heating and cooling, that day may be labeled with a low value for HDD, CDD, or both. This effectively skews energy use per HDD or CDD when using multiple regression models. Temperature normalization based on average daily temperature regression performs marginally better with respect to these temperature variations.

When temperatures increase, increased occupancy (even at lower levels compared to the mid-April peak) drives loads higher. This is clearly illustrated in Figure 8b representing Feeder A. The load events with CDD values between 9 °C and 11 °C required 9.2% more load compared to similar events in this same temperature range in 2018, consistent with the idea of higher home occupancy rates driving higher demand for cooling on hot days. The effect of SIP response can differ for weekend and weekday loads. This is illustrated with the highest heat day in this figure, which has substantially less load than the second highest load event: note that this day falls on a weekend (for which occupancy shifts due to SIP should be reduced) versus higher impacts on adjacent weekdays during this extended extreme heat event. Mixing weekdays and weekends for analysis results in model variance challenges due to substantially different activities for these two day types. This is especially true during typical, non-SIP periods such as the baseline. Clearly, increased occupancy drives up cooling requirements during extreme heat events. Capturing a representative spectrum of temperatures and loads for each month while occupancy was varying due to SIP response to allow direct calculation is challenging. For example, as illustrated in Figure 8a, the high heat events observed in July 2018 and 2019 were not replicated in July 2020, which weakens any comparison across these months to assess 2020 SIP response effects on energy use.

Daily energy use patterns were strongly impacted early in the pandemic. Compared to the counterfactual model, energy use was slower to rise in the early morning and was higher during mid-day hours, with a moderate increase in daily peak energy use across all feeders (see Figure 9). Assessed with restricted temperature analysis, the impact of these features decreased with a slow resumption toward baseline energy use as SIP response reduced.

Energy usage impacts for large multi-family apartment complexes is likely different from that for the single family and small multi-family residences studied above. Figure 10 shows results for an additional example, Feeder E, representing a large apartment complex. During the early pandemic period, energy use for this case largely tracked other residential loads. By summer, the shutdown of many shared-use areas within these buildings to reduce potential community spread of COVID-19 reduced the cooling burden to these buildings, resulting in a net drop compared to the baseline during the period when the cooling burden is the highest (mid-summer). This effect, plus the centralization of cooling and heating, are likely substantial divergence points comparing large apartment complexes and high-rises to low- and medium-density homes and low-rise apartments, which have limited shared facilities and individual heating and cooling supplies.

Commercial energy use, a major component represented in the NPL figure, is illustrated by a single mid-rise building source (see Feeder F in Figure 10). This example is included as a contrast to the residential feeders analyzed above, as an approximate indicator for impacts of SIP on non-essential business activity (jewelry manufacture and distribution). For this commercial feeder, a major drop in energy use occurred during week 12 of 2020 (16–22 March), corresponding to the initiation of SIP restrictions, which is when residential energy use increased. By mid-June (Week 21) energy use in the commercial feeder had greatly increased. This follows a weakening of SIP response, previously discussed. The second-wave restrictions did not substantially reverse the increase in energy use, which showed continued growth until early fall. The lower energy use in November and December of 2020 compared to the 2018–2019 counterfactual composite baseline may reflect the reduction of typical high-intensity holiday shopping during those months, including extended hours.

Figure 10. Direct change in energy use (non-normalized for temperature) for Feeder E (large apartment complex) and Feeder F (commercial building).

The current findings show limited evidence of a higher increase in non-temperature sensitive load over the COVID-19 period for lower income areas than higher income areas. The expected effect of SIP response on energy by income is not clear, as various factors predict mixed results. For instance, more highly educated, higher-income professionals were more likely to be able to shift to working from home, while less-educated workers were more likely to either continue working outside the home (e.g., in essential service or manufacturing) or lose their jobs. Lower-income households tend to have more members to use devices if everyone is at home, but higher-income households have more square footage and more devices to be used per person; furthermore, lower-income households spend less money (and time) on entertainment and dining outside the home than higher-income households normally, and would thus experience less change. Residential use of portable space heaters and window AC units, more common among older housing stock in lower-income areas, also adds to electricity use. Given the limited number of neighborhoods sampled here and the small observed effect, this result is considered questionable, but suggestive of further consideration; additional research would be required to clearly ascertain the income differences in effect of SIP response on energy use. However, it is worth noting that even the same or lower increase in energy use is a greater hardship for lower-income households, as they already experience a significantly higher energy burden (that is, the proportion of their income spent on energy bills) and have little or no discretionary income to cover unexpected expenses.

Overall, SIP compliance was initially strong, but this effect was temporary. Approximately one month elapsed between the rapid ramp-up of SIP response and a long-term decrease and eventually leveling out of SIP compliance. This occurred even with daily briefings from health experts and government officials reporting increasing caseloads in the LA area. Energy models considering change in occupancy must expect a ramp-up, peak, and an extended dynamic equilibrium for general change in occupancy. Considering the near future of the COVID-19 timeline, stay-at-home rates will continue to subside into an extended equilibrium that is likely higher than pre-pandemic levels. This suggests that increased telecommuting from home will continue to raise the energy burden during high heat events. Mid-day energy use, compared to a pre-COVID-19 baseline has had a modest increase–this can help offset the increasing glut in solar energy mid-day during normal conditions. However peak conditions, especially in late afternoon when solar is switching to spinning reserves can still impact energy supplies during this critical ramp-up and source switching period.

5. Conclusions

This research adds to the growing body of knowledge on how the COVID-19 pandemic has affected human behavior and the resulting impact on energy usage. Increased residential occupancy has impact on energy use. Over the course of the 2020 pandemic period, fatigue with SIP compliance led to a rebound toward earlier pre-pandemic occupancy rates (reduced SIP response) and a substantial rise in regional COVID-19 active cases. It is reasonable to assume that in future pandemic events, similar behaviors are to be expected. The potential for extended SIP activity for extended periods has limits. The timing of an SIP period can strongly affect energy use change. During temperate periods, limited heating or cooling impact will likely be observed with a constant increased non-temperature sensitive load increase. Even with occupancy patterns trending more toward normal, impacts on energy used for cooling during heat events was observed. As the current analysis examined only electricity, and space heating in this region is largely fueled by natural gas, observed stay-at-home impacts on heating were minimal. However, as electrification development continues, increased reliance on electric heating should be reflected in larger impacts of residential occupancy on electrical energy use. As long-term work at home activity continues, increased residential energy use during weekdays will continue for applicable households. Modeling this change is outside the scope of this study but relevant to future expected household energy change and population impacts. The results suggest the possibility of a higher impact of stay-at-home behavior on energy change for communities with lower median income level, however, evidence is weak and further research would be necessary to confirm such a relationship.

Continued efficiency measures for miscellaneous electric loads can help reduce non-temperature sensitive loads. Focus on reducing wasteful energy use (i.e., devices not properly entering low-power mode when not in use) is a major potential area of research. The analysis this study has provided on residences is also applicable to businesses, to highlight opportunities for better managing plug and process loads, especially while not in use, and may be a fruitful area for follow-up study. Follow-up studies using similar approach methodology with data for areas with substantial heating and cooling loads would help draw the maximum impact of stay-at-home behaviors when considering temperature sensitive loads as a major energy load contributor.

Supplementary Materials: The following are available online at https://www.mdpi.com/article/10.3390/app11104476/s1, Numeric tools, calculation scripts, and calculation workbooks as well as extended data subsets used within this report are available at: https://github.com/CalPlug/CovidResEnergyAnalysis2021. Supplementary Report Contents: Table S1: Extended feeder summary table showing solar and EV installation penetration within the bounding ZCTA for each feeder, Table S2: Change in energy use for 2020 compared to a counterfactual 2018–2019 comparison baseline showing change in energy usage by communities, Figure S1: Monthly temperature summary for Los Angeles across all periods used for analysis, Table S3: Summary of regression segments (80% CI) and temperature change point values for both comparison and pandemic evaluation periods. Table S4: Specific dates for the ISO weeks for 2018, 2019, and 2020. Figure S1: Monthly temperature summary for Los Angeles across all periods used for analysis. Data sourced from KCQT (Downtown LA), NWS. Figure S2: Comparison model of the effect of ambient temperature on feeder energy use—MST normalized to 100%. Figure S3: Non-temperature normalized energy use for all feeders presented on a weekly basis. Table S5: Report abbreviation and acronym list.

Author Contributions: Conceptualization, G.-P.L., M.J.K. and A.T.; methodology, M.J.K., J.E.P. and A.S.; software, M.J.K.; validation, M.J.K. and J.E.P.; formal analysis, M.J.K. and J.E.P.; inves-tigation, M.J.K. and J.E.P.; resources, A.T., G.-P.L. and D.J.; data curation, M.J.K., A.S. and J.E.P.; writing—original draft preparation, M.J.K., A.S. and J.E.P.; writing—review M.J.K. and J.E.P.; visualization, M.J.K.; supervision, M.J.K.; project administration, G.-P.L., A.T. and D.J. All authors have read and agreed to the published version of the manuscript.

Funding: This research received no external funding.

Institutional Review Board Statement: Not applicable.

Data Availability Statement: Not applicable.

Acknowledgments: The authors wish to thank Miguel Malabanan, Mehdi Shafaghi, Jeremiah Valera, and Luke Sun from LADWP for technical support and Sabine Kunrath, Mahejabeen Kauser, and Katie Gladych from CalPlug for their support with background research, data preparation, and manuscript formatting/copy-editing assistance.

Conflicts of Interest: The authors declare no conflict of interest.

References

1. Le Quéré, C.; Jackson, R.B.; Jones, M.W.; Smith, A.J.P.; Abernethy, S.; Andrew, R.M.; De-Gol, A.J.; Willis, D.R.; Shan, Y.; Canadell, J.G.; et al. Temporary reduction in daily global CO_2 emissions during the COVID-19 forced confinement. *Nat. Clim. Chang.* **2020**, *10*, 647–653. [CrossRef]
2. Gillingham, K.T.; Knittel, C.R.; Li, J.; Ovaere, M.; Reguant, M. The Short-run and Long-run Effects of Covid-19 on Energy and the Environment. *Joule* **2020**, *4*, 1337–1341. [CrossRef] [PubMed]
3. Forster, P.M.; Forster, H.I.; Evans, M.J.; Gidden, M.J.; Jones, C.D.; Keller, C.A.; LambolliD, R.D.; Le Quéré, C.; Rogelj, J.; Rosen, D.; et al. Current and future global climate impacts resulting from COVID-19. *Nat. Clim. Chang.* **2020**, *10*, 913–919. [CrossRef]
4. Heyd, T. Covid-19 and climate change in the times of the Anthropocene. *Anthr. Rev.* **2020**. [CrossRef]
5. Cicala, S.; Holland, S.P.; Mansur, E.T.; Muller, N.Z.; Yates, A. Expected Health Effects of Reduced Air Pollution from COVID-19 Social Distancing. *SSRN Electron. J.* **2020**. [CrossRef]
6. He, C.; Yang, L.; Cai, B.; Ruan, Q.; Hong, S.; Wang, Z. Impacts of the COVID-19 event on the NOx emissions of key polluting enterprises in China. *Appl. Energy* **2021**, *281*, 116042. [CrossRef] [PubMed]
7. Today in Energy: Daily Electricity Demand Impacts from COVID-19 Mitigations Efforts Differ by Region. Available online: https://www.eia.gov/todayinenergy/detail.php?id=43636 (accessed on 3 June 2020).
8. Prol, J.L.; Sungmin, O. Impact of COVID-19 measures on short-term electricity consumption in the most affected EU countries and USA states. *Iscience* **2020**, *23*, 101639. [CrossRef] [PubMed]
9. Hauser, P.; Schönheit, D.; Scharf, H.; Anke, C.-P.; Möst, D. Covid-19's Impact on European Power Sectors: An Econometric Analysis. *Energies* **2021**, *14*, 1639. [CrossRef]
10. New York ISO. COVID-19 Related Updates. Available online: https://www.nyiso.com/covid (accessed on 2 June 2020).
11. California ISO. COVID-19 Impacts to California ISO Load & Markets: Market Analysis and Forecasting 17 March–26 July 2020. Available online: http://www.caiso.com/documents/covid-19-impacts-isoloadforecast-presentation.pdf (accessed on 2 August 2020).
12. Ruan, G.; Wu, D.; Zheng, X.; Zhong, H.; Kang, C.; Dahleh, M.A.; Sivaranjani, S.; Xie, L. A cross-domain approach to analyzing the short-run impact of COVID-19 on the US electricity sector. *Joule* **2020**, *4*, 2322–2337. [CrossRef] [PubMed]
13. Hinson, S. COVID-19 is Changing Residential Electricity Demand. Available online: https://www.pecanstreet.org/2020/05/covid/ (accessed on 8 June 2020).
14. Hinson, S. COVID Five Months in: A Sustained Increase in Residential Electricity. Available online: https://www.pecanstreet.org/2020/08/covid-summer-update/ (accessed on 1 September 2020).
15. U.S. Energy Information Administration. Annual Energy Outlook 2020: Electricity. Available online: https://www.eia.gov/outlooks/aeo/pdf/AEO2020%20Full%20Report.pdf (accessed on 15 April 2020).
16. Mainzer, E.; Batjer, M.; Hochschild, D. *Preliminary Root Cause Analysis Mid-August 2020 Heat Storm*; California ISO, California Public Utilities Commission, and California Energy Commission: Sacramento, CA, USA, 2020.
17. Roth, S. What caused California's rolling blackouts? Climate change and poor planning. *Los Angeles Times*, 6 October 2020.
18. U.S. Energy Information Administration. Short-Term Energy Outlook. Available online: https://www.eia.gov/outlooks/steo/archives/dec20.pdf (accessed on 29 January 2021).
19. Chen, M.; Sanders, K.T.; Ban-Weiss, G.A. A new method utilizing smart meter data for identifying the existence of air conditioning in residential homes. *Environ. Res. Lett.* **2019**, *14*, 094004. [CrossRef]
20. Chen, M.; Ban-Weiss, G.A.; Sanders, K.T. Utilizing smart-meter data to project impacts of urban warming on residential electricity use for vulnerable populations in Southern California. *Environ. Res. Lett.* **2020**, *15*, 064001. [CrossRef]

21. U.S. Geographic Responses to Shelter in Place Orders (Los Angeles County, California). Available online: https://www.safegraph.com/data-examples/covid19-shelter-in-place (accessed on 11 January 2021).
22. Saad, L. Americans Less Amenable to Another COVID-19 Lockdown. Available online: https://news.gallup.com/poll/324146/americans-less-amenable-covid-lockdown.aspx (accessed on 1 December 2020).
23. Hickman, A.; Saad, L. Reviewing Remote Work in the U.S. Under COVID-19. Available online: https://news.gallup.com/poll/311375/reviewing-remote-work-covid.aspx (accessed on 2 June 2020).
24. Unemployment Rate in Los Angeles County, CA (CALOSA7URN). Available online: https://fred.stlouisfed.org/series/CALOSA7URN (accessed on 13 January 2021).
25. Burns, D. How the coronavirus job cuts played out by sector and demographics. *Reboot-Live*, 4 April 2020.
26. Cohn, D.V. About a Fifth of U.S. Adults Moved Due to COVID-19 or Know Someone Who Did. Available online: https://www.pewresearch.org/fact-tank/2020/07/06/about-a-fifth-of-u-s-adults-moved-due-to-covid-19-or-know-someone-who-did/ (accessed on 4 August 2020).
27. Chen, M.; Ban-Weiss, G.A.; Sanders, K.T. The role of household level electricity data in improving estimates of the impacts of climate on building electricity use. *Energy Build.* **2018**, *180*, 146–158. [CrossRef]
28. SBW Consulting Inc. *Uncertainty Approaches and Analyses for Regression Models and ECAM*; Prepared for the Bonneville Power Administration: Bellevue, WA, USA, 2017.
29. United States Census Bureau. *QuickFacts: California*; United States Census Bureau: Houston, MA, USA, 2019.
30. LA County Daily COVID-19 Data. Available online: http://publichealth.lacounty.gov/media/coronavirus/data/index.htm (accessed on 11 January 2021).
31. COVID-19: Keeping Los Angeles Safe. Available online: https://corona-virus.la/ (accessed on 10 February 2021).
32. Newsom, G. *Executive Order N-33-20*; State of California Executive Department: Sacramento, CA, USA, 2020. Available online: https://ca-hwi.org/public/uploads/pdfs/3.19.20-attested-EO-N-33-20-COVID-19-HEALTH-ORDER_.pdf (accessed on 20 November 2020).
33. Danny, T. Climate Zone Tool, Maps, and Information Supporting the California Energy Code. Available online: https://www.energy.ca.gov/programs-and-topics/programs/building-energy-efficiency-standards/climate-zone-tool-maps-and (accessed on 19 November 2020).
34. MesoWest. Available online: https://mesowest.utah.edu/cgi-bin/droman/download_api2.cgi?stn=KLAX (accessed on 5 October 2020).
35. Universal Translator 3. Available online: http://utonline.org/cms/ (accessed on 1 December 2020).
36. Bryce, A. Easy Data Transform (v. 1.x). Available online: https://www.easydatatransform.com/ (accessed on 10 June 2020).
37. Research Into Action Inc.; Quantum Energy Services & Technologies Inc. (QuEST); Stetz Consulting LLC.; Kolderup Consulting; Warren Energy Engineering LLC.; Left Fork Energy Inc.; Schiller Consulting Inc. *Regression for M&V: Reference Guide: Version 1.1*; Bonneville Power Administration: Portland, OR, USA, 2012.
38. Kissock, J.K.; Haberl, J.S.; Claridge, D.E. *Development of a Toolkit for Calculating Linear, Change-Point Linear and Multiple-Linear Inverse Building Energy Analysis Models, ASHRAE Research Project 1050-RP, Final Report*; Energy Systems Laboratory, Texas A&M University: College Station, TX, USA, 2002.
39. Koran, B.; Hicks, G. *User's Guide to ECAM Version 6: Energy Charting and Metrics Version 6*; SBW Consulting, Inc.: Bellevue, WA, USA, 2018.
40. Sun, Y.; Baltazar, J.-C. Analysis and Improvement on the Estimation of Building Energy Savings Uncertainty. *ASHRAE Trans.* **2013**, *119*, 1–8.
41. COVID-19 Time-Series Metrics by County and State: Statewide COVID-19 Cases Deaths. Available online: https://data.ca.gov/dataset/covid-19-time-series-metrics-by-county-and-state1 (accessed on 28 March 2021).
42. Brasted, C. Why cooking and baking fill a void. *British Broadcasting Corporation (BBC): Worklife*, 2 February 2021.
43. Alphabet Corporation. *Google Trends*; Alphabet Corporation: Mountain View, CA, USA, 2021.

applied sciences

MDPI

Article

Solar Energy in Urban Planning: Lesson Learned and Recommendations from Six Italian Case Studies

Matteo Formolli [1], Silvia Croce [2,3], Daniele Vettorato [2], Rossana Paparella [3], Alessandra Scognamiglio [4], Andrea Giovanni Mainini [5] and Gabriele Lobaccaro [6,*]

1 Department of Architecture and Technology, Norwegian University of Science and Technology, Alfred Getz vei 3, 7491 Trondheim, Norway; matteo.formolli@ntnu.no
2 EURAC Research, Institute for Renewable Energy, Via Druso 1, 39100 Bolzano, Italy; silvia.croce@eurac.edu (S.C.); daniele.vettorato@eurac.edu (D.V.)
3 Department of Civil, Environmental and Architectural Engineering, Università degli Studi di Padova, Via Marzolo 9, 35131 Padua, Italy; rossana.paparella@unipd.it
4 Italian National Agency for New Technologies, Energy and Sustainable Economic Development (ENEA), Largo Enrico Fermi 1, 80055 Portici, Italy; alessandra.scognamiglio@enea.it
5 Department of Architecture Built Environment and Construction Engineering, Politecnico di Milano, Via G. Ponzio 31, 20133 Milan, Italy; andreagiovanni.mainini@polimi.it
6 Department of Civil and Environmental Engineering, Norwegian University of Science and Technology, Høgskoleringen 7a, 7491 Trondheim, Norway
* Correspondence: gabriele.lobaccaro@ntnu.no

Abstract: This paper presents the results of the analysis conducted on six case studies related to solar energy integration in urban and rural environments located on the Italian territory. The analysis has been carried out within the Subtask C—Case Studies and Action Research of the International Energy Agency Solar Heating and Cooling Program Task 51 "Solar Energy in Urban Planning". Three different environments hosting active and passive solar energy systems (existing urban areas, new urban areas, and agricultural/rural areas) have been investigated to attain lessons learned and recommendations. Findings suggest that (a) it is important to consider solar energy from the early stages of the design process onwards to achieve satisfactory levels of integration; (b) a higher level of awareness regarding solar potential at the beginning of a project permits acting on its morphology, achieving the best solution in terms of active and passive solar gains; (c) when properly designed, photovoltaic systems can act as characterizing elements and as a distinctive architectural material that is able to valorize the aesthetic of the entire urban intervention; (d) further significant outcomes include the importance of supporting the decision strategies with quantitative and qualitative analyses, the institution of coordinating bodies to facilitate the discussion between stakeholders, and the need for deep renovation projects to fully impact existing buildings' stock; (e) when large solar installations are planned at the ground level, a landscape design approach should be chosen, while the ecological impact should be reduced by carefully planning the adoption of alternative solutions (e.g., agrivoltaics) compatible with the existing land use.

Keywords: solar energy; urban planning; active solar systems; passive solar strategies; Italian case studies; agrivoltaics

Citation: Formolli, M.; Croce, S.; Vettorato, D.; Paparella, R.; Scognamiglio, A.; Mainini, A.G.; Lobaccaro, G. Solar Energy in Urban Planning: Lesson Learned and Recommendations from Six Italian Case Studies. *Appl. Sci.* **2022**, *12*, 2950. https://doi.org/10.3390/app12062950

Academic Editor: Salvatore Vasta

Received: 13 February 2022
Accepted: 9 March 2022
Published: 14 March 2022

1. Introduction

Cities represent the place where 80% of the gross domestic product is generated and where around half of the world population lives, with expected growth to two-thirds by the middle of the current century [1]. The combination of these two factors demonstrates why urban areas are considered responsible for up to 75% of the global energy consumption and more than 70% of energy-related carbon dioxide (CO_2) emissions [2].

In this context, the decarbonization of the energy sector, in line with the Paris Agreement's goals to keep the temperature increase below 2 °C, requires a substantial contribution from cities. A growing number of cities, mainly located in the European Union (EU) and North America, have set renewable energy targets to support the green energy transition [3]. At the municipal level, the adoption of regulations, economic incentives, and the direct financial involvement of municipalities and public authorities, combined with the storytelling of successful practices, proved to be crucial in raising public awareness and in steering private developers toward sustainable strategies [4,5]. Similarly, in the past decades, national level measures constituted a strong enabler for the exploitation of renewable energy sources (RES) [6].

Solar photovoltaic (PV) and solar thermal (ST) had respectively the first (36%) and the fourth (10.5%) highest expansion rate among the RES in the past 30 years [7]. In one of the preconfigured future scenarios, the International Energy Agency (IEA) is expecting that PVs would cover almost one-third of the new electricity demand within 2030, with an average annual growth rate of 13% [8]. Utility-scale ground-mounted systems dominate the market [9]; however, several potential issues related to their impact on the ecosystem, landscape, and competition with agricultural land are emerging [10,11].

The major alternative is distributed rooftop systems (i.e., applied and integrated systems), representing a market share of 40% in 2020 [12]. This technology has the highest potential use in the built environment, especially for photovoltaic systems fully integrated into the building envelopes (i.e., facades and roofs), which perform the double function of energy producer and building cladding. The advantages of building-integrated photovoltaics (BIPV) and solar thermal (BIST) are manifold: unused surfaces can be turned into active energy generators [13], losses associated with transmission and distribution of electricity are reduced thanks to on-site production [14], and higher energy flexibility toward extreme weather conditions is guaranteed [3]. On the other hand, the installation of such systems in urban areas, especially on vertical surfaces (i.e., facades), requires considering complex phenomena such as inter-building solar reflections and overshadowing effects [15–17], and other energy and climate-related issues, such as high surface temperature [18] and fire hazard [19]. The visibility and the socio-cultural sensitivity impact of the systems should also be considered [20–23]. In Europe, BIPV accounted for a cumulative installed capacity of 6.9 GWp at the end of 2019, with Italy covering about 38% of the total, due to the early campaign of statal incentives to boost the adoption of PV technology [24].

Other two viable alternatives, representing small market niches in rapid expansion, are floating solar and agrivoltaics [12]. These technologies allow to reduce land use for solar installations and through appropriate design can provide a series of benefits. Floating solar has the advantage of reducing the evaporation of water reservoirs. Furthermore, the presence of water helps to keep the panels' temperature low, stabilizing the efficiency of the system [25,26]. Similarly, agricultural PV can improve crop yield, limit evaporation, provide shade to livestock or crops, and protect against extreme weather conditions and soil erosion [12,27]. The potential for agrivoltaic systems in Europe is immense; if solar would be deployed on 1% of the arable land, its technical capacity would amount to over 900 GW, which is more than six times the current installed PV capacity in the EU [28].

Framework and Aim of the Work

This paper presents part of the findings related to Italian case studies within the *Subtask C—Case Studies and Action Research* (STC) framed in the International Energy Agency (IEA) Solar Heating and Cooling Program (SHC) Task 51 "*Solar Energy in Urban Planning*" [29–31]. During the whole duration (2013–2017), the international experts working in Task 51 (i.e., architects, urban planners, public authorities, researchers, etc.) promoted through their work the integration of active and passive solar solutions in the built environment. Furthermore, the collection of international case studies through a common template carried out within STC allowed to create an overview of solar energy in urban planning by highlighting potentialities and fragilities and by summarizing lessons learned for urban stakeholders, public authorities, and researchers. The developed template is here introduced and used to present six Italian case studies, which make use of solar energy in different ways and are characterized by different built environments. Five of them deal with consolidated and new urban areas, with photovoltaic or solar thermal panels fully integrated or applied to the buildings' envelopes, while one case provides an example of an agricultural PV system and its relationship with the landscape. Different from [31] where an international overview is given, in this paper, the Italian case studies are presented more in detail, and specific conclusions are drawn for solar energy implementation in the Italian context. Additionally, a comparison between the different built environments through similarities and differences is provided.

In the next section, an overview of the Italian energy planning legislation is given, followed by the presentation of the case studies and their discussion. Finally, the limitations of the study are presented, together with a conclusive summary of the main lessons learned and recommendations.

2. Background: Italian Legislative Framework

The framework of Italian legislation on urban and energy planning is characterized by a hierarchical approach stretching from the national level down to the regional, provincial, and municipal levels. According to the Constitution, urban planning and energy-related topics are a shared task between the State, the Regions, and the autonomous Provinces. Consequently, regional authorities may implement autonomous legislations as long as they do not contradict the general principles and requirements provided by national and EU regulations.

The Italian national legislation promoting energy efficiency and the diffusion of RES has been developed as an implementation of the major European directives. In 2011, the National Renewable Energy Action Plan [32] transposed the Directive 2009/28/EC setting the targets for renewable energy production by 2020. The plan also defined a minimum quota of production from RES for all new buildings and buildings subjected to major renovation. The minimum share of RES quota for domestic hot water (DHW) was set to 50% of the primary energy consumption. Furthermore, a calendar with a progressive higher RES share was established for the sum of primary energy consumptions for DHW, heating, and cooling. In December 2018, the EU targets have been revised by the European Directive 2018/2001/EC on renewable energy [33] and by the Regulation 2018/1999/EU on the governance of the energy union and climate action [34]. These regulations were part of the "Clean Energy for all Europeans Package", which promoted a 40% reduction in greenhouse gas (GHG) emissions, a goal of 32% final consumption from RES, and an energy efficiency target of 32.5%. Finally, at the beginning of 2020, the European Parliament adopted the European Green Deal, which set the objective "to increase the EU's GHG reductions target for 2030 to at least 50% and towards 55% compared with 1990 levels in a responsible way" and to achieve climate neutrality for the continent by 2050 [35]. Building on the EU legislation, and in line with the European Regulation 2018/1999/EU, which required all EU countries to develop 10-year National Energy and Climate Plans (NECPs) for the period 2021–2030, the Italian Integrated National Plan for Energy and Climate (INECP) was adopted in January 2020. The INECP set clear targets to 2030: (i) 30% of energy from RES in the final gross energy consumption, (ii) 43% reduction in primary energy consumption compared to the Price-Induced Market Equilibrium System (PRIMES) 2007 scenario, and (iii) 38% overall reduction in GHG emissions compared to 1990 [36].

2.1. National Standards

The INECP sets some growth targets for power and thermal energy from RES at the national level. In the case of solar energy, targets are set to 28,550 MW for 2025 and 52,000 MW for 2030 (in 2017, the production amounted to 19,682 MW).

With regard to the thermal sector, the targets are 590 ktoe in 2025 and 751 ktoe in 2030 [36]. Furthermore, the "Clean Energy for all Europeans Package" was implemented in

Italy by the "Renewables Decree" (D.Lgs 28/2011 [32]), which also led to the definition of guidelines for energy performance certification of buildings (D.M. 26 June 2015. Specifically, Article 11 of Decree D.Lgs 28/2011 introduces the obligation to integrate renewable energy sources in new buildings or buildings subject to major renovations. Indeed, in case of new construction, refurbishment, or demolition and reconstruction, RES should cover 50% of DHW consumption, and 50% of the sum of consumptions for DHW, space heating, and cooling. For public buildings, the share is increased to 55%; while in the case of private buildings located in historic city centers, the share is reduced to 25% (27.5% for public buildings) [32]. In 2021, the share of RES has been further increased, with the introduction of the D.Lgs 199/2021. From June 2022, in case of new construction and major renovations, RES should now cover 60% of the consumptions for DHW, space heating, and cooling. Conversely, a 65% coverage should be demonstrated for public buildings [37]. Concerning passive solar, minimum daylight levels in buildings are regulated by national laws (D.M. 190/1975, C.M. LL. PP. 22 November 1974 n. 1301 and D.M. n. 26/1975). These set minimum levels of the daylight factor depending on the type of building:

- Residential buildings: $\geq$2% and a window-to-floor ratio $\geq 1/8$ is required.
- Hospitals and schools: $\geq$3% (rooms and labs), $\geq$2% (gyms and canteen), $\geq$1% (offices and other service rooms).

The Italian National Status of Ground-Mounted PV

The national energy target for 2030 foresees an increase in ground-mounted photovoltaics of about 35 GW. However, the actual installation trend of 1 GW/yr is far below the 6.5 GW/yr needed to fulfill the target on time [38]. The installation of PVs on the ground is currently hampered by barriers along the authorization process due to landscape preservation and land use concerns. The adoption of innovative solutions such as agricultural PV aims at overcoming these barriers by combining a dual use of land and establishing a synergy between energy production and agriculture, which can be seen as a driver. In that regard, the current legislation is in constant evolution. In April 2021, the National Recovery and Resilience Plan (RRP), part of the Next Generation EU (NGEU) program, allocated 1.1 billion euros investment for the "agrivoltaic development" with the specific objective of installing 2 GW of agrivoltaic capacity and improving the competitiveness of the agricultural sector [39]. Additional elements regarding the definition of agrivoltaics have been introduced by the National Law 29 July 2021 [40] where these systems are described as "innovative integrative solutions, with PV modules raised from the ground, also including tracking systems, which do not compromise the continuity of the agricultural activities on ground". This law introduces simplifications in the process, which would lead to a rationalization of the authorizations for PV toward the national energy targets. Despite a specific definition for agrivoltaics not being currently available in Italy, an official document is expected in a short time. Meanwhile, the National Agency for New Technologies, Energy and Sustainable Economic Development (ENEA) has launched the National Network for Sustainable Agrivoltaics [41], while the Italian Electrotechnical Committee (CEI) has initiated a working group on the topic.

2.2. Municipal Standards

The targets set at the national level are pursued by Italian Regions, which can enhance the minimum quota or specify the renewable energy systems to use. Therefore, depending on regional laws, municipalities can set their own energy-related regulations in the building code. Generally, the building codes consider two major categories of built environment: historical city centers and non-historical urban areas [42].

Historical city centers are subjected to strict regulations about solar system applications due to the need to preserve the visual effect of their views and panoramas. Concerning this category, buildings codes set either prohibition of installing solar systems or restrictions on the visual effect and location of the systems (e.g., modules have to be located on roof surfaces not visible from main streets and to be building integrated). In addition to these restrictions, the approval of the Superintendence for Architectural Heritage and Landscape is required for installing solar systems on constructions subjected to landscape or heritage protection [43]. In the second category—non-historical urban areas—solar systems are generally permitted as building-integrated. For new buildings or buildings subjected to major renovations, it is required that PV or ST plants are installed integrated or adherent to the roofs, following their same orientation and inclination.

Further regulations on solar energy systems are sometimes included in municipal building codes, deriving from national laws on building energy efficiency and the promotion of renewable energy production. However, the standards regard only the design of buildings at the architectural scale. Energy performance certification is mandatory for new constructions and for selling or renting a dwelling in existing buildings [44].

3. Materials and Methods

This section illustrates the methodology used to analyze the case studies. The method was developed within the *Subtask C—Case Studies and Action Research* of the IEA SHC Task 51 to provide experts in the field of urban planning with an exhaustive set of information and examples on the integration of solar energy in heterogeneous environments. A total of six cases located on the Italian territory are presented here.

3.1. Classification of the Environments

As a first step, the case studies were divided according to the three types of environments presented in Table 1.

Table 1. Classification of the environments.

Environment	Description
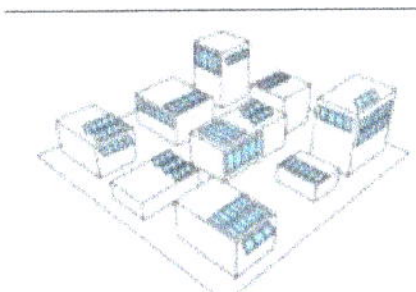	*Existing urban areas.* The first environment, represented by two case studies, includes fill-ins and densification processes, new buildings within a consolidated built environment, or the refurbishment of existing buildings. A scale larger than a single building is considered to assess the impact of a project on its surrounding.
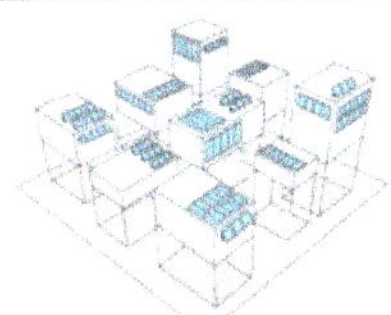	*New urban areas.* The second environment includes three case studies, and it is characterized by projects where completely new infrastructures and detailed development plans are required. The involvement of urban planners since the beginning of the process can play a significant role in the successful integration of solar energy.
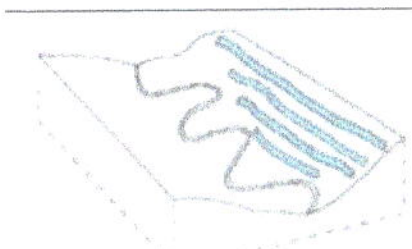	*Landscape.* The third environment includes one case study, and it investigates the impacts of large solar installations in the landscape.

3.2. Template Definition and Description

The second step consisted of the creation of a template to systematically organize the information of the case studies within a homogeneous framework. Despite the adoption of a standard structure, the template has a certain degree of flexibility. A total of ten sections are available, six of which are common among all the investigated cases, while the others can be compiled according to the type of environment described and the available information. A schematic representation of the template structure is visible in Figure 1, followed by the sections' description.

Figure 1. Representation of the different sections of the template used. In full color, the sections compiled for all the cases, while the optional ones are shown in white.

1. *Overview*. It briefly describes the context in which the case study has been developed. Particular attention is posed on the adopted solar energy strategies, local and national regulations, and future planning development.
2. *Challenges, issues, and decision strategies*. Issue and challenges encountered during the realization of the project are presented with an emphasis on relevant features of energy characterization (i.e., integrated panels, overshadowing effects).
3. *The planning process*. This section presents the timeline of the project and the different spatial scales investigated during the planning. In addition, the involvement of stakeholders and researchers and the most influential decisions taken during the planning process are presented.
4. *Energy concept*. The focus is on energy technologies, PV, ST, and passive solar gains. The energy needs and the adopted solar strategies are described.
5. *Architecture, Visibility, Sensitivity, and Quality*. This section investigates the critical issues of implementing active solar strategies into a built environment. This is completed using a methodology developed by EPFL researchers within the framework of IEA SHC Task 41 "*Solar Energy and Architecture*", which is now integrated into the tool LESO—QSV [45–47]. As a first step, the solar system installation is evaluated according to the architectural integration quality. To be considered successful, the integration should be coherent with the entire building design logic regarding system geometry (i.e., size, position), system materiality (i.e., visible materials, surface textures, colors), and system modular pattern (i.e., module shape, size, joints). The coherency of these three aspects is evaluated using a three-level scale (fully—partly—not coherent), corresponding to three colors (green—yellow—red). The second step is to assess the criticality of the surface where the solar system is installed, which is depending on its close or remote visibility from the public space (low—medium—high) and from the sensitivity (low—medium—high) of the urban context in which is located. Exemplary of a high sensitivity context is a historical city center, a medium one is a post-war residential development, and a low one is an industrial district. Regarding new realization, the evaluation should be based on the future vocation of the urban area. Figure 2 summarizes the approach.

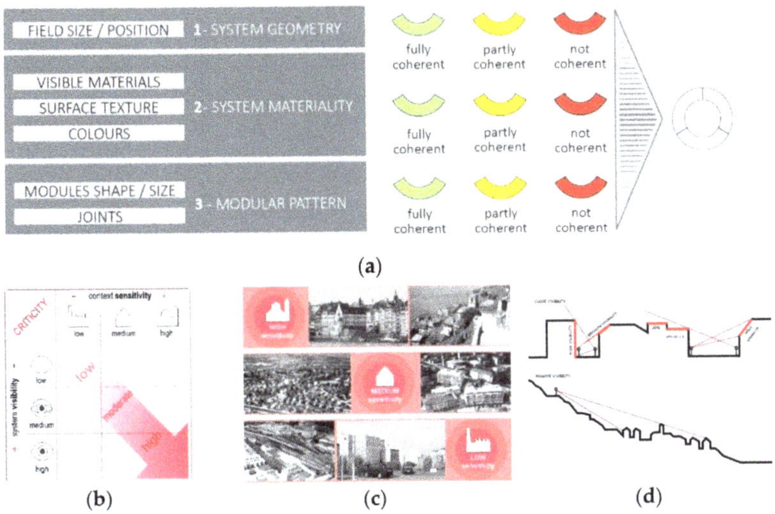

Figure 2. (**a**) Architectural integration quality matrix; (**b**) Criticality matrix developed in relation to (**c**) context sensitivity and (**d**) system visibility.

6. *Solar landscape*. This section is specific to the landscape environment. It includes the definition of the solar system used, its functional features (i.e., modules' pattern, presence of edges—Figure 3), and the energy production. This is defined according to the classification proposed by ENEA on solar landscape plan [10].

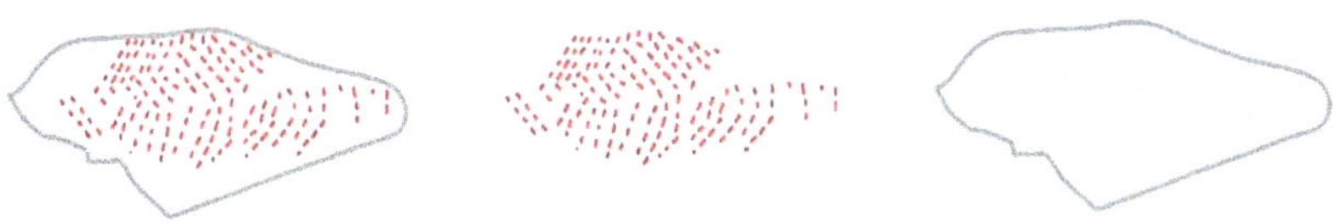

Figure 3. The spatial system as a whole (pattern) (on the **left**), the photovoltaic space (in the **middle**), and the "pore" space (on the **right**).

7. *Site potential*. It refers again to the landscape environment following the methodology developed by ENEA, defining the site potential. The factors influencing the suitability of the site for the installation of PV or ST are here identified and qualitatively assessed.
8. *Environmental, economic, and social impacts*. The impact of the project under the social, economic, and environmental aspects is analyzed, with particular attention to the relation between the solar system and the surrounding environment.
9. *Approaches, methods, and tools*. It contains information about methods (i.e., procedures to assess and evaluate solar in relation to other aspects of urban planning), tools (e.g., rule of thumb, software, etc.), and tested approaches utilized during the planning process and in the integration of solar energy.
10. *Lessons learned and recommendations*. The final section presents the outcomes of the project together with potential solutions and recommendations gained from it.

An overview of the six analyzed case studies is visible in Table 2.

Table 2. Overview of the analyzed case studies. (CCHP: Combined Heating and Cooling; DH: District Heating; DC: District Cooling).

Location	Picture	Name	Classification	Area [m^2]	Energy Strategies	Highlights
Alessandria		Photovoltaic Village	Existing urban area/ Refurbishment	72,000	PV roof PV façade PV facilities	• The first example of an Italian public residential building project where solar energy is applied at an urban scale. • Photovoltaics cover 100% of electricity consumption for common areas and up to 70% in apartments.
Bolzano		SINFONIA	Existing urban area/ Refurbishment	1750	PV roof ST roof Geothermal	• European project aiming to transform Bolzano into a smart-city and to retrofit 34,000 m^2 of social housing • A successful example of the use of an Integrated Design Process (IDP)
Trento		Le Albere	New urban area	116,000	PV roof PV façade PV shading Geothermal CCHP plant	• Regeneration and reconnection to the city's fabric of a large brown-field • Building-integrated photovoltaic modules give identity and enforce the aesthetic of the project
Brescia		Violino District	New urban area	48,450	PV roof DH	• Social housing project developed according to bioclimatic design principles • Volumetric composition and orientation of buildings designed to maximize the solar exposure
Bolzano		Casanova District	New urban area	100,000	PV roof ST roof Geothermal DH-DC	• Attention to mobility and social aspects • Shape, orientation and reciprocal position of buildings are designed according to solar and wind

Table 2. *Cont.*

Location	Picture	Name	Classification	Area [m^2]	Energy Strategies	Highlights
Monticelli d'Ongina		Agrovoltaico	Landscape	21,000	PV on two axes trackers	• Double use of land combining energy and food production ad different layers. • The photovoltaic system is designed as a landscape, minimizing its ecological impact.

4. Results and Discussion

In this section, the six case studies are presented and discussed, starting from the ones in existing urban areas, followed by new urban areas, and concluding with the integration of solar technologies into the landscape.

4.1. Photovoltaic Village in Alessandria

4.1.1. Overview

The case study is a residential neighborhood constituted of several multistorey buildings, for a total of 192 flats, located in the southwest fringe of Alessandria (Piemonte). The project was developed within the context of a complex initiative undertaken by the Alessandria Municipal Council after the flood of 1994. It aims to regenerate a social housing community using a sustainable, environmental, and social approach through the realization of new buildings, urban facilities (i.e., a community center, recreational areas, sheltered sits, and parking lots) and the refurbishment of existing residential blocks (Figure 4) [48]. The extensive use of photovoltaic technology characterizes the project.

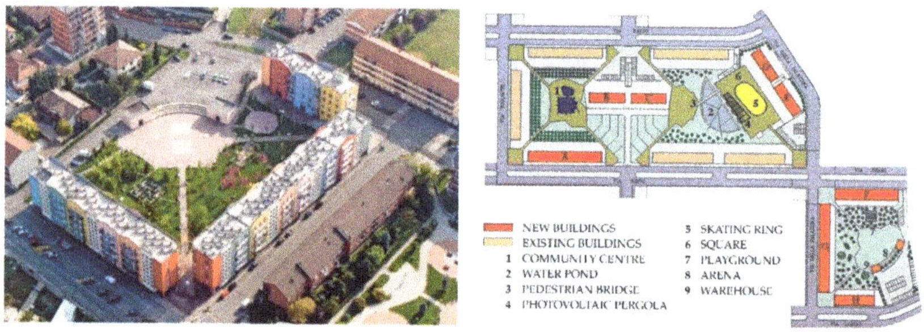

(a) (b)

Figure 4. The (**a**) aerial view of new buildings in the eastern part of the Photovoltaic Village (source: © Municipality of Alessandria) and a site plan of the entire area indicating the position of facilities, new and existing buildings (**b**) (source: © Municipality of Alessandria).

4.1.2. Challenges, Issues, and Decision Strategies

The challenge encountered during the realization of the project was the definition of an innovative planning process to meet the interests of both public and private actors and the promotion of energy-efficient solutions and RES. In addition, the existing urban plan was substituted due to the adoption of new legislative procedures. The decision strategy was based on three principles: (i) active participation and collaboration between the actors; (ii) a pivotal role of the municipal authorities in the whole design process; (iii) the project conception as a pilot for town planning and suburban requalification.

4.1.3. The Planning Process

The project began in 1996 with the enforcement at the regional level of the Environmental, Building and Urban Requalification Program [49], with energy saving as one of the main objectives for buildings and urban projects. This stimulated the adoption of solar energy and sustainable practices and led to create the Building Operators Council in 1997 to coordinate the actors involved in the process. The construction phase lasted five years (2000–2005). The most important phases were the large scale of comprehensive/strategic planning, represented by the Environmental, Building and Urban Requalification Program, and the architectural design stage of the existing buildings. The Municipal Town Council, the Building Operators Council constituted by private and cooperative builders from the Province of Alessandria, architects, and various other private stakeholders took part in developing this initiative. Furthermore, researchers from Politecnico di Torino contributed to the design and monitoring of solar systems [48].

4.1.4. Energy Concept

The buildings are designed to reduce their impact on the environment through energy-saving measures, and the use of RES. PV modules, with an efficiency of up to 15%, are installed on roofs with a tilt angle of 30°, on the southern facades in front of the stairways of two buildings in the community center (indicated with letters B and C in Figure 4b) and as part of a photovoltaic pergola for the public space (Figure 5). PV systems cover a total area of 3000 m^2 (i.e., 1600 m^2 panel net surface). The overall power is 163 kW with estimated energy production of 674–830 kWh/kWp per year. This is expected to cover 100% of the electricity consumption for common areas and up to 70% for the flats [48].

(a) (b)

Figure 5. Ground (**a**) and aerial (**b**) view of the pergola equipped with PV panels in the public area (source: © Municipality of Alessandria).

4.1.5. Architecture, Visibility, Sensitivity, and Quality

The LESO–QSV method illustrated in the methodology is used in this section to evaluate the integration quality of the installed solar systems. As visible from the colored segments of the ring of the matrix in Figure 6, the integration has been judged partly coherent for all the three investigated aspects. Since the modules do not perform the double function of replacing a building component and simultaneously producing energy, being just overlaid to the roofs and facades, they cannot be considered as fully integrated into the building envelope from an architectural point of view. Moreover, the blue color of the solar cells contrasts with the finishing color of the facades. The systems' visibility from a close point of view is high, especially for the facades' elements, and it remains medium from a remote distance. Nevertheless, the context can be classified as of medium sensitivity due to the low historical value and the absence of any relevant elements or monuments.

(a) (b)

Figure 6. Architectural integration matrix for the PV systems in the Photovoltaic Village of Alessandria (left); (**a**) PV panels installed on facade in front of the stairways (source: © PierFranco Robotti); (**b**) aerial view of the systems installed on the flat roofs (Source: © Municipality of Alessandria).

4.1.6. Environmental, Economic, and Social Impacts

The project follows eco-sustainability and bio-compatibility principles. Particular attention was reserved to open spaces where green areas, a water pond, and urban furniture were carefully designed to create different bioclimatic zones for the users to stroll, relax, and socialize. The project is financed by a mix of public and private sources. The region financed part of the investments for its environmental sustainability implications, while approximately 70% of the total cost for the installation of photovoltaics was covered by public funds through the program "10,000 Photovoltaic Roofs" coordinated by the Italian Ministry of the Environment [48,50].

4.1.7. Approaches, Methods, and Tools

A monitoring campaign on the PV systems was carried out by researchers of Politecnico di Torino for a period of 12 months (September 2004–August 2005) to measure the electricity generated during the buildings' operation phase and correlate it to the actual energy consumption. Data were collected every 15 min and function curves for daily, weekly, monthly, and annual periods were generated. The analysis shows that PV systems can fulfill the electricity demand for seven months, but an electricity supply through the grid is still needed. In addition, in the absence of regional, national, or community incentives, the investment can be paid off in a long-term period (more than 30 years).

4.1.8. Lesson Learned and Recommendations

The Photovoltaic Village in Alessandria illustrates the limits of retrofitting existing buildings with solar technologies in the absence of a deep renovation opportunity. Despite the overall positive results, the designers were forced to consider a limited range of design options since they had to comply with existing buildings' masses and materials. A different outcome may have been obtained through an in-depth renovation process involving replacing/recladding facades and roofs. Furthermore, the integration of solar systems in public spaces has only been marginally explored. However, the institution of a Building Operators Council as a coordinating body for the different involved actors was a successful practice replicable in similar contexts.

4.2. SINFONIA Bolzano

4.2.1. Overview

The SINFONIA case study is the result of a five-year European project aiming to implement large, scalable, and integrated energy solutions in mid-sized cities. Bolzano, the capital of South-Tyrol province (Trentino Alto Adige), was selected together with Innsbruck (Austria) as a testbed for this initiative to understand the replicability of the proposed solutions in other European cities [51]. The project aimed to achieve the following:

- Obtain between a 40% and 50% reduction in primary energy use and a 20% increase in RES share for the selected districts in the two pilot cities of Bolzano and Innsbruck.
- Prove the feasibility of large-scale energy measures, retrofitting, grid optimization, and district heating and cooling systems in mid-sized European cities.
- Define district typologies and refurbishment models to ensure scalability and replicability. These last two aspects are supported by five "early adopter" cities where to test the initial outcomes of the two pilots.
- Develop a network to support and engage other European cities in strengthening their smart-energy solutions.

In this framework, refurbishment interventions were undertaken for six residential complexes in Bolzano to improve their energy performance and indoor comfort. The initiative is in line with the municipal goal to reduce energy use and increase the share of RES. The presented case study focuses on one of these interventions, consisting of two buildings blocks of 36 flats each, located in the south part of the town (Figure 7).

(a) (b)

Figure 7. North building before (**a**) and after (**b**) the intervention (source: © Ivo Corrà).

4.2.2. Challenges, Issues, and Decision Strategies

The location of the building complex has been one of the main challenges in the development of the project. Its proximity to the side of a mountain represented an accessibility issue during the refurbishment phase. Other challenges included the limited surface for the integration of solar energy systems, which was also due to the proximity to the mountain slope toward the south and the presence of overshadowing due to four stairwell towers above the roof levels. Finally, the intervention should have the minimum possible impact on the tenants, as all refurbishment activities have been carried out with lived-in flats. Regarding the decision strategy, targets for the energy concept were set at the beginning to reach: (i) a final energy balance $\leq$ 22.5 kWh/(m^2 yr); (ii) DHW from renewable energy sources $\geq$ 9 kWh/(m^2 yr); and (iii) PV installation $\geq$ 53 kWp for the entire complex. Furthermore, the optimization of the solar systems and the improvement of interior daylight conditions were also investigated. An Integrated Design Process (IDP) was followed to assure good stakeholders' involvement.

4.2.3. The Planning Process

The refurbishment was carried out in three phases. The first was the design phase, which was developed through an IDP by a multidisciplinary team of experts from the Municipality of Bolzano, the design team, research institutes, and Agenzia Casaclima (the local certification body for energy efficiency in constructions). The two following phases were the construction phase and the monitoring phase, which lasted one year. All the spatial scales were investigated during the five years of the realization (2014–2019).

4.2.4. Energy Concept

The buildings' complex was originally realized in the 1990s without energy-saving measures, apart from a thin insulation layer (4 cm) on the external walls. Hence, the project involved the refurbishment of the existing facades by using timber prefabricated multifunctional elements to improve the energy performance and the indoor thermal comfort as well as aesthetically rehabilitate the buildings. Renewable energy strategies were also adopted with the installation of PV and ST systems on the roof and a geothermal heat pump serving centralized heating. PV modules with 15.5% efficiency, covering an area of 337 m^2, are installed on the roof of the northern building (Figure 8). The system has a power of 20 kW, and the modules are mounted horizontally (0° tilt angle) to avoid mutual overshadowing effects. Conversely, the roof surface of the southern building has been used for the installation of a 477 m^2 ST system with a 10° tilt angle and with an estimated production of 11.2 kWh/(m^2 yr) for DHW. This choice was dictated by the need to comply with the national legislation, requiring to cover at least 35% of the energy need for DHW with RES [32]. Both PV and ST cannot be considered architecturally integrated into the building, since they constitute free-standing elements placed on the roofs.

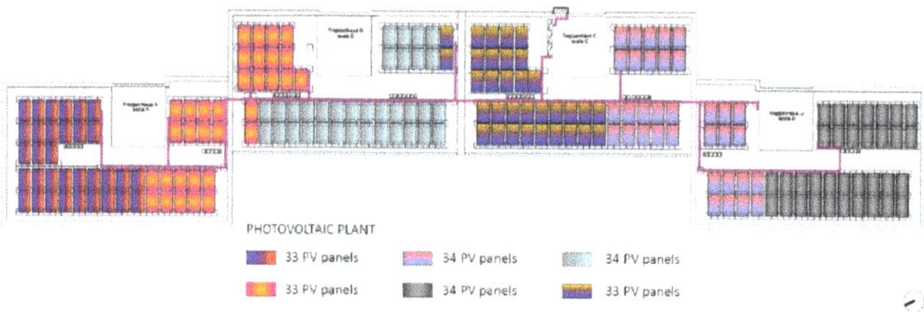

Figure 8. Technical plan of PV system on the roof of the north building (source: © EQ Ingegneria).

4.2.5. Environmental, Economic, and Social Impacts

The SINFONIA project intended to have an impact on the entire municipality of Bolzano by improving the quality of life of the citizens, turning the city into a smart city. The installation of over 100 smart points and three multifunctional interactive totems provides services such as charging electric vehicles, monitoring air quality and weather, and improving the lighting of public spaces. The installation of this diffuse technological network runs in parallel with the refurbishment of six residential housing complexes that enhanced the interior comfort and energy performance of 455 dwellings for a total of around 15 million euros of investments. From a social perspective, the tenants of the refurbished apartments have been directly involved through informative meetings and questionnaire-based surveys during the realization of the project. Additionally, displays have been installed in part of the apartments; data on energy performance and indoor conditions are shown in real-time, enabling the inhabitants to take actions to reduce their consumption. This was completed to raise awareness of the role that their behavior plays in the energy performance of buildings (i.e., ventilation rate, air temperature control, energy use) [52].

4.2.6. Approaches, Methods, and Tools

During the IDP, retrofit solutions were assessed before adopting a mixed approach combining the use of traditional insulation for the lodges and prefabricated multifunctional facades (MFF) for the rest of the building envelope. This choice did not require the use of scaffolding, simplifying and speeding up the construction phase, ensuring a low impact on the site's inhabitants. To define the best solutions for the MFF, prototypes were constructed and tested in laboratory tests. Moreover, the design focused on defining the optimal solution for solar active systems, assessing the potential energy production, and preserving the indoor visual comfort. To this aim, solar potential and daylight analyses were conducted using Diva for Rhino [53] and LadybugTools [54]. To assess the solar potential, close shading elements were modeled in the 3D environment of Rhinoceros [55], while the distant obstacles, such as the mountains, were considered by a horizon line generated from data measured with a Solmetric SunEye. The optimization process was carried out using a genetic algorithm to maximize the total annual and average irradiation and to minimize overshadowing effects. As the installation of the new prefabricated MFF caused a reduction in interior daylight availability, daylight analyses were performed to verify the compliance with the Italian national legislation (i.e., daylight factor above 2%). Details such as the windowsills' materials and loggias' colors were investigated to improve the solar mutual reflection effects. The simulations and the laboratory tests were conducted during the design phase, which allowed reducing the risks and uncertainties during the construction phase.

4.2.7. Lesson Learned and Recommendations

The main lesson learned from the SINFONIA case study is the benefit that an IDP can have on energy-related measures applied to the building.

Fundamental is an early collaboration between the different experts and stakeholders involved in the project (i.e., energy consultants, designers, owners, researchers, technicians). Furthermore, the case study demonstrates the importance of conducting solar potential and daylight analyses during the initial phase of the project by supporting the design team in the most important choices (i.e., finishing materials and colors to improve the interior daylight level, localization of active solar systems to avoid overshadow effect). Finally, it is recommended to consider how the occupants' behavior can influence the energy use of a building and try to engage and inform the users accordingly.

4.3. Le Albere in Trento

4.3.1. Overview

Le Albere is a district in the city of Trento (Trentino Alto Adige) that is constructed on a former industrial site enclosed between the railway and the river Adige. The aim of the project, designed by the Italian architect Renzo Piano, was to reconnect the site with the city center and to re-establish a relationship with its natural context. The result of the intervention is a mixed-use district that includes offices, residential buildings, a science museum (MUSE), a library for the University of Trento, and a park (Figure 9) [56]. The main feature of the project is the extensive integration of active solar systems on roofs and facades of the buildings of the whole complex. In that regard, the choice to adopt this energy strategy was autonomously taken by the actors involved since, at the time of the design, the Italian national legislation on energy and urban planning did not set any specific target for solar energy integration, despite supporting the use of RES.

(a) (b)

Figure 9. (**a**) Aerial view of the Le Albere district (source: © Google Earth) and (**b**) view from the public park (photo: © Silvia Croce).

4.3.2. Challenges, Issues, and Decision Strategies

The main challenge for the case study has been the formal integration of the photovoltaic systems in architectural design. High requirements were set by the architect in terms of the color, appearance, and dimension of PV modules due to the choice of having them as a key and visible element of the project. The strategy adopted by the municipality of Trento was to create a high-quality district, showcasing the best practices in terms of energy efficiency and sustainability, with extensive integration of RES.

4.3.3. The Planning Process

The planning process encompassed different spatial scales, spanning from the provincial down to municipal, local, and detail scales. Le Albere was a former industrial area occupied since 1937 by the tire manufacturing company Michelin. When the factory ceased its activity in 1998, a group of public and private investors, under the name of "Initiative Urbane", purchased the site to regenerate and reconnect it to the rest of the city.

In 2000, the Province of Trento included the renewal of the district into its Program for Urban Regeneration and Sustainable Development of the Territory and into the Trento Strategic Masterplan. Two years later, the architectural firm Renzo Piano Building Workshop was designated to realize the project, and a final master plan was approved in 2004.

The construction phase started in 2008 and lasted for five years when the MUSE, the mixed-use district, and the public park were inaugurated. Conversely, the university library located in the southern part of the parcel opened at the end of 2016.

4.3.4. Energy Concept

Several RES were used in the Le Albere district, such as BIPV systems integrated on many of the buildings that are present on site. The systems have a nominal power of 279 kW and cover a total area of approximately 3200 m^2. Tedlar–glass and glass–glass modules with different orientations and inclinations were specifically designed depending on their location. They are integrated into the buildings' facades and roofs or used as PV shading devices. The electricity produced by the BIPV modules covers the electrical demand of offices, common areas, pump rooms, and the lighting in the basement. The MUSE is served by a geothermal plant while a combined cooling, heating, and power plant, outside of the district's boundary, covers the heating and cooling demand for the whole area.

4.3.5. Architecture, Visibility, Sensitivity, and Quality

Following the LESO–QSV methodology, the integration of PV modules has been judged as fully coherent for all three categories of geometry, materiality, and pattern (Figure 10). Despite the different heights, inclinations, and shapes of the buildings, PVs act as unifying elements for the entire complex, further enhancing and characterizing it. The level of criticality, as a function of context sensitivity and system visibility, can be considered as medium. In fact, the case study is bordered by urban areas without any particular architectural or historical value, with the only exception of a renaissance villa/fortress on the north side. Regarding the systems' visibility, it is high from internal streets, open areas of the park, and roads bordering the site when PVs are installed on facades or integrated into glass surfaces constituting the tilted roofs. On the contrary, low perception levels for an observer located in the urban canyons of the district are achieved when the system is integrated on the opaque galvanized roofs. From an observer in the courtyards or in the park, the photovoltaics are perceived as a geometrical pattern. Finally, the remote visibility is low due to the overall height and dimensional scale of the complex, which is comparable with those of the historical center and the existing industrial structures.

(a) (b)

Figure 10. Architectural integration matrix for the PV systems in the Le Albere district in Trento (left); (**a**) the two typologies of BIPV systems: tedlar–glass modules on the galvanized roof and the glass–glass modules in the background (photo: © Silvia Croce); (**b**) detail of the glass–glass PV modules installed on the facades (photo: © Silvia Croce).

4.3.6. Environmental, Economic, and Social Impacts

Sustainability was central in the Le Albere project since the first phases of the design, starting from the construction materials sorted locally whenever possible to limit the environmental impact. The master plan was developed to preserve the natural features of the area by concentrating all the new constructions toward the east side of the plot and creating a large public park on the west side. The existing street, running parallel to the river, was transformed into an underground passage to allow the continuity of the park and directly interact with the river Adige. Regarding mobility, the realization of two-story underground parking greatly reduces the number of cars on the surface, while the traffic in the district is limited to residents, taxis, and public transport. Great attention was also placed on the design of numerous walkways and bikeways. Furthermore, the MUSE and the university library have obtained the LEED Gold certification, while residential buildings and offices, built according to passive standards, received the level B label of CasaClima (i.e., energy-efficiency classification), corresponding to a heating energy requirement below 50 kWh/(m^2 yr). The project carries a high social value thanks to the restitution of the area to the public realm with a new identity. The district has now a strong cultural and recreational vocation due to the presence of two social attractors (MUSE and university library) at the borders of the plot and a five-hectare park with residential and commercial buildings in the middle. The renewal intervention, which cost around 350 million euros, was entirely financed by a pool of banks, insurance companies, and industry associations in collaboration with the Municipality and the Province of Trento.

4.3.7. Approaches, Methods, and Tools

The approach used for the design of the case study has been strongly oriented toward sustainability. Building information modeling (BIM) tools were used along the entire process to integrate the buildings' design with energy modeling and considerations on internal and external comfort conditions.

4.3.8. Lesson Learned and Recommendations

The Le Albere case study strongly highlights the potential of solar systems integration not only as energy generators but also as architecturally valuable elements that can provide identity and enforce the aesthetic of the urban intervention. This is realized by declaring the presence of the photovoltaic modules and giving them high visibility instead of hiding or camouflaging them. When this approach is pursued, the quality of the system's integration becomes of primary importance and has to be carefully designed. In addition, the project poses sustainability, under its various declination of social, economic, and environmental, as the pivotal element to guide the design process, successfully resewing a large portion of the city with the rest of the urban fabric.

4.4. Violino District in Brescia

4.4.1. Overview

The Violino district is a new residential area located in the outskirt of the city of Brescia (Lombardia). It is constituted of 112 terraced houses and 2 multi-family houses five floors high (143 housing units), which are organized in lots defined by an orthogonal grid (Figure 11). It is the result of a new urban plan for social housing followed by an architectural competition announced by the Brescia Municipal Council [57] to design a district according to bioclimatic principles and with the extensive use of solar strategies. The area required a northeast/southwest oriented grid, which was judged to be not optimal for the full exploitation of the thermal and energy potential of the sun. The issue was overcome by adapting the terraced house' typology to the rigid grid with a partial rotation of the buildings' masses, ensuring adequate solar accessibility to all accommodations.

(a) (b)

Figure 11. (**a**) Aerial view of the terraced houses of Violino District (source: © BAMSphoto—Basilio) and (**b**) the two multi-family houses (photo: © Fabio Cattabiani).

4.4.2. Challenges, Issues, and Decision Strategies

The main challenges were the need for an adequate urban quality in social housing, elevated technical performances to reach sustainability and energy targets, and the rigid urban grid. Most of these challenges were highlighted in the call for tenders, and the participants were asked to submit proposals with quantifiable quality requirements. This allowed adopting a decision strategy based on a scoring system and conveying information on the conformity of each proposal to performance and quality requirements (i.e., thermal, acoustic, hygrometric insulation, daylight, energy production) as well as understanding the economic impact of these measures compared to standard social housing.

4.4.3. The Planning Process

The origin of the district dates back to 1980 with the Piano Regolatore Generale (PRG) of Brescia drawn up by Secchi, Viganò and Scarsato, and it emerged in its current configuration in 2002 when the Municipality of Brescia purchased the land and issued a tender for the development of a new social housing intervention in the southwest sector. Architects, installers, and consultants collaborated in the planning process, while the construction phase (from 2004 to 2006) was carried out by local cooperatives and construction companies. The intervention was financed with funding from the Lombardia Region and from Azienda Lombarda Edilizia Residenziale (ALER). All the spatial scales were investigated during the realization of the project.

4.4.4. Energy Concept

The entire district has been designed according to bioclimatic principles to benefit from climatic conditions. Therefore, each building is oriented to maximize daylight and passive solar gain as well as the generation of electricity through photovoltaic systems. A characteristic element is represented by the greenhouses integrated into the terraced houses and acting as solar collectors and thermal buffers. The sun rays that penetrate the windows are transformed into heat in contact with a massive surface: the heat is retained, thus counteracting the excessive temperature ranges. The terraced houses are all equipped with photovoltaic systems of 1.3 kWp mounted flat or with a 30° southwest inclination, while multi-family houses have systems ranging from 5 to 20 kWp with a 30° angle.

4.4.5. Architecture, Visibility, Sensitivity, and Quality

The integration of PV modules in the terraced houses can be described as fully coherent regarding system geometry and materiality, while they can be considered partially coherent for the modular pattern (Figure 12). Many gaps are visible between the panels, giving the overall impression of an uneven surface. Regarding the systems installed on the two multi-family houses, they cannot be considered as coherent with any of the listed aspects, since they are mounted inclined on a flat roof, using a metal structure. The area is characterized by a medium sensitivity level due to the absence of historical buildings, monuments, or meaningful elements. The close visibility of PV panels is also classifiable as medium, being the systems either installed on tilted roofs, visible from the street canyons, or on flat roofs. Concerning the remote visibility, the plain terrain of the site does not provide any vantage point from where the PV systems are visible, with the only exception represented by the top stories of the two multi-family houses.

Figure 12. Architectural integration matrix for the PV systems at Violino District in Brescia (left); (**a**) system modular pattern (photo: © Fabio Cattabiani); (**b**) view from the roof of one of the multi-family houses (photo: © Fabio Cattabiani).

4.4.6. Environmental, Economic, and Social Impacts

The environmental dimension of the project resides in the bioclimatic approach utilized for the design of the entire intervention, where great attention was placed on the exploitation of solar radiation. From the economic and social point of view, the project was fully financed by public funds with the scope to provide an affordable and high-standard social housing solution. Some accommodations were reserved for specific age groups to create a more heterogeneous and diversified environment. For instance, 26 dwellings are for elderly people, while 12 are for young couples. The rationalization of local roads, with the interposition of three public pathways, provides a protected connection to the main area of the neighborhood and fosters public aggregation and socialization.

4.4.7. Approaches, Methods, and Tools

The approach adopted for designing the terraced houses aims to guarantee the "right of sun" to each residential unit. The interior distribution of spaces, reflected in the exterior composition of volumes, is based on their function. The south orientation is dedicated to the most used spaces, namely the living room on the ground floor and individual rooms at the upper level. Services such as bathrooms and vertical distribution are on the north side, while the west side is occupied by the kitchen and the master bedroom. The greenhouse, painted in dark hues to further enhance heat gains in winter, is integrated into the south facade. Colors have also played a significant role in the project, as a study has been carried out to diminish the perception of the repetitiveness of the terraced houses' typology using different shades of colors. Furthermore, this choice highlights the volumetric composition of the residential units and guarantees different perceptions during the day according to the various light conditions (Figure 13). Since 2007, the project was also subjected to a monitoring campaign to evaluate the performance of PV systems, their energy production, and the amount of electricity fed into the grid. Measurements show that the PV systems installed in housing units produce, on average, 4.54 kWh per day. Moreover, between 2014 and 2015, the Smart Domo Grid Project [58,59] was carried out by the energy company A2A S.p.A., the Politecnico di Milano, and the company Whirlpool.

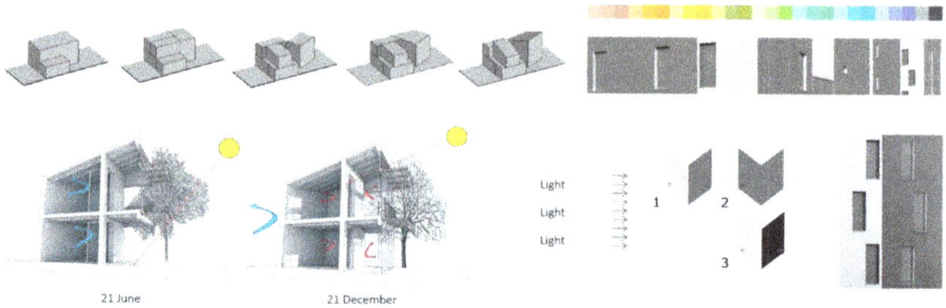

Figure 13. Volumes' composition (**top left**); section of terrace house with passive strategies (**bottom left**); color plan by Jorrit Tornquist (**right**) (source: © Boschi + Serboli Architetti Associati).

The project involved 21 families and consisted of testing a smart system to enhance energy efficiency and reduce electricity costs for the users. Using an energy management application that measures consumption and provides insight on how to minimize energy electricity costs, users could avoid overload and plan the use of appliances in the most convenient moment of the day.

4.4.8. Lesson Learned and Recommendations

The Violino District underlined the importance to work simultaneously on volume composition, internal space distribution, and technical systems to maximize the exploitation of passive strategies and achieve hygrothermal comfort conditions. The sun was used as an important bioclimatic element in the design process to determine buildings' orientation and facades' exposure. The greenhouses were used as natural temperature regulators to trap and slowly release solar heat gains, especially in winter.

4.5. CasaNova District in Bolzano

4.5.1. Overview

CasaNova is a large social housing project in Bolzano (Trentino Alto Adige). It has been developed by the municipality to provide affordable rental solutions while guaranteeing high living standards and a low ecological footprint. It is located in the southern parcel of the city, on an area of 10,000 m^2 delimited on the northern side by an existing neighborhood and the countryside, and on the southern side by the railway and the Isarco river. Groups of three to four buildings each form eight urban blocks arranged around green courtyards for a total of 950 apartments (Figure 14), which mostly comply with the highest level of CasaClima energy classification.

(a) (b)

Figure 14. (**a**) Aerial view of CasaNova district showcasing the buildings' arrangement around courtyards to form urban blocks (source: © altoadige.it); (**b**) view of the block designed by CDM Architetti Associati in the north part of the plot (photo: © Andrea Martiradonna).

4.5.2. Challenges, Issues, and Decision Strategies

Despite being the project established on a greenfield (former agricultural area), the municipality chose the lot for its position in continuity with an already urbanized area to avoid excessive decentralization. The focus was placed on creating a strong relationship between the new development and the countryside through extensive use of greenery, both at ground level and on the roofs, as well as in the design of pedestrian paths and buildings' alignments. On the other hand, several issues were encountered during the realization of the district such as the absence of a well-established connection system with the rest of the city, the presence of rural buildings within the plot, and hydrogeological risks related to the proximity to the river and the high level of the water table in the area. Furthermore, the inhabitants of the adjacent neighborhoods and the farmers firmly opposed the project due to its possible future expansion [60].

4.5.3. The Planning Process

The project started in 2001 when the Bolzano municipality acquired 10,000 m^2 of farmland and turned it into buildable land. After an architectural competition held at the European level [61], the realization of a master plan was assigned to the Dutch firm Frits Van Dongen. During the urban design phase, several workshops were carried out together with the political and technical representatives of the municipality, representatives of the owners, the neighborhood council, and other actors involved in the project. Successively, the eight lots were designed by different architectural firms after another competition organized by the Bolzano municipality. The construction phase started in 2007 and terminated in 2014 with the realization of a public space with a mix of functions in the center of the plot. Finally, a cycling path and a train station were built to improve the connectivity of the neighborhood with the rest of the city.

4.5.4. Energy Concept

Before starting the architectural urban design, the energy plan of the neighborhood was developed. It set the compliance with the highest CasaClima certification energy standard (i.e., heating index of 30 to 50 kWh/m^2y depending on the surface/volume ratio of each building) [62]. The energy concept included passive strategies of bioclimatic planning approach and active systems to produce electricity and DHW (Figure 15).

The compact shape of the buildings aims to maximize solar exposure and reduce overshadowing effects by placing the higher buildings in the north part of each block with sloped roofs in a unique direction. The PV and ST systems installed on most of the buildings cover almost completely the DHW need for the neighborhood. Space heating during winter and the pre-heating of the DHW is guaranteed by geothermal heat pumps and by the district heating network, which are connected to a cogeneration plant. Space cooling in summer is provided by a controlled ventilation system where the air is pre-cooled by the geothermal pumps. These solutions allow a reduction of 65% of the energy consumption compared to a traditionally built neighborhood.

4.5.5. Environmental, Economic, and Social Impacts

The environmental impact of the CasaNova district is kept low thanks to the design approach oriented toward a strong reduction in energy consumption. An example is in the construction method, which used reinforced concrete only for the main structure and in the slabs, and prefabricated elements for the external walls. The prefabricated elements allowed to cut the construction time nearly in half, and save electricity and water in situ due to the minor use of cranes and machines. Another environmental feature is the reuse of the rainwater thanks to the installation of four water tanks at the corners of each urban block. The collected water is used to irrigate the green areas at the ground and roof level and for toilets. The outdoor thermal comfort is addressed by arranging the buildings to exploit the fresh breeze in summer and act as a protection against cold drafts in winter and by the extensive presence of green spaces. To reduce the impact of traffic and to guarantee good air quality standards, only one road crosses the districts, and all parking spaces are located underground. Finally, the large offer of affordable rental solutions and the good accessibility achieved after the realization of a new train stop, bus connections, and cycling paths represent a relevant economic and social impact achievement of the project.

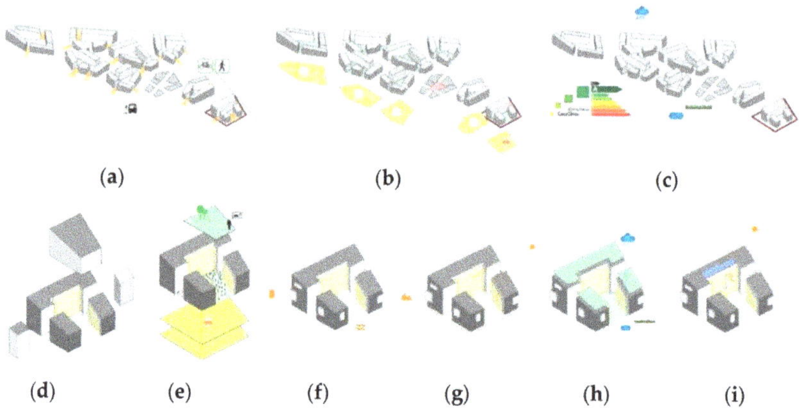

Figure 15. Design concept of CasaNova district. At the neighborhood level: (**a**) urban mobility and accessibility to the area, (**b**) social public spaces organization, (**c**) shadow cast and buildings' morphology to reach the highest energy class according to the Italian energy certificate system. At the building level: (**d**) morphology of the building block, (**e**) internal connections and social spaces, (**f**) study of the views from the building stock, (**g**) study of the solar exposure and orientation, (**h**) green roofs and water management system, and (**i**) use of RES; (Source: © Wienerbeger).

4.5.6. Approaches, Methods, and Tools

The case study aims to reduce energy consumption, maximize solar exposure, and create a permeable urban fabric able to establish a dialogue with the countryside. The architects' approach reflects these objectives in the arrangement of compact-shaped buildings around open courtyards, with taller ones located on the north side of the blocks to avoid overshadowing effects. The same attention is reserved for the composition of the single housing units, where large glazed areas are used to exploit solar heat gains and every apartment has a loggia oriented toward south or west. Finally, in 2009, Eurac Research was appointed as technical and operational support for building monitoring. A campaign was carried out to verify the energy consumption and internal comfort conditions, and to compare them to the design values. The results show that seven out of nine blocks consume more than what was calculated during the design phase. The probable causes have been identified both on technical management and user behavior [62].

4.5.7. Lessons Learned and Recommendations

The shape, orientation, and reciprocal arrangement of the buildings demonstrated how the preliminary planning allows reaching high energy and living standards. This approach contributes to the energy saving of the entire district, and it lasts for its entire lifetime. Aspects such as internal layout, choice of materials, rainwater management, use of district heating and cooling, and integration of active solar systems are considered. It is important for architects and city planners to have the right instruments to evaluate these factors since the early design phases. A continuous process of verification and optimization allows to solve technical and management problems and reach the planned energy performance. The case study shows the importance of infrastructural connections and recreative spaces: the district's attractiveness is guaranteed by a new train stop, bicycle lane, green spaces, and a multifunctional area with services.

4.6. Similarities and Differences among the Different Built Environments

Table 3 presents the similarities and differences among the built environments.

Table 3. Similarities and differences among the different built environments.

Built Env.	Similarities	Differences
Existing-Existing	Social housing projects aiming to improve the life quality of citizens and the energy performance of buildings. Active involvement of municipalities and stakeholders highlights the importance of a dialogue between public and private. Pilot studies proposing replicable solutions and developed within the context of European projects. PV systems not fully integrated into existing buildings. Provincial or Municipal regulatory frameworks to promote energy efficiency. Establishment of a monitoring phase and involvement of researchers.	SINFONIA focuses on the refurbishment of existing buildings. Photovoltaic Village includes the construction of new apartments blocks, public spaces, facilities. Different scales of intervention. SINFONIA has refurbishment projects all over the city, the realization of DH/DC grids, air quality, and smart mobility measures at city scale. In Photovoltaic Village the intervention is limited to the neighborhood scale. Different use of RES. In Bolzano PV, ST and geothermal heat pumps are used. The solar systems are installed only on buildings' roofs. In Photovoltaic Village only PV technology is adopted and the installation encompasses roofs, facades, and public areas.
New-New	Use of public tenders as an instrument to assign the design of the projects. Aim to represent the state of the art of technology in the sustainable building industry and inspire replicability. Morphologies of neighborhoods and buildings shaped to maximize use of sun. Attention in connecting the new development to the rest of the city through the realization of public paths. Use of green spaces as a unifying element for the interventions and as a way to connect it to the natural surroundings. Reduce traffic by limiting the number of roads or restricting access to vehicles. Constant dialogue between public and private stakeholders during the process.	Le Albere and Casanova use energy certification (CasaClima, LEED) to assess the sustainability level. Violino does not. Le Albere and Casanova include mixed functions within the neighborhood, while Violino has only residential buildings. Focus on social housing and affordable rental solution in Violino and Casanova. Violino and Le Albere show a higher PV integration level than Casanova. Adoption of a monitoring campaign carried out by researchers in Casanova and Violino. Le Albere was realized on a brownfield area, while the other two cases are developed on greenfield sites. Stricter urban grid limits planning freedom in Violino compared to the other two cases.

Table 3. *Cont.*

Built Env.	Similarities	Differences
New-Existing	Aim to act as inspiring and replicable projects at national/international level. Social housing constitutes a driving force for the experimentation and adoption of solar strategies at a larger scale. Municipalities play a coordinating and steering role in the projects' development. Monitoring campaign and research involvement to track energy performance. Attention to public spaces to connect the interventions with the surroundings. Use of finishing colors to address different aspects of the project (foster variety In Violino, improve daylight in SINFONIA).	Different level of active solar systems integration and quality, with new intervention generally providing better outcomes thanks to the higher morphological freedom. New areas typically result in land consumption, while acting on existing buildings' stock preserves the territory. More attention to car traffic mitigation, pedestrian-centered area, and city connectivity in new realizations compared to existing projects. Easier implementation of a mix of renewable energy sources in new projects compared to existing urban areas.

4.7. *Agrovoltaico*

4.7.1. Overview

Recent years have seen the development of solutions called "agrivoltaic" combining the dual use of food production at the ground level with electricity generation through PV modules located at an upper layer (Figure 16) [10,63]. The case study presented, the Agrivoltaico ® [64], is a patented system developed by the private company REM (Revolution Energy Maker) and tested for the first time in Pianura Padana (Emilia Romagna), a vast alluvial plain corresponding to the river Po's basin in Northern Italy. The area has a strong agricultural vocation thanks to a capillary irrigational system and a flat homogeneous morphology. These characteristics make the territory suitable for the installation of solar systems, which can however compete with the production of food.

Figure 16. View of the dual use of land with the PV track system on the upper layer and agricultural works at ground level (photo: © REM).

The solution proposed by REM exemplifies how a double use of land can be achieved. The system was built in a period of strong opposition toward the realization of utility-scale photovoltaic systems on agricultural land (Figure 17).

(a) (b)

Figure 17. (**a**) View of the PV track system during the first crop (photo: © REM); (**b**) aerial view of the solar plant in landscape (photo: © REM).

4.7.2. Challenges, Issues, and Decision Strategies

The main challenges encountered during the development of the project were related to the need to overcome the national legislation that did not allow for the installation of large PV systems (above 1 MWp) in agricultural areas [65]. This has been possible thanks to an innovative solution, where the PV modules are suspended 5 m above the ground using a metal structure. Only 2% of the land is occupied by the punctual supports holding the metal structure (in compliance with the regional regulation requiring a soil occupation for PV installation < 10%), leaving great freedom of movement to the harvest machines operating below. To balance the low density of the system compared to a standard one mounted on the ground, double-axis sun-tracking devices were used for the panels. The different configurations of the system during the day and its low density allow a minimal impact on the total radiation falling to the ground. Regarding the decision strategy used for the project, REM collaborated with several industrial partners, a university (the Institute of Agronomy, Genetics and Field crops of the Università Cattolica del Sacro Cuore of Piacenza), and local authorities to assure the design's requirements.

4.7.3. The Planning Process

The project was realized in four phases between 2010 and 2012 and addresses landscape and architectural planning scales. The scope was to overcome legislative and public acceptance barriers for the installation of PV systems in agricultural areas through an innovative solution allowing food and energy production with the regional regulation on land occupation. The private investor REM initially invested 2.5 million euros in the pilot project before replicating it in two other Italian locations for a total of 30 million euros in value including the satellite activities and nearly 700 people involved.

4.7.4. Solar Landscape

The notion of a "solar landscape" was proposed in [10], and it consists of a different approach to the design of PV systems on the ground. The focus is not anymore on the integration of the system in the landscape but rather the design of the system as a landscape. This different perspective requires shifting the focus of the PVs' planning from the sole energy performance to a new set of paradigms, which are more in line with the ecological burden that the creation of a new landscape implies. The pattern of the system (dense or porous) is determined by the spatial arrangement of the modules, the type and grain of the patch, the pattern type, and its edges (Table 4). Other important aspects are the function of the space and its connectivity. The first aspect is related to the porosity of the system that determines the amount of solar radiation reaching the ground and consequently its ecological potential. No changes in the land's function or type of cultivation were needed for Agrivoltaico due to the high porosity of the system. The second aspect depends primarily on the distance of the PV modules from the ground and if any fence or enclosures are present at the boundary of the area. The case study has a great level of connectivity with the surrounding landscape, allowing animals, people, and harvesting machines to easily cross the area. The installed nominal power is 3.2 MW, and it is composed of 11,535 polycrystalline PV modules tiltable on two axes using a wireless control.

Table 4. Formal functional features of the solar landscape for Agrivoltaico.

Patch type	Small Straight borders	 ✓	Large Convoluted borders	✓
Grain type	Small patches	✓	Large patches	
Pattern	Porous	✓	Dense	
Pattern type	Parallel stripes	✓	Non-parallel stripes	
Edge/borders	Continuous	✓	Discontinuous	

4.7.5. Site Potential

The site has a cultural value and requires the preservation of its features, being the product of a long human transformation of a humid area into one suitable for agriculture. Nowadays, the area is intensively exploited for the cultivation of crops thanks to the efficient network of irrigation and flat morphology. Despite the PV system suspended above the ground does not perform any other function than producing energy, the site represents an example of double use of land (i.e., agriculture/food production and energy production from PV system).

Moreover, the PV double-axis tracking system controlled by a specific algorithm allows for optimal control of the dynamic shading on the crops, aiming at optimizing agricultural production. The assessment of the sensitivity (low or high) of several selected landscape factors is visible in Table 5.

Table 5. Sensitivity of landscape factors for Agrivoltaico.

Sensitivity	**Low**	**High**
Landform	✓	
Landscape pattern and complexity	✓	
Land use		✓
Land cover		✓
Settlement and manmade influence		✓
Historic character		✓
Historic features		✓
Inter-visibility with adjacent landscape		✓
Sense of remoteness		✓
Sense of openness		✓

4.7.6. Environmental, Economic, and Social Impacts

The environmental impact of the system can be considered low, since the original use of the land is preserved and the system does not constitute a barrier in the territory. Furthermore, it is entirely made of safe, non-polluting, and recyclable materials (as the recycled aluminum that the trackers are made of) and its construction technique makes it easy to disassemble at the end of the lifecycle (25–30 years) [66]. Despite the visibility of the system changes during the day because of the tracking, no mitigation strategy was required due to the porous pattern of the modules. Finally, the collaboration established with local authorities since the early phases of the planning process helped to achieve an optimal design and raised public awareness and participation.

4.7.7. Approaches, Methods, and Tools

The company REM Tec has patented the Agrivoltaico technology worldwide. Two solutions of sun-tracking systems are currently available, having in common the length (12 m) and the height (4 to 5 m) of each tracker but differing in the system's power and the number of modules. In the first one, ten panels are installed on each tracker for a peak power spanning between 2.5 and 4.35 KW, while the second solution presents a denser configuration with 32 panels and peak power from 8.64 to 10.46 KW. Furthermore, the cooperation with the Università Cattolica del Sacro Cuore of Piacenza permitted us to understand the impact of the system on different crop species and develop optimal design solutions combining energy and food production.

4.7.8. Lesson Learned and Recommendations

The Agrivoltaico case study demonstrates that the ecological impact of large PV systems installed on the ground can be greatly reduced with the adoption of innovative solutions allowing a double use of land. Barriers posed by legislation and regulations can often be overcome through a conscious design approach and collaboration with local authorities, industrial, and scientific partners. In the final analysis, the installation of large PV systems has to be considered as a matter of landscape design, where the paradigm is shifted from the sole energy production toward a more holistic approach, taking into account the various ecological aspects of a landscape transformation.

5. Limitations of the Study

This study does not pretend to be seen as an exhaustive illustration of the status of solar energy in the Italian urban context but it rather wishes to give a satisfactory overview of it through exemplary cases selected and analyzed by experts in the field. The selection of the six case studies reflects the knowledge and available information provided by the experts during the IEA SHC Task 51 "Solar Energy in Urban Planning". The geographical distribution of the cases, all located in the northern part of Italy, can be considered as the major limitation of the study. The inclusion of projects from southern regions could have further enriched the study and investigated different climatic conditions. Furthermore, three out of six cases are located in a single region (i.e., Trentino Alto Adige), which has special administrative conditions compared to the rest of the country. More landscape PV case studies should be analyzed to provide a more complete overview of this environment as well as a critical analysis of the similarities and differences. It is also worth mentioning that IEA SHC Task 51 was terminated in 2017; therefore, potentially relevant cases realized after that year were not included in this paper. The ongoing IEA SHC Task 63 "Solar Neighborhood Planning" [67,68] will partially address this issue.

6. Conclusions

In this paper, six case studies located in the Italian territory were presented and discussed following a common template developed within Subtask C of the IEA-SHC Task 51 "*Solar Energy in Urban Planning*". The integration of solar energy in three different environments—namely existing urban areas, new urban areas, and landscape—has been investigated. General considerations can be outlined, allowing for a partial generalization of the results for the entire country or at least for the administrative territory in which the cases are located. The main lesson learned is the importance of including solar energy since the beginning of the design process to achieve a higher level of integration.

When this basic rule is not observed and a lack of communication between experts and stakeholders occurs, the risk of encountering pitfalls during the following phases exponentially increases, and only sub-optimal results can be obtained. An additional advantage of the early analysis of solar energy potential is the possibility to optimize the urban morphology and the buildings' form to maximize active and passive solar strategies. Several design aspects can be controlled in the early design phases, although they tend to crystallize rapidly as the project progress. Furthermore, an optimized design of buildings' masses is a passive measure that can have positive and long-lasting impacts. This aspect is particularly visible when dealing with existing buildings, often offering limited design solutions if not subject to deep renovation interventions. Finally, an important lesson is the need to consider solar systems beyond their electricity generation function. This is true when the integration occurs into buildings as well as with utility-scale landscape systems. A set of recommendations specific for the three different analyzed environments can be summarized as follows.

I. *Existing urban areas (refurbishment):*
    - Deep renovation processes involving morphological and material changes represent potential successful solutions for solar systems integration as demonstrated by the case studies SINFONIA Bolzano and Photovoltaic Village in Alessandria.
    - The use of solar systems in public areas (e.g., shading devices on pergola as in the Photovoltaic Village in Alessandria) has a high unexploited potential.
    - The institution of coordinating bodies and the adoption of an integrated design process can have a significant role in the application of energy-related measures while improving interdisciplinarity and collaboration among stakeholders as happened in both SINFONIA Bolzano and Photovoltaic Village in Alessandria.
    - It is important to use simulation software since the early-design phases and to focus on final-user behavior as in SINFONIA Bolzano.

II. *New urban areas:*
    - Photovoltaics can be utilized as a distinctive architectural element and material, which can enforce the identity and aesthetic of urban interventions as in the case studies of the Le Albere district, Violino District in Brescia, and CasaNova.
    - The shape, orientation, reciprocal arrangement of volumes, materials, internal layout, and opening distribution are long-term passive strategies that can maximize the contribution of solar energy to building efficiency and comfort, as visible in the case studies of the Violino District in Brescia and CasaNova in Bolzano.
    - The last two points listed in the existing urban areas are also applied for new urban areas as demonstrated in all the presented case studies such as Le Albere district, Violino District in Brescia, and CasaNova.

III. *Landscape*
    - The ecological impact of ground-mounted PV can be greatly reduced using innovative solutions that combine a dual use of land.
    - Barriers represented by regulations and legislation can be often overcome with a conscious design and an early collaboration between the different involved actors.
    - PV installation at the ground level should be considered as a landscape design matter, where the pattern, patch, grain, and borders of the system are carefully planned.

Author Contributions: Conceptualization G.L., M.F. and S.C.; methodology G.L., M.F. and S.C.; investigation M.F.; resources G.L., S.C., D.V., R.P. and A.S.; writing—original draft preparation G.L., M.F. and S.C.; writing—review and editing M.F., G.L., S.C., D.V., R.P., A.S. and A.G.M.; supervision G.L. All authors have read and agreed to the published version of the manuscript.

Funding: This research was funded by The Research Council of Norway (NFR) and the Norwegian University of Science and Technology (NTNU) through the Zero Emission Neighbourhoods in Smart Cities Research Centre (FME ZEN) (project No. 257660) and the research project HELIOS—enHancing optimal ExpLoitatIOn of Solar energy in Nordic cities through digitalization of built environment (FRIPRO FRINATEK) (project No. 324243), the Department of Civil, Environmental, Architectural Engineering—University of Padua, the Department of Architecture Built Environment and Construction Engineering—Politecnico di Milano, the Italian Ministry of Economic Development in the framework of the Operating Agreement with ENEA for Research on the Electric System, and the European Union's Seventh Programme for Research, Technological Development and Demonstration (grant agreement No. 609019). The European Union is not liable for any use that may be made of the information contained in this document.

Acknowledgments: The authors wish to thank all the experts within the research project IEA SHC Task 51 "Solar Energy in Urban Planning", who contributed to the successful collection and analysis of the Italian case studies within the subtask C "Case studies and actions research" and for the copyright of graphics, images, tables, and figures. The authors also wish to express their gratitude to the IEA SHC Executive Committee for supporting Task 51 and in this way strengthening the international collaboration among researchers and practitioners. This work has been written within the FME ZEN Research Centre and the research project HELIOS. The authors M.F. and G.L. gratefully acknowledge the support from the FME ZEN Research Centre, its partners, and the NFR. The author R.P. wishes to thank the Department of Civil, Environmental, Architectural Engineering—University of Padua. The authors S.C. and D.V. acknowledge the European Union's Seventh Programme for Research, Technological Development and Demonstration. The author A.S. thanks for the support received by the Italian Ministry of Economic Development in the framework of the Operating Agreement with ENEA for Research on the Electric System.

Conflicts of Interest: The authors declare no conflict of interest.

References

1. *UN Population Division The World's Cities in 2018*; United Nations: New York, NY, USA, 2018; ISBN 978-92-1-148306-2.
2. Edenhofer, O.; Pichs-Madruga, R.; Sokona, Y.; Kadner, S.; Minx, J.C.; Brunner, S.; Agrawala, S.; Baiocchi, G.; Bashmakov, I.A.; Blanco, G.; et al. 2014: Technical Summary. In *Climate Change 2014: Mitigation of Climate Change*; Contribution of Working Group III to the Fifth Assessment Report of the Intergovernmental Panel on Climate Change; Cambridge University Press: Cambridge, UK; New York, NY, USA, 2014.
3. *IRENA Rise of Renewables in Cities: Energy Solutions for the Urban Future*; International Renewable Energy Agency: Abu Dhabi, United Arab Emirates, 2020.
4. Formolli, M.; Lobaccaro, G.; Kanters, J. Solar Energy in the Nordic Built Environment: Challenges, Opportunities and Barriers. *Energies* **2021**, *14*, 8410. [CrossRef]
5. Worldwatch Institute Reducing the Environmental Footprint of Buildings. In *State of the World*; Island Press: Washington, DC, USA, 2016; pp. 115–133. ISBN 978-1-61091-569-4.
6. *IRENA Renewable Energy Target Setting*; International Renewable Energy Agency: Abu Dhabi, United Arab Emirates, 2015.
7. *IEA Renewables Information: Overview*; IEA: Paris, France, 2021.
8. *IEA World Energy Outlook 2020*; IEA: Paris, France, 2020; p. 214.
9. Pasqualetti, M.J. Reading the Changing Energy Landscape. In *Sustainable Energy Landscapes*; CRC Press: Baca Raton, FL, USA, 2012; ISBN 978-0-429-18484-0.
10. Scognamiglio, A. 'Photovoltaic Landscapes': Design and Assessment. A Critical Review for a New Transdisciplinary Design Vision. *Renew. Sustain. Energy Rev.* **2016**, *55*, 629–661. [CrossRef]
11. *SolarPower Europe Solar Sustainability Best Practices Benchmark*; SolarPower Europe: Brussels, Belgium, 2021.
12. *REN21 Secretariat Renewables 2021 Global Status Report*; REN21: Paris, France, 2021.
13. Croce, S.; Vettorato, D. Urban Surface Uses for Climate Resilient and Sustainable Cities: A Catalogue of Solutions. *Sustain. Cities Soc.* **2021**, *75*, 103313. [CrossRef]
14. Paatero, J.V.; Lund, P.D. Effects of Large-Scale Photovoltaic Power Integration on Electricity Distribution Networks. *Renew. Energy* **2007**, *32*, 216–234. [CrossRef]
15. Lobaccaro, G.; Carlucci, S.; Croce, S.; Paparella, R.; Finocchiaro, L. Boosting Solar Accessibility and Potential of Urban Districts in the Nordic Climate: A Case Study in Trondheim. *Sol. Energy* **2017**, *149*, 347–369. [CrossRef]
16. Lobaccaro, G.; Lisowska, M.M.; Saretta, E.; Bonomo, P.; Frontini, F. A Methodological Analysis Approach to Assess Solar Energy Potential at the Neighborhood Scale. *Energies* **2019**, *12*, 3554. [CrossRef]
17. Good, C.S.; Lobaccaro, G.; Hårklau, S. Optimization of Solar Energy Potential for Buildings in Urban Areas—A Norwegian Case Study. *Energy Procedia* **2014**, *58*, 166–171. [CrossRef]
18. Lobaccaro, G.; Croce, S.; Vettorato, D.; Carlucci, S. A Holistic Approach to Assess the Exploitation of Renewable Energy Sources for Design Interventions in the Early Design Phases. *Energy Build.* **2018**, *175*, 235–256. [CrossRef]
19. Mazziotti, L.; Cancelliere, P.; Paduano, G.; Setti, P.; Sassi, S. Fire Risk Related to the Use of PV Systems in Building Facades. *MATEC Web Conf.* **2016**, *46*, 05001. [CrossRef]
20. Probst, M.M.; Roecker, C. Criteria for Architectural Integration of Active Solar Systems IEA Task 41, Subtask A. *Energy Procedia* **2012**, *30*, 1195–1204. [CrossRef]
21. Florio, P.; Peronato, G.; Perera, A.T.D.; Di Blasi, A.; Poon, K.H.; Kämpf, J.H. Designing and Assessing Solar Energy Neighborhoods from Visual Impact. *Sustain. Cities Soc.* **2021**, *71*, 102959. [CrossRef]
22. Florio, P.; Munari Probst, M.; Schüler, A.; Roecker, C.; Scartezzini, J.-L. Assessing Visibility in Multi-Scale Urban Planning: A Contribution to a Method Enhancing Social Acceptability of Solar Energy in Cities. *Solar Energy* **2018**, *173*, 97–109. [CrossRef]

23. Lingfors, D.; Johansson, T.; Widén, J.; Broström, T. Target-Based Visibility Assessment on Building Envelopes: Applications to PV and Cultural-Heritage Values. *Energy Build.* **2019**, *204*, 109483. [CrossRef]
24. Corti, P.; Bonomo, P.; Frontini, F.; Mace, P.; Bosch, E. *BIPV Status Report 2020. Building Integrated Photovoltaics: A Practical Handbook for Solar Buildings' Stakeholders*; SUPSI–Swiss BIPV Competence Centre and Becquerel Institute: Brussels, Belgium, 2020.
25. Sahu, A.; Yadav, N.; Sudhakar, K. Floating Photovoltaic Power Plant: A Review. *Renew. Sustain. Energy Rev.* **2016**, *66*, 815–824. [CrossRef]
26. World Bank Group; ESMAP; SERIS. *Where Sun Meets Water: Floating Solar Market Report*; World Bank: Washington, DC, USA, 2019.
27. Weselek, A.; Ehmann, A.; Zikeli, S.; Lewandowski, I.; Schindele, S.; Högy, P. Agrophotovoltaic Systems: Applications, Challenges, and Opportunities. A Review. *Agron. Sustain. Dev.* **2019**, *39*, 35. [CrossRef]
28. *SolarPower Europe Agrisolar Best Practices Guidelines*; Version 1.0; SolarPower Europe: Brussels, Belgium, 2021.
29. Wall, M.; Snow, M.; Lundgren, M.; Dahlberg, J.; Wyckmans, A.; Siems, T.; Voss, K.; Simon, K. Annex Plan Task 51: Solar Energy in Urban Planning. Available online: https://task51.iea-shc.org/data/sites/1/publications/IEA-SHC-Task51-Annex.pdf (accessed on 8 March 2022).
30. Jørgensen, O.B.; Croce, S.; Dahlberg, J.; Delmas, A.; Garde, F.; Giostra, S.; He, J.; Horvat, M.; Kanters, J.; Kappel, K.; et al. *National and International Comparison of Case Studies on Solar Energy in Urban Planning*; Lund University: Lund, Sweden, 2018. [CrossRef]
31. Lobaccaro, G.; Croce, S.; Lindkvist, C.; Munari Probst, M.C.; Scognamiglio, A.; Dahlberg, J.; Lundgren, M.; Wall, M. A Cross-Country Perspective on Solar Energy in Urban Planning: Lessons Learned from International Case Studies. *Renew. Sustain. Energy Rev.* **2019**, *108*, 209–237. [CrossRef]
32. Italian Republic. *Legislative Decree No. 28 of 3/3/2011 "Attuazione Della Direttiva 2009/28/CE Sulla Promozione Dell'uso Dell'energia Da Fonti Rinnovabili, Recante Modifica e Successiva Abrogazione Delle Direttive 2001/77/CE e 2003/30/CE"*; OJ of the Italian Republic: Rome, Italy, 2011.
33. *European Parliament Directive 2018/2001 of the European Parliament and of the Council of 11 December 2018 on the Promotion of the Use of Energy from Renewable Sources*; Official Journal of the European Union L 328/82; EU: Brussels, Belgium, 2018.
34. *European Parliament Regulation (EU) 2018/1999 of the European Parliament and of the Council of 11 December 2018 on the Governance of the Energy Union and Climate Action*; Official Journal of the European Union L 328/1; EU: Brussels, Belgium, 2018.
35. *European Commission The Europen Green Deal*; European Council: Brussels, Belgium, 2020.
36. Integrated National Energy and Climate Plan 2030. Available online: https://energy.ec.europa.eu/system/files/2020-02/it_final_necp_main_en_0.pdf (accessed on 8 March 2022).
37. Italian Republic. *Legislative Decree No. 199 of 8/11/2021 "Attuazione Della Direttiva (UE) 2018/2001 Del Parlamento Europeo e Del Consiglio Dell'11 Dicembre 2018 Sulla Promozione Dell'uso Dell'energia Da Fonti Rinnovabili"*; OJ of the Italian Republic: Rome, Italy, 2021.
38. Re Rebaudengo, A. PNRR Il Piano Necessario per Accedere Ai Fondi Del Next Generation EU: Una Grande Opportunità per l'Italia. Available online: http://www.elettricitafutura.it/public/editor/Position_Paper/2021/PARLAMENTO/2021.02.02_PNRR_POST_Rev0402.pdf (accessed on 8 March 2022).
39. Italian Republic Piano Nazionale Di Ripresa e Resilienza (PNRR). Available online: https://www.governo.it/sites/governo.it/files/PNRR.pdf (accessed on 8 March 2022).
40. Italian Republic. *Legislative Decree No. 108 of 29/7/2021 "Conversione in Legge, Con Modificazioni, Del Decreto-Legge 31 Maggio 2021, n. 77, Recante Governance Del Piano Nazionale Di Ripresa e Resilienza e Prime Misure Di Rafforzamento Delle Strutture Amministrative e Di Accelerazione e Snellimento Delle Procedure"*; OJ of the Italian Republic: Rome, Italy, 2021.
41. ENEA Agrivoltaico Sostenibile. Available online: https://www.agrivoltaicosostenibile.com/ (accessed on 8 March 2022).
42. Paparella, R.; Saretta, E.; Scognamiglio, A. Renewable Energy Systems Application into Italian Historic City Centres: Review of Building Codes. In Proceedings of the GDC 2013, Sarajevo, Bosnia and Herzegovina, 3 October 2013; pp. 46–51.
43. Mazzarella, L. Energy Retrofit of Historic and Existing Buildings. The Legislative and Regulatory Point of View. *Energy Build.* **2015**, *95*, 23–31. [CrossRef]
44. Ministero Dello Sviluppo Economico. *Legislative Decree of 26/6/2015 "Applicazione Delle Metodologie Di Calcolo Delle Prestazioni Energetiche e Definizione Delle Prescrizioni e Dei Requisiti Minimi Degli Edifici"*; OJ of the Italian Republic: Rome, Italy, 2015.
45. Munari Probst, M.; Roecker, C. Solar Energy Promotion & Urban Context Protection: Leso-Qsv (Quality-Site-Visibility) Method. In Proceedings of the PLEA 2015, Bologna, Italy, 9 September 2015.
46. Munari Probst, M.C.; Roecker, C. Criteria and Policies to Master the Visual Impact of Solar Systems in Urban Environments: The LESO-QSV Method. *Solar Energy* **2019**, *184*, 672–687. [CrossRef]
47. Devetaković, M.; Djordjević, D.; Radojević, M.; Krstić-Furundžić, A.; Burduhos, B.-G.; Martinopoulos, G.; Neagoe, M.; Lobaccaro, G. Photovoltaics on Landmark Buildings with Distinctive Geometries. *Appl. Sci.* **2020**, *10*, 6696. [CrossRef]
48. Comune di Alessandria. *Progettare Vivere Costruire Alla Luce Del Sole. Il Villaggio Fotovoltaico Di Alessandria*; Internal Report; Comune di Alessandria: Alessandria, Italy, 2008.
49. *Regione Piemonte Regional Law No. 18 of 9/4/1996 "Integrated Intervention Programme of the Environmental, Building and Urban Requaification Enacting the Article 16 of Law 17 February 1992, n. 179*; Regione Piemonte: Piemonte, Italy, 1996.
50. *Italian Republic Legislative Decree No. 74 of 16/3/2001 "Programma Tetti Fotovoltaici"*; OJ of the Italian Republic: Rome, Italy, 2001; pp. 55–58.

51. SINFONIA. Low Carbon Cities for Better Living. Available online: http://www.sinfonia-smartcities.eu./ (accessed on 8 March 2022).
52. DellaValle, N.; Bisello, A.; Balest, J. In Search of Behavioural and Social Levers for Effective Social Housing Retrofit Programs. *Energy Build.* **2018**, *172*, 517–524. [CrossRef]
53. Jakubiec, J.; Reinhart, C. DIVA 2.0: Integrating Daylight and Thermal Simulations Using Rhinoceros 3D, DAYSIM and EnergyPlus. In Proceedings of the 12th Conference of Internation Building Performance Simulation Association, Sydney, Australia, 14 November 2011.
54. Sadeghipour Roudsari, M.; Pak, M. Ladybug: A Parametric Environmental Plugin for Grasshopper to Help Designers Create an Environmentally-Conscious Design. In Proceedings of the BS 2013, Chambéry, France, 26–28 August 2013; pp. 3128–3135.
55. McNeel, R. *Others Rhinoceros 3D*; Robert McNeel & Associates: Seattle, WA, USA, 2010.
56. Piano, R.; Frampton, K. *Renzo Piano—The Complete Logbook, 1966–2016*; Thames & Hudson: London, UK, 2016; ISBN 978-0-500-34310-4.
57. Scognamiglio, A.; Palumbo, M. Forms of Energy #5. Available online: https://www.domusweb.it/en/architecture/2010/11/18/forms-of-energy-5.html (accessed on 8 March 2022).
58. Smart Domo Grid: Brescia Smart City del Futuro. Available online: https://www.darioflaccovio.it/blog/informazione-tecnica/smart-domo-grid-brescia-smart-city-del-futuro (accessed on 8 March 2022).
59. Midas I Quaderni Di Edilio n.9—Housing Sociale e Sostenibilità. Available online: https://issuu.com/edilio/docs/09progettiefficienzaenergetica (accessed on 8 March 2022).
60. Gelsomino, L.; Marinoni, O.; Bohigas, O. *Territori Europei Dell'Abitare: 1999–2010*; Compositori: Bologna, Italy, 2009; ISBN 978-88-7794-648-5.
61. *IPES Concorso Ipes Casanova Bolzano: 4 Progetti per una Città che Cresce*; Istituto per L'edilizia Sociale della Provincia Autonoma di Bolzano: Bolzano, Italy, 2006.
62. Castagna, M.; Antonucci, D.; Lollini, R. Monitoring of CasaNova Low Energy District: Result and Discussion. *Energy Procedia* **2016**, *96*, 895–906. [CrossRef]
63. Toledo, C.; Scognamiglio, A. Agrivoltaic Systems Design and Assessment: A Critical Review, and a Descriptive Model towards a Sustainable Landscape Vision (Three-Dimensional Agrivoltaic Patterns). *Sustainability* **2021**, *13*, 6871. [CrossRef]
64. REM Tec—La Soluzione per Il Fotovoltaico Legata All'Agricoltura. Available online: https://remtec.energy/en/agrovoltaico (accessed on 8 March 2022).
65. *Regione Emilia Romagna Delibera No. 28 of 6/12/2010 "Prima Individuazione Delle Aree e Dei Siti per l'installazione Di Impianti Di Produzione Di Energia Elettrica Mediante l'utilizzo Della Fonte Energetica Rinnovabile Solare Fotovoltaica"*; Regione Emilia Romagna: Emilia Romagna, Italy, 2010.
66. Agostini, A.; Colauzzi, M.; Amaducci, S. Innovative Agrivoltaic Systems to Produce Sustainable Energy: An Economic and Environmental Assessment. *Appl. Energy* **2021**, *281*, 116102. [CrossRef]
67. Wall, M.; Hachem-Vermette, C.; Vettorato, D.; Croce, S.; Kanters, J.; Florio, P. Annex Plan Task 63: Solar Neighborhood Planning. Available online: http://task63.iea-shc.org/Data/Sites/1/publications/IEA-SHC-Task63--Solar-Neighborhood-Planning--Annex.pdf (accessed on 8 March 2022).
68. IEA SHC | | Task 63 | | Solar Neighborhood Planning. Available online: https://task63.iea-shc.org/ (accessed on 8 March 2022).

applied sciences

MDPI

Article

Numerical Validation of the Radiative Model for the Solar Cadaster Developed for Greater Geneva

Benjamin Govehovitch [1,*], Martin Thebault [2], Karine Bouty [2], Stéphanie Giroux-Julien [1], Éric Peyrol [3], Victor Guillot [4], Christophe Ménézo [2] and Gilles Desthieux [4]

1 Centre d'Énergétique et de Thermique de Lyon, University Claude Bernard Lyon 1, CNRS, CETHIL, UMR5008, 69622 Villeurbanne, France; stephanie.giroux@univ-lyon1.fr
2 LOCIE/FRESBE, University Savoie Mont Blanc, F-74944 Annecy-le-Vieux, France; martin.thebault@univ-smb.fr (M.T.); karine.bouty@univ-smb.fr (K.B.); christophe.menezo@univ-smb.fr (C.M.)
3 Département de Chimie-Biochimie, University Claude Bernard Lyon 1, BioDyMIA EA3733, 01000 Bourg en Bresse, France; eric.peyrol@univ-lyon1.fr
4 Haute école du Paysage d'Ingénierie et d'Architecture de Genève (Hepia), Institute for Landscaping Architecture Construction and Territory (inPACT), University of Applied Sciences Western Switzerland, CH 1202 Geneva, Switzerland; victor.guillot@hesge.ch (V.G.); gilles.desthieux@hesge.ch (G.D.)
* Correspondence: benjamin.govehovitch@univ-lyon1.fr

Abstract: The achievement of the targets for reducing greenhouse gas emissions set by the Paris Agreements and the Swiss federal law on the reduction of greenhouse gas emissions (CO_2 law) requires massive use of renewable energies, which cannot be achieved without their adoption by the general public. The solar cadaster developed as part of the INTERREG G2 Solar project is intended to assess the solar potential of buildings at the scale of Greater Geneva—for both industrial buildings and for individual residential buildings—at a resolution of 1 m. The new version of the solar cadaster is intended to assess the solar potential of roofs, as well as that of vertical facades. The study presented here aims to validate this new version through a comparison with results obtained with two other simulation tools that are widely used and validated by the scientific community. The good accordance with the results obtained with ENVI-met and DIVA-for-Rhino demonstrates the capability of the radiative model developed for the solar cadaster of Greater Geneva to accurately predict the radiation levels of building facades in configurations with randomly distributed buildings (horizontally or vertically).

Keywords: solar cadaster; solar potential modeling; numerical validation

Citation: Govehovitch, B.; Thebault, M.; Bouty, K.; Giroux-Julien, S.; Peyrol, É.; Guillot, V.; Ménézo, C.; Desthieux, G. Numerical Validation of the Radiative Model for the Solar Cadaster Developed for Greater Geneva. *Appl. Sci.* **2021**, *11*, 8086. https://doi.org/10.3390/app11178086

Academic Editors: Tiziana Poli, Andrea Giovanni Mainini, Gabriele Lobaccaro, Mitja Košir, Juan Diego Blanco Cadena and Constantinos A. Balaras

Received: 24 June 2021
Accepted: 25 August 2021
Published: 31 August 2021

1. Introduction

The massive use of renewable energies and, more particularly, solar energy is necessary in order to meet the objectives of the Paris Agreements and the Swiss federal law on the reduction of greenhouse gas emissions (CO_2 law) [1]. As buildings are some of the main contributors to climate change, one of the main challenges is, therefore, turning buildings from energy consumers into energy producers. This cannot be done without the development of tools dedicated to the evaluation of solar potential.

This massive deployment involves the evaluation of solar potential on a large scale. In addition, the urban environment has many specific features that are to be taken into consideration when evaluating this potential. This can be achieved with solar cadasters.

However, these tools are usually limited to the evaluation of the solar potential on rooftops [2–5]. This limitation can be an obstacle to the development of the use of solar energy because the potential of vertical surfaces in an urban environment can prove to be predominant [6–9].

Thus, there is a need for large-scale tools that are able to take vertical facades into account, in addition to roofs. From a physical point of view, the effects of surrounding surfaces (the ground or surrounding buildings) on the vertical radiative balance are not well known and may appear to be non-negligible. From the point of view of solar production potential, the issues of the variability linked to shading effects are added. Tools that are able to consider these issues are being developed and need validation.

A solar cadaster tool is being developed at the scale of Greater Geneva, which is a trans-border agglomeration around the city of Geneva between France and Switzerland, and it totals an area of 2000 km^2. A first solar cadaster was developed for the Canton of Geneva only (280 km^2), as presented by Desthieux et al. [10]. The goal now is to update and extend this solar cadaster for the whole of the Greater Geneva area in the framework of INTERREG G2 Solar. Its aim is to intensify solar energy production in the agglomeration by mobilizing a large panel of stakeholders from public institutions, municipalities, energy providers, the private sector, and universities. The difficulty lies in the fact that the solar cadaster must cover a large area (2000 km^2 for Greater Geneva) while providing accurate information at a small spatial resolution and with a reasonable computation time.

The main objective of this study is to validate a solar cadaster tool by using a comparison with the results from other numerical tools. To this purpose, the results from the solar cadaster are compared with those of two widely used tools: DIVA-for-Rhino and ENVI-met. The latter, thanks to its holistic model, makes it possible to deepen the assessment of the buildings' photovoltaic potential by evaluating their surface temperatures.

The validation of the solar cadaster involves three different tools based on different models and methods and with a varied field of applications. Thus, these tools are presented first, prior to introducing the methodology used. The results in terms of predicted shortwave irradiance on both the roofs and the facades are then detailed and discussed.

2. Materials and Methods

2.1. Modeling Approaches for Radiative Transfers

Two main methods are commonly used for the evaluation of radiative transfers: ray-tracing-based models or radiosity methods. Their principles, as well as their advantages and drawbacks, are described in the following.

2.1.1. Ray-Tracing Methods

Ray-tracing methods consist of following the paths taken by electromagnetic rays. There are two types of ray-tracing approaches. Forward ray-tracing methods, also called light tracing, consist of launching rays from light sources in a set of directions. Backward ray-tracing methods consist of following the light path backward. In this latter case, rays are launched in a set of directions from the element of interest (the eye of an observer or an irradiated surface). A comparison between forward and backward ray tracing is illustrated in Figure 1.

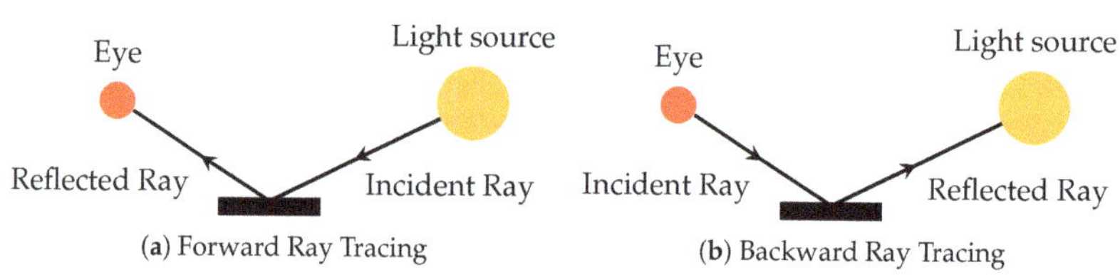

Figure 1. Illustration of backward and forward ray tracing.

For both of these methods, the directions can be random (probabilistic approach) or set (deterministic approach). After rays have been launched, an intersection test is carried out between each of these rays and the objects in the scene; then, new rays are cast from the points of impact while considering the properties of the materials encountered (reflection, transmission, absorption). This method makes it possible to take the effects of light into account, such as refraction of light (for example, in the case of a scene comprising glasses). The behavior of a ray at the point of intersection then depends on the refractive properties of the material encountered. This feature is particularly useful in the context of image synthesis or the study of transparent media with different refractive indexes.

This method also has the advantage of making the calculation of specular reflections possible (see Figure 2). On the other hand, it is not suitable for a calculation of diffuse reflections (which are usually the case for urban surfaces), as these require a significant calculation time because of the number of rays to be launched (in all directions) at each new intersection. This method has to be coupled with a diffuse radiation model.

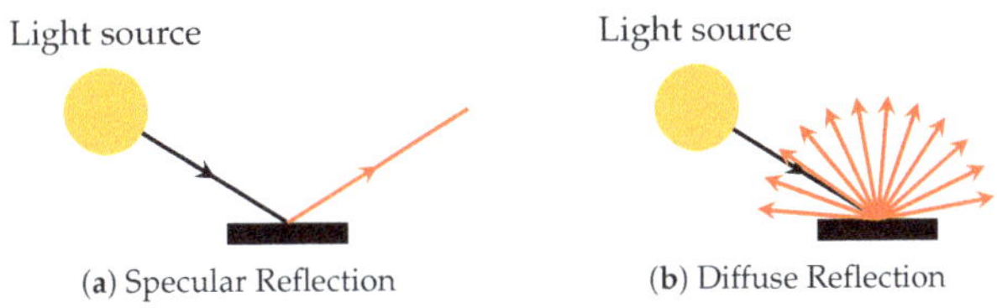

Figure 2. Illustration of specular and diffuse reflection.

2.1.2. Radiosity Method

The radiosity method is based on the calculation of the balance of radiosity, which is the total radiative flux exiting a surface (W m^{-2}). This method allows the calculation of diffuse and reflected components of the irradiation only by solving a system of equations involving all the objects of a scene. Taking specular reflections (see Figure 2) into account is not directly possible with this method. However, unlike ray tracing, the calculations are independent of the position of the possible observer (surface of interest)—the calculation is generalized over the whole scene.

The radiosity method is based on solving the equation introduced by Kajiya [11] and detailed in the work of Sillion and Puech [12], where the exiting power is expressed as:

$$B(x) = E(x) + \rho \int_{\Omega} L_i(x,\theta,\phi) \cos\theta d\omega \quad (1)$$

where x is a point in space, $B(x)$ is the exiting power (emitted, reflected, and transmitted) at point x in space (in W m^{-2}), $E(x)$ is the power emitted by point x (source term, in W m^{-2}), ρ is the coefficient of reflection of the surface, $L_i(x,\theta,\phi)$ is the incident power at point x from the solid angle determined by the angles θ and ϕ, Ω is the set of directions (θ,ϕ) of the hemisphere, and $d\omega$ is the elementary solid angle (sr).

This equation can be discretized:

$$B_i = E_i + \rho_i \sum_{j=1}^{N} F_{ij} B_j \quad (2)$$

where B_i and B_j are the power exiting the i-th cell and j-th cell, respectively, or the radiosity of the i-th or j-th cell (in W m^{-2}); this term takes into account both the term emitted and the effect of the reflections. E_i is the power initially emitted by the i-th cell, ρ_i is the reflection coefficient of the i-th cell, i.e., the fraction of the energy received that the cell returns (unitless value between 0 and 1), N is the number of meshes of which the scene is composed, and F_{ij} is the form factor between the i-th and the j-th cells.

Solving the radiosity equation requires one to first know the source term E_i (known if the source emits directly), the reflexivity ρ_i, which is generally known as one of the characteristics of the materials, and the matrix of the form factors F_{ij}, which must be calculated because it is dependent on the studied configuration.

2.2. Tools Considered

Three different tools were considered in this study: the solar cadaster (CadSol) for Greater Geneva (which is the one to be validated) and two widely validated tools—DIVA-for-Rhino (DIVA) and ENVI-met (EM). Their models and applications are described in the following.

2.2.1. Solar Cadaster

As summarized by Desthieux et al. [10] and Freitas et al. [13], the solar cadaster tool is a geographic information system (GIS) tool. Such tools enable one to process large amounts of data and spatial analyses. These tools also provide automatic or systematic environmental analyses of urban areas, such as solar radiation calculations. These are different from tools that are classified in the category of computer-aided design (CAD). The latter process more accurate spatial and weather data in terms of spatial and time scales, but they require much more computing time and, thus, address the local scale (limited sets of buildings).

The irradiance received from the direct component I_b on an element of a surface at location x at a given time t is given by:

$$I_b(x,t) = BNI(t) \times r_b(x,t) \times S_b(x,t) \quad (3)$$

in which BNI corresponds to the direct normal component of the irradiance (also called the beam), r_b is the transposition factor, and S_b corresponds to the shadow cast from a neighborhood building ($S_b(x,t) = 0$ if, at time t, the surface located at x is shaded by another building, or $S_b(x,t) = 1$ otherwise). The transposition factor, r_b, depends on the solar elevation h and the slope of the considered surface β ($\beta = 0$ corresponds to a horizontal surface and $\beta = 90^\circ$ corresponds to a vertical surface; see Figure 3). It is calculated as $r_b(x,t) = sin(h(t))/cos(\beta(x))$. The irradiance over the element of the surface is then considered as the average of the calculated irradiances at its edges. The calculation differences between the three tools, as described below, are illustrated in Figure 4.

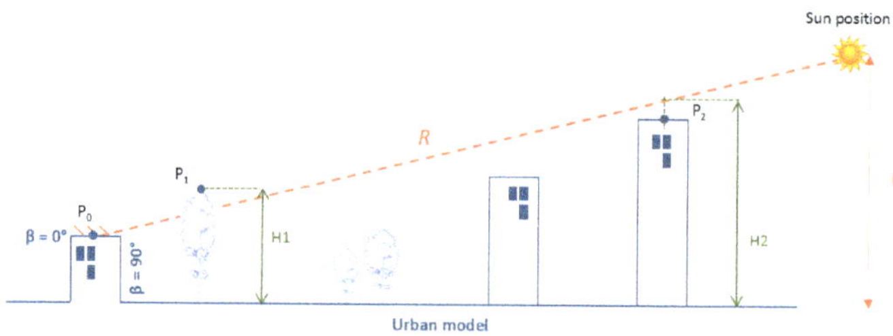

Figure 3. Shadow casting of the solar cadaster, adapted with permission from [10,14].

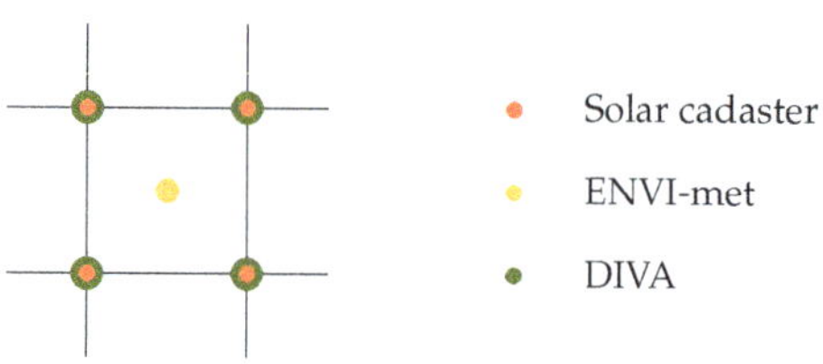

Figure 4. Differences in terms of irradiance calculations for a mesh.

In order to model the contribution of the diffuse component to the received irradiance, the Hay model is used. This model considers two components: a circumsolar (anisotropic) and an isotropic component. Similarly to the direct component, the circumsolar component is calculated at each time step by considering the sun's position and the shadowing. For the isotropic component, the sky-view factor is computed. In the solar cadaster, a sky model of 580 light sources is used.

The diffuse component, I_d, of the solar radiation is then calculated as follows:

$$I_d(x,t) = DHI(t) \times \left(\frac{GHI(t) - DHI(t)}{I_0} r_b(x,t) + SVF(x) \left(\frac{GHI(t) - DHI(t)}{I_0} \right) \right) \quad (4)$$

where GHI is the global irradiation on a horizontal surface, DHI is the diffuse irradiation on a horizontal surface, I_0 is the hourly extraterrestrial irradiation, r_b is the transposition factor as defined in (3), and SVF is the sky-view factor.

Finally, the reflected component is simply considered at the current stage as isotropic and is estimated based on Iqbal (1983) [15] as follows:

$$I_r(x,t) = 0.5 \times GHI(t) \times \rho(1 - \cos\beta(x)) \tag{5}$$

where GHI is as defined in (4), ρ is the coefficient of reflection of the surface, as defined in (1), and β is the slope of the surface.

More details about the calculation method and the computing tools used are available in [10,16].

2.2.2. DIVA-for-Rhino

DIVA-for-Rhino is a highly optimized daylighting and energy modeling plug-in for Rhinoceros. This software uses ray-tracing and light-backwards algorithms based on the physical behavior of light in a 3D volumetric model. For hourly solar radiation, the Daysim interface is used. Daysim has been validated by several studies to be accurate in modeling visible-wavelength natural light for multiple sky conditions [17].

The daylight coefficient approach and the all-weather sky luminance model according to [18] are used here. In this approach, the irradiance received on an element of surface x is calculated as the sum of all sky segments visible from this element of the surface. For the diffuse component, 145 sky segments are used [19] concomitantly with three ground segments [20]. For the direct component, Daysim uses 65 sun positions; therefore, at a specific time t, Daysim will use one of the 65 positions that is closest to the real sun position at time t.

To calculate the complete set of daylight coefficients, two ray-tracing runs are performed:

- One for each of the 148 diffuse and ground divisions;
- The second ray-tracing run uses 65 direct solar positions that are distributed along the annual solar path to calculate the contribution of the direct component. For that, 65 angular light sources with a solar cone opening of 0.53 are assigned for the building model. The positions of the light sources correspond to the representative sun positions of the building site.

The default Daysim simulation parameters were chosen. Up to two reflections from direct solar irradiation and one reflection from diffuse sky irradiation from the environment were considered.

Similarly to the solar cadaster, the irradiance calculated by Diva-for-Rhino for each surface element corresponds to the average of those calculated at its vertices.

2.2.3. ENVI-Met

ENVI-met is a software aiming at simulating the urban microclimate by taking into consideration all of the phenomena that occur in an urban environment. It is based on coupled balance equations (including those of mass, momentum, and energy). This involves taking the built and natural environment into account.

The ENVI-met model has been widely used in numerous studies dealing with different issues, including the evaluation of outdoor thermal comfort [21,22], mitigation of the urban heat island effect [22,23], or assessment of buildings' solar and photovoltaic potential [24].

Regarding radiative transfers, ENVI-met uses a hybrid approach based on a radiosity method (see Section 2.1.2) for the evaluation of the irradiance with a deterministic ray-tracing method for the calculation of the view factors. As ENVI-met is under a proprietary license, access to its code is limited. However, some information is available. Regarding the calculation of the diffuse solar radiation, the model is isotropic. This means that there is no difference between the different sky segments and no dependency on the actual position of the sun. Concerning the sky, it is divided into 414 segments (207 upward and 207 downward) [25].

Unlike the two other tools, the irradiances predicted by ENVI-met are given as that calculated at the center of the mesh.

2.3. Case Study

The present study focuses on two fictitious districts composed of three rows and three columns of buildings, i.e., a total of nine buildings. This ability of this kind of fictitious district has already been proven with respect to the study of the solar potential of neighborhoods [26–28]. In addition, it makes it possible to study the same district at different location, depending on the weather conditions used as input for the simulations.

These two districts, which are called the homogeneous and the heterogeneous district, have different arrangements, but they share some common urban morphology indicators, which are listed in Table 1.

Table 1. Urban morphology indicators of the homogeneous and heterogeneous districts.

Indicator	Definition	Value
Shape Factor	Ratio between the external building envelope surface and the building volume	0.23
Floor Area Factor	Ratio between the building gross floor area and the site area	2.5
Site Coverage	Ratio between the building footprint and the site area	0.25
Average Building Height	Average height (or rise of height) of buildings in an urban model	30 m
Resolution		1 m
Number of meshes per roof		400

Regarding the thermo-radiative properties of the different surfaces, the coefficient of reflection is set to 0.2 for both the ground's and the buildings' envelopes (vertical facades and roofs). These values are taken as a compromise between the typical values for concrete and soil.

2.3.1. Study of the Influence of the Mesh Resolution on the Results

The mesh sensitivity was tested on a simple configuration—presented in Figure 5—which consisted of two buildings that were 25 and 35 m high and separated by 10 m. Three grid sizes of 0.5, 1, and 2 m square meshes were chosen. The daily cumulative irradiation on the south facade of the highest building is given in Figure 6. It appears that the cumulative irradiation predicted by the solar cadaster increased along with the coarseness of the resolution.

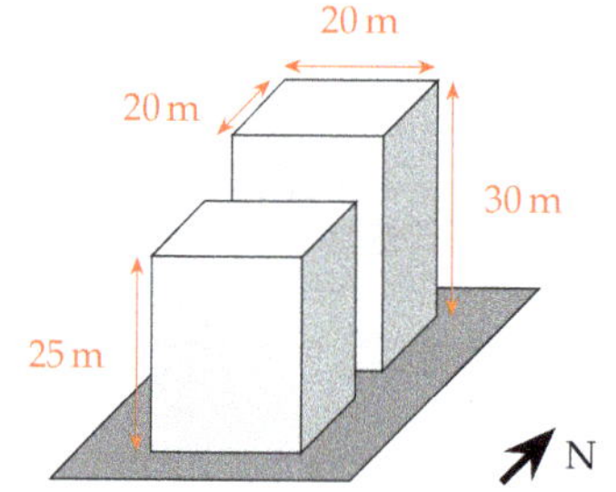

Figure 5. Geometry used for the study of the influence of the mesh.

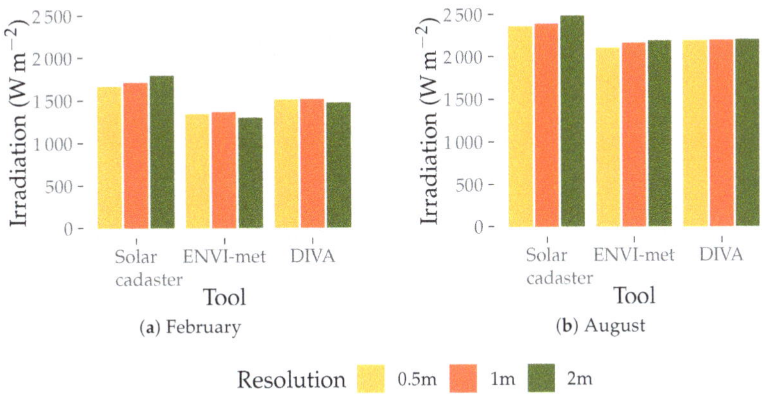

Figure 6. Cumulative irradiation for different mesh resolutions.

The influence of the mesh resolution on the results is detailed in Table 2 for the months of February and August. The mesh resolution had an influence on the irradiation predicted for every tool considered. Indeed, the coarser the resolution, the higher the discrepancy in terms of cumulative irradiation. Nonetheless, DIVA, which is based on a ray-tracing method, was less influenced by the mesh resolution, unlike ENVI-met and the solar cadaster, which are based on radiosity methods.

Table 2. Divergence of the cumulative irradiation according to the resolution of the mesh.

Resolutions Compared	February			August		
	Solar Cadaster	ENVI-Met	DIVA	Solar Cadaster	ENVI-Met	DIVA
0.5 m–1 m	2.81%	1.33%	0.22%	1.11%	1.20%	0.07%
1 m–2 m	4.56%	−3.93%	−2.80%	3.87%	2.63%	0.3%
0.5 m–2 m	7.23%	−5.33%	−2.57%	4.94%	4.80%	0.4%

On the other hand, the influence of the mesh resolution was more important in February than in August. This was due to the sun's lower course, which led to more shadowing.

In the particular case of the solar cadaster tool, a resolution of 2 m is, therefore, coarse and is not recommended for an analysis of irradiation in an urban environment. A resolution of 1 m remains acceptable for large areas, and the deviation from 0.5 m is small. This difference is explained by the fact that the obstacles' mutual shadows between the two plots will be better detected and considered in every facade point with a finer resolution.

A mesh composed of square cells with a resolution of 1 m was retained for the following study. This constituted a good compromise between the accuracy of the results and the computation time.

2.3.2. Homogeneous Neighborhood

The homogeneous neighborhood was composed of nine identical buildings, which were 20 m wide and 30 m high, and each was separated by 20 m, as shown in Figure 7. In this case, the widths of the streets were the same, and so was the height-to-width ratio (Table 3) for all of the buildings. This configuration also made it possible to have the same sky-view factor regardless of the orientation of the vertical facade, as well as a sky-view factor that was equal to 0.5 for the roofs. The homogeneous neighborhood provided a simple configuration and made it easier to study the different radiative phenomena that occurred on the buildings' facades.

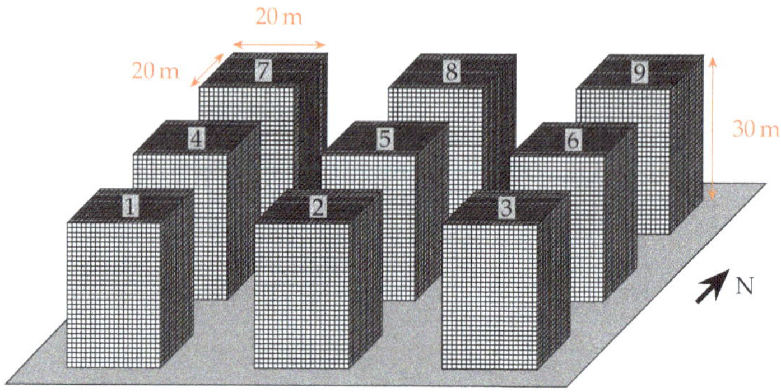

Figure 7. Sketch of the homogeneous neighborhood.

Table 3. Urban morphology indicators specific to the homogeneous district.

Indicator	Definition	Value
Building Height		30 m
Street Width		20 m
Height-to-Width Ratio	Ratio between the building height and the width of the distance between buildings	1.5
Number of meshes per facade		600

2.3.3. Heterogeneous Neighborhood

As a counterpart used to make the study of radiative phenomena easier, the homogeneous neighborhood (see Figure 7) could not be fully representative of an actual district, as actual districts are generally composed of buildings of different heights or random positions. To bridge this gap, a heterogeneous neighborhood was considered as well. Its layout is given in Figure 8.

This district is representative of a common type of actual district because of the non-homogeneous spatial distribution of the buildings, as well as their different heights. The buildings' footprints and their total volume were kept constant, as were the urban morphology indicators given in Table 1. The urban morphology indicators specific to the heterogeneous district are given in Table 4.

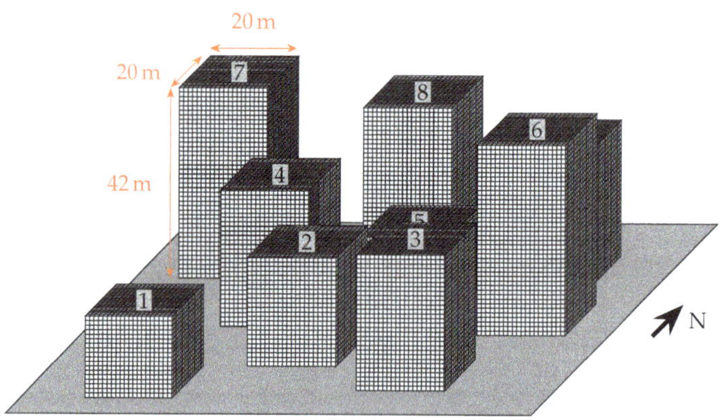

Figure 8. Sketch of the heterogeneous neighborhood.

Table 4. Urban morphology indicators specific to the heterogeneous district.

Indicator	Definition	Value
Building Height		18 m to 42 m
Street Width		5 m to 25 m
Height-to-Width Ratio	Ratio between the building height and the width of the distance between buildings	0.72 to 8.4
Number of meshes per facade		360 to 840

2.4. Weather Conditions for the Study

The present study focused on two different days: 15 February, which had a short daylight period, and 16 August, which had a longer one. These two days are representative of a winter and a summer day for a location such as Geneva (cold and not very sunny for the first, hot and more sunny for the second). Furthermore, the sun's path corresponding to these two days was halfway between the solstices and the equinoxes.

The two days were actually those for which the sun's path was closest to the mean paths for the months. Indeed, in order to reduce the number of simulated days (two days to be simulated for the months of February and August instead of 59) while avoiding the specific conditions of a given day, this study considered two representative average days for the two considered months [10]. A representative average day (RAD) is defined, according to Equation (6), as a day for which the meteorological conditions (including irradiation level, temperature, and wind velocity and direction) for each hour are equal to the average of these conditions over all days of the month (N_{day}). Carrying out the study for two different months allowed us to evaluate the influence of the irradiation level on the accuracy of the predicted results.

$$X_{RAD}(t) = < X_i(t) >_{N_{day}} \tag{6}$$

where $< \cdots >$ stands for the time-average operator.

The two considered months had a double advantage. First, they allowed the validation of the values predicted by the solar cadaster under low and high irradiation levels (February and August, respectively). Second, they made it possible to evaluate the influence of the evolution of the sun's path over the year on the irradiance profiles.

Since the solar cadaster was developed for Greater Geneva, the fictitious districts studied were located at this place. The data used as input for the simulations came from the METEONORM® version 7.3 database for this city. The hourly averaged values issued from the METEONORM database were then averaged by month according to Equation (6) in order to reduce the computation time. Since the values used as input for the simulations were averaged by hour, the level of irradiation was considered constant between two consecutive hours for both the input and output data.

In this study, only the shortwave radiation (SW) was considered, as it is the most important energy content in the solar spectrum. Thus, the waveband considered was 2500 nm. The daily profiles of the different components of the solar radiation over Geneva for the representative days of the months of February and August are given in Figure 9. The direct shortwave radiation represents the amount of energy that comes straight from the sun, while the diffuse part is the amount of solar radiation reflected by the atmosphere prior to hitting the ground. The total shortwave radiation is the sum of the direct and the diffuse parts.

Although the overall evolution of the shortwave radiation is similar between February and August, differences are noticeable. Indeed, the maximum in terms of GHI is two times higher in August. This is mainly due to the direct part of the solar radiation, which exceeds $350\,\mathrm{W\,m^{-2}}$ in August, in comparison with $150\,\mathrm{W\,m^{-2}}$ in February.

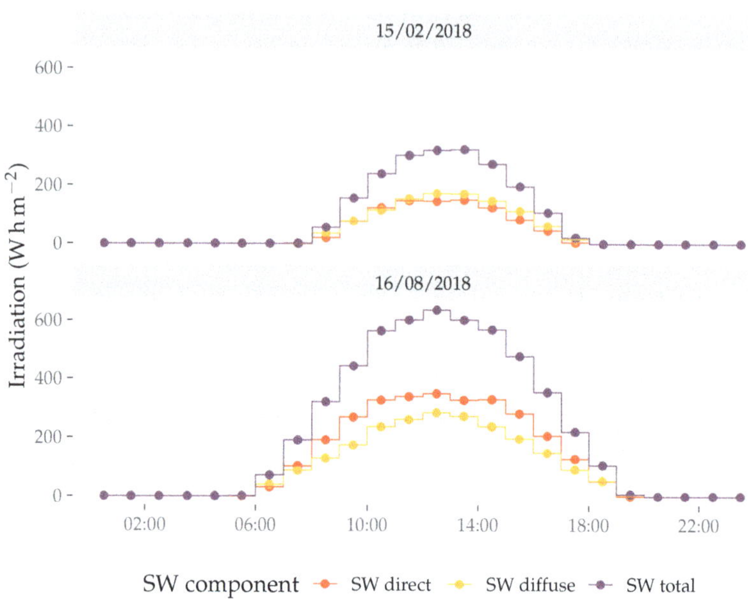

Figure 9. Shortwave radiation used as input.

3. Results

The validation of the solar cadaster radiation model is a multi-step process. In the first step, the focus is on the horizontal surfaces of the homogeneous district because they are subject to a smaller number of phenomena. In the second step, the mean irradiance values over the vertical facades of the central building (see building 5 in Figures 7 and 8) are studied. Finally, the spatial distribution of the irradiance is analyzed.

3.1. Mean Irradiance over the Unshaded Horizontal Surface

The mean irradiance level over the roofs is given in Figure 10. For the homogeneous district (see Figure 7), the irradiance level is the same for the roofs of the nine buildings, since they are not subject to shading effects. Regarding the heterogeneous district, the irradiance level is the that of buildings 6 and 7 (see Figure 8), which are the highest buildings.

This comparison demonstrates the ability of the solar cadaster to accurately reproduce the basics of the solar conditions (including the beam horizontal irradiance (BHI) and the diffuse horizontal irradiance (DHI)). This first preliminary analysis allows us to ensure that the results of the modeling of the global horizontal irradiance (GHI) are the same for all tools. In other words, the difference that will be observed in what follows will be the consequence of the presence of shading and the inclination of the facades.

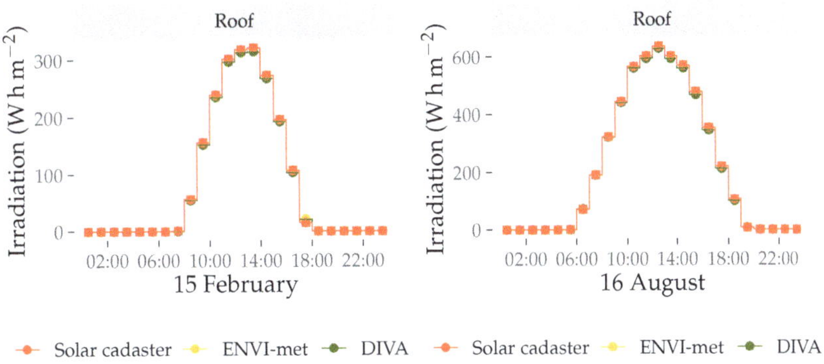

(a) February (b) August

Figure 10. Spatial average of the irradiance received on the unshaded roofs (homogeneous neighborhood or buildings 6 and 7 of the heterogeneous neighborhood).

3.2. Mean Irradiance over the Vertical Facades

The goal of this study is to demonstrate the ability of the solar cadaster to provide accurate results for vertical facades. The results for the homogeneous district are given in Figures 11 and 12, while the results for the heterogeneous district are given in Figures 13 and 14. It appears that the values predicted by the solar cadaster are in good accordance with the values predicted by DIVA-for-Rhino and ENVI-met.

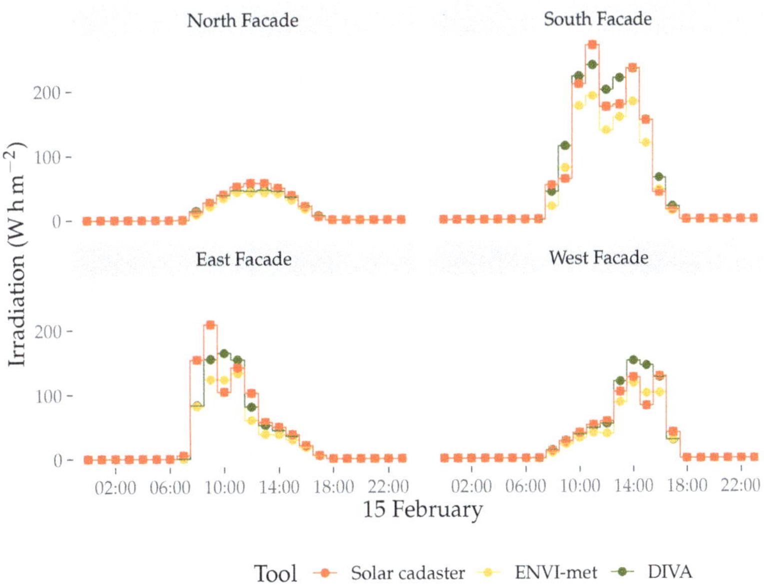

Figure 11. Spatial average irradiance received on the facades for the homogeneous district in February.

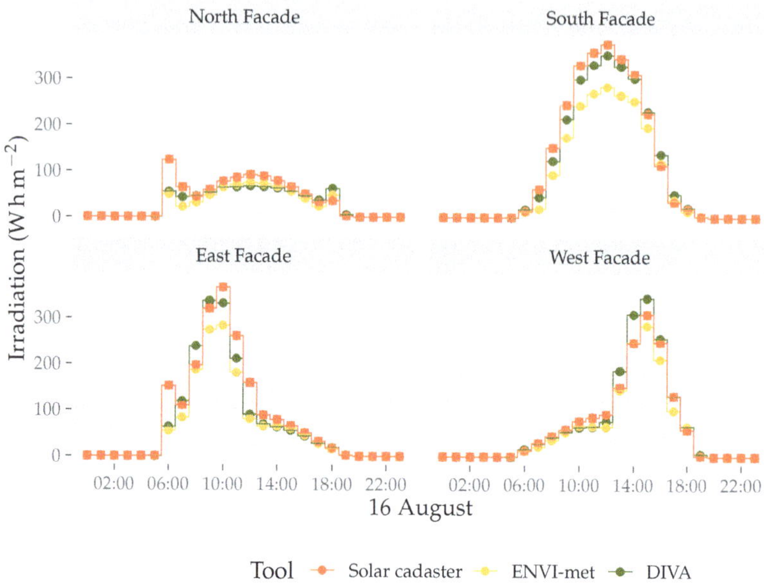

Figure 12. Spatial average irradiance received on the facades for the homogeneous district in August.

The different orientations of the facades make it possible to evaluate the prediction of the irradiance level for different solar conditions. Indeed, the north facade is mainly irradiated by the diffuse and reflected parts of the solar radiation, while the proportion of direct radiation is higher for the south facade. The east facade is directly irradiated in the morning, while the west facade faces the sun later in the day.

Regarding the level of irradiation, it does not have an influence on the concordance of the predicted values. Indeed, the results for the summer day are, in general, as accurate as those for the winter day. The differences between the results of the tools—quantified here with the normalized root mean squared error (NRMSE)—are not significantly impacted by the month of the year considered (see Table 5). The NRMSE values are calculated from differences between the daily profiles of the spatial average irradiance received on the facades and those predicted by the solar cadaster and ENVI-met on the one hand or by the solar cadaster and DIVA-for-Rhino on the other hand.

In both cases, the NRMSE remains lower for the roofs (less than 1 % for the roofs versus 4 % to 25 % for the vertical facades; see Table 5). This demonstrates the good accordance between the three tools with respect to the evaluation of the incident radiation on the roof, which are without a mask or reflection here. The observed differences for the vertical facades, which result in an increase in the NRMSE in the last four lines in Table 5, are then due to a difference in the consideration of masks and inter-building reflections.

Table 5. Normalized root mean squared error of the predicted level of irradiance (%) for the homogeneous district.

Facade	February		August	
	ENVI-Met	DIVA	ENVI-Met	DIVA
Roof	0.37	0.97	$1.18 \cdot 10^{-5}$	0.85
North Facade	9.45	6.59	15.72	14.72
South Facade	16.76	24.59	13.58	10.08
East Facade	12.06	10.81	10.06	7.70
West Facade	7.29	9.53	4.41	5.07

Nevertheless, the time of the year does not influence only the level of irradiance, but also the sun's path. Indeed, regarding the north facade (see Figures 11–14), the results from the solar cadaster are close to the those from DIVA-for-Rhino and ENVI-met (although a bit overestimated). Nonetheless, one peak can be observed in the results at 6:00 a.m. in the month of August (see Figure 12), which is not present for the same facade in February (see Figure 11). This difference in terms of the profile over time for the irradiance level is due to the evolution of the sun's path over the year. Indeed, during the month of August, the sun rises in the north-east and sets in the north-west, while it rises in the east and sets in the west in February.

The sun's path has an influence on the level of irradiation as well. Indeed, in the month of February (Figure 11), the south facade shows a slight decrease around noon, which is not present in the GHI (see Figure 9). This decrease is actually due to the shadow cast by building 2 on the central building (see Figures 7 and 8). This does not occur in August because the sun's path is high enough for the shading effect not to occur.

The results for the heterogeneous district make it possible to evaluate the ability of the solar cadaster to accurately predict the irradiation level for a more complex city. Indeed, the buildings are randomly located in this case (see Figure 8). The results for the two months considered are given in Figures 13 and 14.

It appears that the morphology of the neighborhood, although it is more random and complex than that of the homogeneous neighborhood, does not have a significant negative impact on the accuracy of the results predicted by the solar cadaster, which are in good accordance with those of DIVA-for-Rhino and ENVI-met (see Tables 5 and 6). Nonetheless, the level of solar irradiance over the different facades is lower in the case of the heterogeneous district than in the case of the homogeneous one. This can be explained by the fact that the central building is smaller than those around it and is thus subject to more of a shading effect.

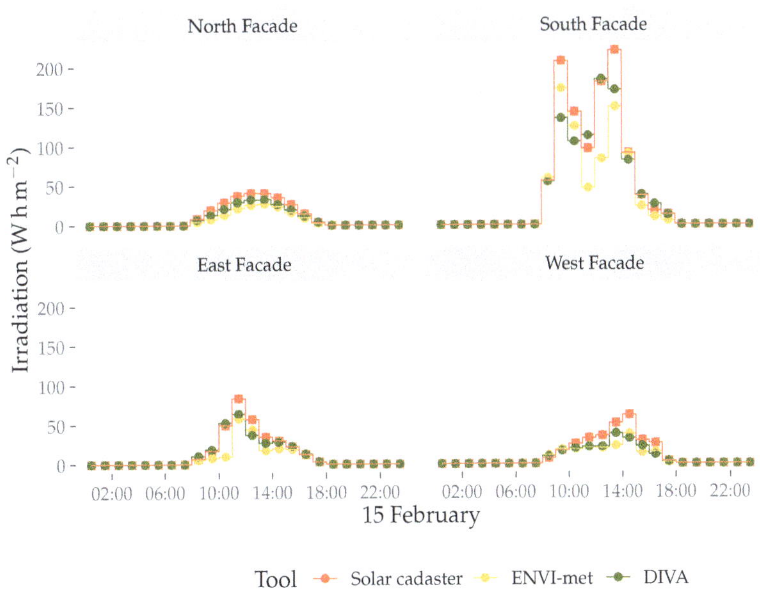

Figure 13. Spatial average irradiance received on the facades in the heterogeneous district in February.

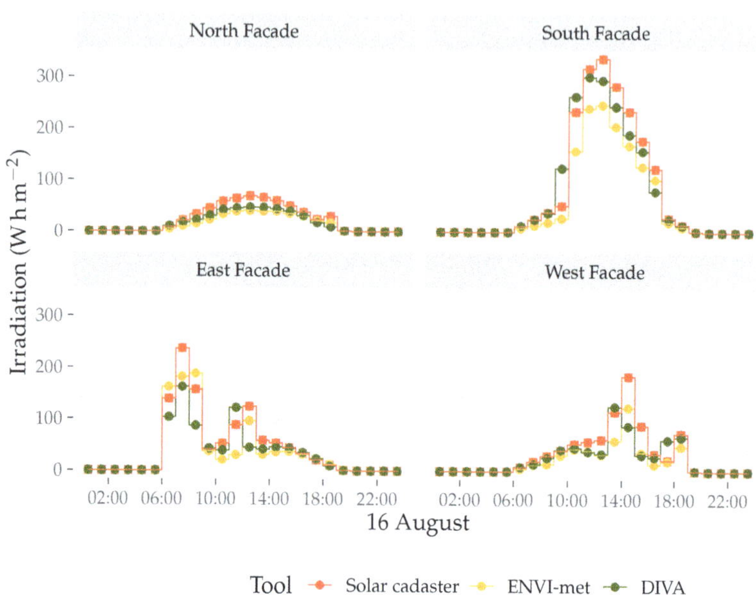

Figure 14. Spatial average irradiance received on the facades in the heterogeneous district in August.

Table 6. Normalized root mean squared error of the predicted level of irradiance (%) in the heterogeneous district.

Facade	February		August	
	ENVI-Met	DIVA	ENVI-Met	DIVA
North Facade	32.46	28.00	36.30	32.98
South Facade	12.11	10.09	12.45	7.76
East Facade	22.46	18.72	18.66	21.54
West Facade	26.69	25.30	21.22	21.01

3.3. *Irradiance Maps*

Although the results presented so far show a good accordance between the three tools, a consideration of the spatial mean value alone is not sufficient. Indeed, as shown in Figures 15 and 16, the range of the level of irradiance received over the facade can be very important.

The range of the predicted values may be due to the method of modeling radiative phenomena (ray-tracing or radiosity) or the shading effect. The latter may have an important impact on the level of irradiance. Thus, the study of the mean irradiance over the facades needs to be complemented with the study of irradiance maps.

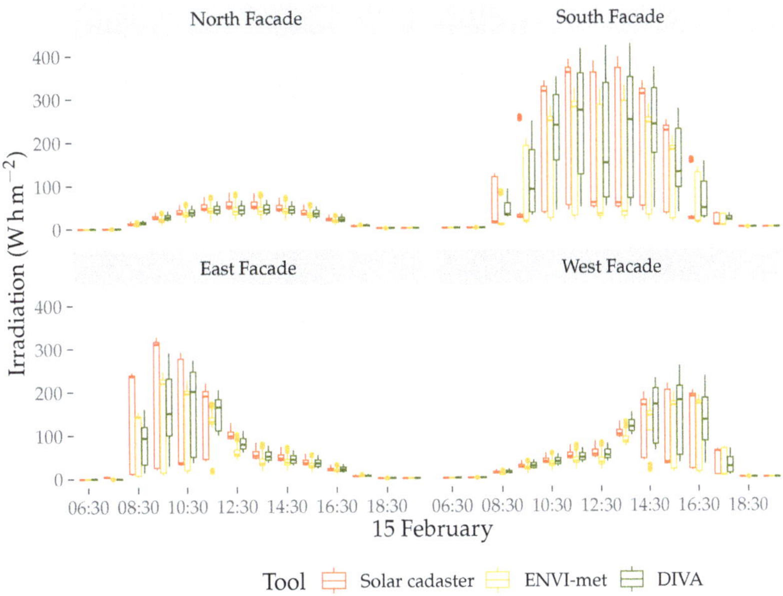

Figure 15. Distribution of the irradiance over the vertical facades for the homogeneous district in February.

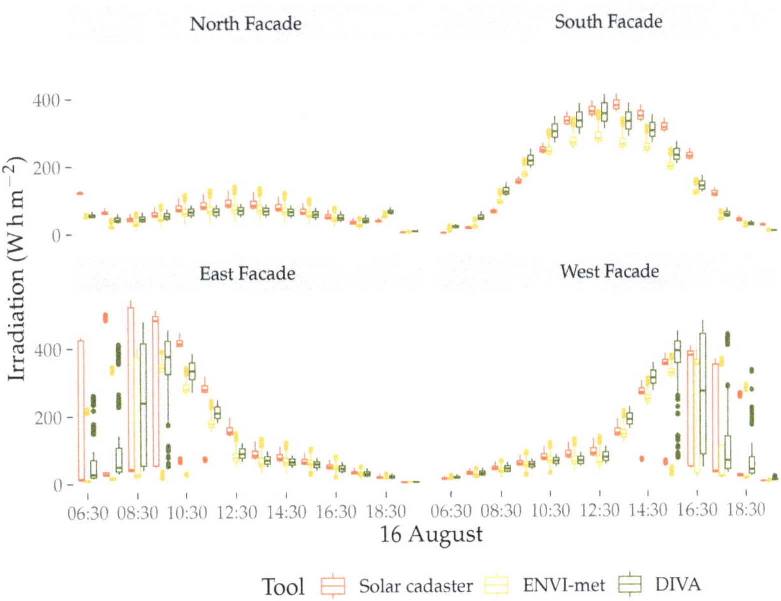

Figure 16. Distribution of the irradiance over the vertical facades for the homogeneous district in August.

The level of irradiance on unshaded roofs appears to be perfectly homogeneous and equal for the three tools considered, and it corresponds to the sum of the BHI and the DHI when given as the input of the simulation.

Nevertheless, the level of irradiance on the vertical facades depends on the orientation of the facade, the time of the day, and the time of the year. The levels of irradiance for the south facade of the central building of the homogeneous district are given in Figure 17. Figure 17a–c show the results for the month of February, while Figure 17d–f show the results for the month of August.

Regarding the data, not all are used for comparisons. All of the facades of the buildings are 20 m wide. However, the data at the edges of facades were not analyzed. Indeed, these data are difficult to compare because of the difference in terms of the calculation method for the three tools considered (depending on the tool, irradiance is considered as that at the center of the mesh or as the average over it); see Figure 4. This source of difference, due to the schemes specific to the tools, does not lead to a significant difference in the results, except at the edges of the facades.

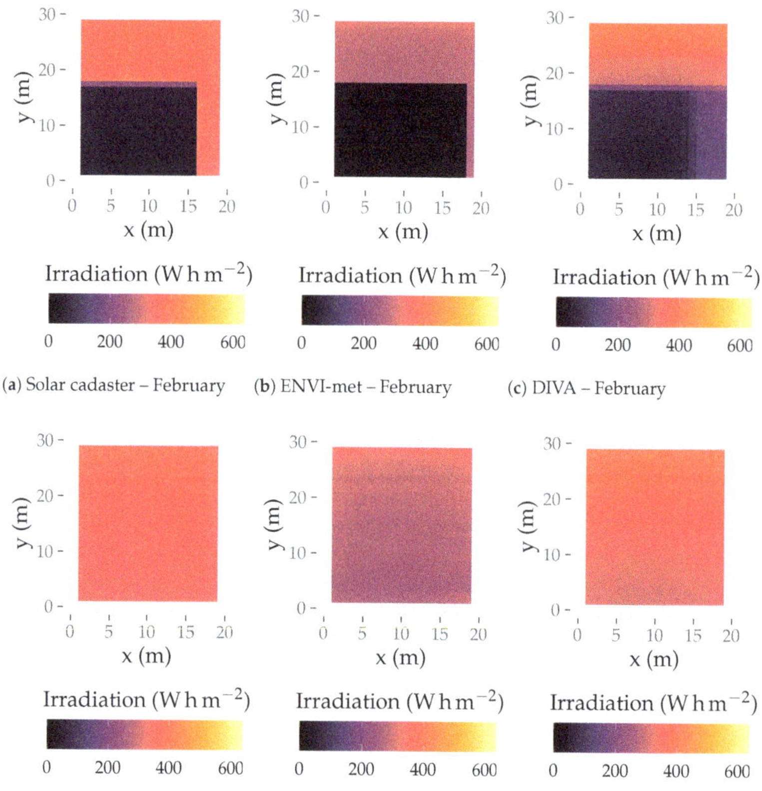

(a) Solar cadaster – February (b) ENVI-met – February (c) DIVA – February

(d) Solar cadaster – August (e) ENVI-met – August (f) DIVA – August

Figure 17. Irradiance map of the homogeneous district for the south facade over 12 h.

The irradiance levels appear to be more homogeneous on the south facade for the month of August than for the month of February. This can be explained by the sun's higher path at this time of year. The greater range in terms of value (see Figures 15 and 16) is then due to the shading effect, which is more intense in February and leads to a range of irradiance of 50 $W\,m^{-2}$ over the south facade in August in comparison with 350 $W\,m^{-2}$ in February.

Regarding the level of irradiance predicted by the solar cadaster, which is shown in Figure 17a,d, the values are similar to those predicted by ENVI-met and DIVA-for-Rhino.

Nevertheless, although the predicted irradiance levels are close to each other, there are two particular points. First, the shape induced by the shading is not exactly the same for the solar cadaster and ENVI-met on the one hand and for DIVA-for-Rhino on the other hand. The latter predicts a smoother evolution of the predicted irradiance level. This smoothness is due to the method used for the evaluation of the radiative phenomena; DIVA-for-Rhino uses ray-tracing while ENVI-met and the solar cadaster use the radiosity method. The second point is how the irradiance is evaluated. ENVI-met considers the irradiance to be that at center of the mesh, while DIVA-for-Rhino and the solar cadaster consider it to be the average of the irradiance over the mesh. In the case of DIVA-for-Rhino, it reinforces the smoothness of the evolution of the irradiance over the facade. Concerning the solar cadaster, this has an impact at the edge of the shade (the horizontal purple line at the height of 17 m in Figure 17a).

4. Discussion

The results presented so far show a general good accordance between the three tools considered in terms of both the mean values the irradiance maps. Nevertheless, two points must be noted:

1. The prediction of the irradiance is highly sensitive around sunrise and sunset;
2. A slight time offset between the results of the different tools seems to be present.

Regarding the sensitivity of the predicted irradiance, the highest discrepancies between the results of the tools appear at the first or the last hour of sunshine (see the north and the west facades in Figure 12). These discrepancies are due to a lack of precision between the calculation of the sun's path and the level of irradiation. This shift has only a slight incidence on the prediction of the irradiance during the day, but sees its influence strongly increase when the sun is close to the horizon. A sunrise that is predicted too early in the morning results in an overestimation of the predicted irradiance at this time of day, while a sunset predicted too late in the evening results in a peak of the predicted irradiance before the dusk.

Figure 17 shows the existence of a time lag between the three tools studied. Indeed, the shapes of the shadow on the south facade are different among the results of the solar cadaster (Figure 17a), ENVI-met (Figure 17b), and DIVA-for-Rhino (Figure 17c). This offset has only a slight influence on the results. Nonetheless, this impact may increase incrementally, especially early in the morning or late in the evening, as mentioned before (see the north facade in Figure 12).

5. Conclusions

The solar cadaster has proven its ability to provide accurate results in terms of the level of irradiance on both roofs and vertical facades. Indeed, the results predicted by the new version of the solar cadaster developed for Greater Geneva are in good accordance with those predicted by ENVI-met and DIVA-for-Rhino.

Regarding the concordance of the results, the non-shaded horizontal surfaces match almost perfectly. This is not the case for the shaded facades and vertical facades, but the results are very satisfactory. The differences observed between the tools come from the differences between the models in terms of diffuse radiation, the evaluation of reflections, or shading.

Finally, the solar cadaster provides precise results at a low spatial resolution while keeping the computation time low. It ranges from a few hours for the solar cadaster and DIVA-for-Rhino to a few days for ENVI-met. It can therefore be used on a large scale and can prove to be a reliable tool for promoting and intensifying the use of solar energy on the scale of Greater Geneva. However, work is in progress to improve the modeling of reflected components and, thus, the reliability of the tool for more complex geometries.

Author Contributions: Conceptualization, B.G., M.T., G.D., C.M. and S.G.-J.; Methodology, B.G., M.T. and G.D.; Software, V.G. and G.D.; Resources, G.D.; Writing—original draft preparation, B.G.; Writing—review and editing, B.G, M.T., K.B., É.P., S.G.-J., C.M. and G.D.; Visualization, B.G.; Supervision, C.M. and G.D.; Funding acquisition, C.M. and G.D.; All authors have read and agreed to the published version of the manuscript.

Funding: This research was funded by EU-INTERREG V France-Suisse program (G2-SOLAIRE Ref: 4610), and received support by the French National Research Agency, through Investments for Future Program (ref. ANR-18-EURE-0016—Solar Academy).

Institutional Review Board Statement: Not applicable.

Informed Consent Statement: Not applicable.

Data Availability Statement: The data that supports the findings of this study are available from the corresponding autho upon reasonable request.

Acknowledgments: The authors would like to thank the INTERREG V Suisse–France program for providing financial support for conducting this study in the framework of the G2 Solar project, which aims to extend the solar cadaster to Greater Geneva and intensify solar energy production at this level. This work was supported by the French National Research Agency through the Investments for Future Program (ref. ANR-18-EURE-0016—Solar Academy). The research units LOCIE and FRESBE are members of the INES Solar Academy Research Center.

Conflicts of Interest: The authors declare no conflict of interest. The funders had no role in the design of the study; in the collection, analyses, or interpretation of data; in the writing of the manuscript, or in the decision to publish the results.

References

1. Loi Fédérale sur la réDuction des éMissions de Gaz à Effet de Serre. 2018. Available online: https://fedlex.data.admin.ch/eli/fga/2018/108 (accessed on 24 June 2021).
2. Wiginton, L.; Nguyen, H.; Pearce, J. Quantifying rooftop solar photovoltaic potential for regional renewable energy policy. *Comput. Environ. Urban Syst.* **2010**, *34*, 345–357. [CrossRef]
3. Cellura, M.; Di Gangi, A.; Longo, S.; Orioli, A. Photovoltaic electricity scenario analysis in urban contests: An Italian case study. *Renew. Sustain. Energy Rev.* **2012**, *16*, 2041–2052. [CrossRef]
4. Hachem, C.; Athienitis, A.; Fazio, P. Design of roofs for increased solar potential BIPV/T systems and their applications to housing units. *ASHRAE Trans.* **2012**, *118*, 660–676.
5. Orioli, A.; Di Gangi, A. Load mismatch of grid-connected photovoltaic systems: Review of the effects and analysis in an urban context. *Renew. Sustain. Energy Rev.* **2013**, *21*, 13–28. [CrossRef]
6. Díez-Mediavilla, M.; Rodríguez-Amigo, M.C.; Dieste-Velasco, M.I.; García-Calderón, T.; Alonso-Tristán, C. The PV potential of vertical façades: A classic approach using experimental data from Burgos, Spain. *Sol. Energy* **2019**, *177*, 192–199. [CrossRef]
7. Hachem, C.; Athienitis, A.; Fazio, P. Energy performance enhancement in multistory residential buildings. *Appl. Energy* **2014**, *116*, 9–19. [CrossRef]
8. Hsieh, C.M.; Chen, Y.A.; Tan, H.; Lo, P.F. Potential for installing photovoltaic systems on vertical and horizontal building surfaces in urban areas. *Sol. Energy* **2013**, *93*, 312–321. [CrossRef]
9. Vulkan, A.; Kloog, I.; Dorman, M.; Erell, E. Modeling the potential for PV installation in residential buildings in dense urban areas. *Energy Build.* **2018**, *169*, 97–109. [CrossRef]
10. Desthieux, G.; Carneiro, C.; Camponovo, R.; Ineichen, P.; Morello, E.; Boulmier, A.; Abdennadher, N.; Dervey, S.; Ellert, C. Solar Energy Potential Assessment on Rooftops and Facades in Large Built Environments Based on LiDAR Data, Image Processing, and Cloud Computing. Methodological Background, Application, and Validation in Geneva (Solar Cadaster). *Front. Built Environ.* **2018**, *4*, 14. [CrossRef]
11. Kajiya, J.T. The Rendering Equation. *SIGGRAPH Comput. Graph.* **1986**, *20*, 143–150. [CrossRef]
12. Sillion, F.X.; Puech, C. *Radiosity and Global Illumination*; Morgan Kaufmann Publishers Inc.: San Francisco, CA, USA, 1994.
13. Freitas, S.; Catita, C.; Redweik, P.; Brito, M. Modelling solar potential in the urban environment: State-of-the-art review. *Renew. Sustain. Energy Rev.* **2015**, *41*, 915–931. [CrossRef]
14. Desthieux, G.; Carneiro, C.; Susini, A.; Abdennadher, N.; Boulmier, A.; Dubois, A.; Camponovo, R.; Beni, D.; Bach, M.; Leverington, P.; et al. Solar cadaster of Geneva: A decision support system for sustainable energy management. In *From Science to Society*; Springer: Berlin/Heidelberg, Germany, 2018; pp. 129–137.
15. Iqbal, M. *An Introduction to Solar Radiation*; Academic Press: Toronto, ON, Canada; New York, NY, USA, 1983.
16. Stendardo, N.; Desthieux, G.; Abdennadher, N.; Gallinelli, P. GPU-Enabled Shadow Casting for Solar Potential Estimation in Large Urban Areas. Application to the Solar Cadaster of Greater Geneva. *Appl. Sci.* **2020**, *10*, 5361. [CrossRef]

17. Reinhart, C.F.; Andersen, M. Development and validation of a Radiance model for a translucent panel. *Energy Build.* **2006**, *38*, 890–904. [CrossRef]
18. Perez, R.; Seals, R.; Michalsky, J. All-weather model for sky luminance distribution—Preliminary configuration and validation. *Sol. Energy* **1993**, *50*, 235–245. [CrossRef]
19. Tregenza, P.R. Subdivision of the sky hemisphere for luminance measurements. *Light. Res. Technol.* **1987**, *19*, 13–14. [CrossRef]
20. Reinhart, C.F.; Herkel, S. The simulation of annual daylight illuminance distributions—A state-of-the-art comparison of six RADIANCE-based methods. *Energy Build.* **2000**, *32*, 167–187. [CrossRef]
21. Perini, K.; Chokhachian, A.; Dong, S.; Auer, T. Modeling and simulating urban outdoor comfort: Coupling ENVI-Met and TRNSYS by grasshopper. *Energy Build.* **2017**, *152*, 373–384. [CrossRef]
22. Kolokotsa, D.D.; Giannariakis, G.; Gobakis, K.; Giannarakis, G.; Synnefa, A.; Santamouris, M. Cool roofs and cool pavements application in Acharnes, Greece. *Sustain. Cities Soc.* **2018**, *37*, 466–474. [CrossRef]
23. Farhadi, H.; Faizi, M.; Sanaieian, H. Mitigating the urban heat island in a residential area in Tehran: Investigating the role of vegetation, materials, and orientation of buildings. *Sustain. Cities Soc.* **2019**, *46*, 101448. [CrossRef]
24. Govehovitch, B.; Giroux-Julien, S.; Gaillard, L.; Peyrol, E.; Ménézo, C. Numerical and experimental study of two PV power generation models for buildings façades. Solar World Congress/ISES Conference Proceedings. 2019. Available online: http://proceedings.ises.org/paper/swc2019/swc2019-0182-Govehovitch.pdf (accessed on 24 June 2021).
25. Bruse, M.; The ENVI-Met Team. Advances in Simulating Radiative Transfer in Complex Environments. Available online: https://www.envi-met.com/wp-content/uploads/2020/11/201119-530_06_201116-CaseStudy_IVS-1-1.pdf (accessed on 24 June 2021).
26. Cheng, V.; Steemers, K.; Montavon, M.; Compagnon, R. Urban Form, Density and Solar Potential. In Proceedings of the 23rd Conference on Passive and Low Energy Architecture, Geneva, Switzerland, 6–8 September 2006.
27. Natanian, J.; Aleksandrowicz, O.; Auer, T. A parametric approach to optimizing urban form, energy balance and environmental quality: The case of Mediterranean districts. *Appl. Energy* **2019**, *254*, 113637. [CrossRef]
28. Lobaccaro, G.; Fiorito, F.; Masera, G.; Poli, T. District Geometry Simulation: A Study for the Optimization of Solar Façades in Urban Canopy Layers. *Energy Procedia* **2012**, *30*, 1163–1172. [CrossRef]

applied sciences

MDPI

Article

Identification of Cost-Optimal Measures for Energy Renovation of Thermal Envelopes in Different Types of Public School Buildings in the City of Valencia

María Esther Liébana-Durán [1,*], Begoña Serrano-Lanzarote [1] and Leticia Ortega-Madrigal [2]

[1] Department of Continuum Mechanics and Theory of Structures, School of Architecture, Universidad Politécnica de Valencia, 46022 Valencia, Spain; apserlan@mes.upv.es

[2] Instituto Valenciano de la Edificación, 46018 Valencia, Spain; lortega@five.es

* Correspondence: eliebanad@gmail.com

Abstract: In order to achieve the EU emission reduction goals, it is essential to renovate the building stock, by improving energy efficiency and promoting total decarbonisation. According to the 2018/844/EU Directive, 3% of Public Administration buildings should be renovated every year. So as to identify the measures to be applied in those buildings and obtain the greatest reduction in energy consumption at the lowest cost, the Directive 2010/31/EU proposed a cost-optimisation-based methodology. The implementation of this allowed to carry out studies in detail in actual scenarios for the energy renovation of thermal envelopes of public schools in the city of Valencia. First, primary school buildings were analysed and classified into three representative types. For each type, 21 sets of measures for improving building thermal envelopes were proposed, considering the global cost, in order to learn about the savings obtained, the repayment term for the investment made, the percentage reduction in energy consumption and the level of compliance with regulatory requirements. The result and conclusions will help Public Administration in Valencia to draw up an energy renovation plan for public building schools in the city.

Keywords: public school buildings; energy efficiency; optimal cost; energy renovation; public buildings

Citation: Liébana-Durán, M.E.; Serrano-Lanzarote, B.; Ortega-Madrigal, L. Identification of Cost-Optimal Measures for Energy Renovation of Thermal Envelopes in Different Types of Public School Buildings in the City of Valencia. *Appl. Sci.* **2021**, *11*, 5108. https://doi.org/10.3390/app11115108

Academic Editors: Tiziana Poli, Andrea Giovanni Mainini, Mitja Košir, Juan Diego Blanco Cadena and Gabriele Lobaccaro

Received: 7 April 2021
Accepted: 28 May 2021
Published: 31 May 2021

1. Introduction

It is a fact that the European Union is embarked on a path towards the conversion of economy and society with the aim of locating both of them in a more sustainable territory. A strategic framework is determined to promote a thriving, modern, competitive and climate-neutral economy. Among long-term objectives, a reduction of 90% of emissions by 2050 is included, compared to the levels in 1990 [1]. Currently, 36% of the EU's CO_2 gas emissions comes from the building stock, and almost 50% of final energy consumption is used for heating and cooling [2]. Therefore, to achieve these goals it is essential to renovate the building stock, by improving the energy efficiency and fostering total decarbonisation.

The current rate of building renovation is between 0.4% and 1.2%. This means that, in order to reach long-term European targets by 2050, it is necessary to double the rate of interventions in existing buildings [3]. Europe is driving a wave of renovation, prioritizing the improvement of the worst energy-efficient buildings, including schools and hospitals [4].

According to the 2018/844/EU Directive, 3% of public administration buildings should be renovated every year. However, the large number of properties, the lack of financing, information and planning are some of the obstacles found.

After checking some interventions in Spanish schools, the aforementioned drawbacks make performances consequently be carried out in two ways: a comprehensive renovation of each building or a phased renovation. The latter allows simultaneous performance in several buildings by improving a specific element, for example, windows, facades, heating systems, lighting, etc., or by installing renewable energy equipment. Based on the mentioned points, a simultaneous enhancement in several buildings makes it possible to jointly promote performances, to save time in project preparation and processing, as well as favour a further provision of financial assistance, since the projects are promoted by the public sector.

The main issue is that public administrations do not always have enough data or studies on school buildings or potential improvement scenarios, so there are no results that could be obtained in terms of energy saving. This means that, for example, the same renovation measure is implemented in all types of buildings, without knowing that whether in some of them a reduction of the energy demand before renovating heating systems could be necessary. Moreover, another example is those building types in which improvements in facade insulation could be suitable instead of window replacement.

So as to establish those performances with the greatest reduction in energy consumption at the lowest cost, the Directive 2010/31/EU proposed a cost optimisation methodology. Regarding its implementation in schools, the cost-optimal reports in the EU countries during 2018 [5] show that, whereas for the residential building sector some reference buildings have been established in all countries, for those buildings in the tertiary sector, in particular school buildings, not all countries have drawn up reports on them. Moreover, in those studied, there is no building classification, for example, they are grouped as "educational buildings" or "schools," as in the case of Slovakia and Germany, in which only a single building type is studied. Another example can be found in the Czech Republic with a "nursery school" or in the United Kingdom, with a "secondary school." Ferrara et al. [6] made a review on 88 scientific works based on the implementation of optimal-cost based analysis of calculation methods for designing and optimizing nearly zero-energy buildings in Europe. They show that only 4% of the papers studied include school buildings as case study.

The implementation of this methodology in school buildings shows great potential of growth. In addition, it provides local authorities with specific data on energy saving, maintenance costs, interventions, repayment terms, etc.

Furthermore, some studies on energy renovation in school buildings are worth mentioning. Several of them propose renovation measures for thermal envelopes, heating and lighting systems, use of renewable energy sources, etc. For example, Stocker et al. [7] use a calculation method focused on a standard energy demand with life cycle cost methods. Their results show that the optimal performance according to costs represents a value around 50 to 60 kW h/m2p.a regarding heating and cooling energy demand. Likewise, Dalla et al. [8] implemented cost-optimal methodology in some existing school buildings located in the north-east of Italy. They propose 120 sets of measures, including interventions in thermal envelopes, in systems (photovoltaic system and lighting replacement) and replacement of thermal generators (condensing boiler, biomass boiler or electrical heat pump).

Other authors tried to identify measures that enable to reach nearly zero energy building (nZEB) through an analysis from the cost-benefit perspective, as in the case of Lou et al. [9] who look into energy saving and electricity production schemes in a specific school building by using the building energy set eQUEST. The results show that improvement measures such as high-performance in building thermal envelopes, energy-efficient air-conditioning systems and lighting fixtures, as well as building-integrated photovoltaic panels (BIPV), allow to obtain zero energy buildings. Gaitani et al. [10] analysed some school buildings in terms of energy efficiency and cost optimisation, and designed a comprehensive action plan for renovation, a Technical and Financial Toolkit. Likewise, this study is framed within the European project ZEMedS, focused on the renovation of schools in the Mediterranean area to reach nZEB. With the aim of upgrading school buildings and turn them into nZEB, Ferrari et al. [11] focused their research on criteria laid down for the intervention on historical school buildings officially protected by the Italian Cultural Heritage. They assessed an Italian historical school building, and proved that the nZEB goals could be reached by retrofitting the building itself through measures compatible with the constraint arising from the protection of cultural heritage, and significantly reducing primary energy consumption. Marrone et al. [12] state that a large number of the Italian school building stock has implemented energy retrofitting measures, but the strategies suggested are often taken according to the best and most common practices (considering average energy saving), but not supported by a proper energy research. They evaluated 80 Italian school buildings by using cluster analysis, so as to provide a methodology capable of identifying the best energy retrofitting measures from the cost-benefit viewpoint. Mora et al. [13] state that a large number of Italian schools were built before the entry into force of energy and seismic regulations. Therefore, they simultaneously studied energy retrofitting and seismic upgrading in one school building.

The European project SHERPA (Share knowledge for Energy Renovation in buildings by Public Administrations) [14], is aimed at strengthening the abilities of public administrations at regional and local level to improve energy efficiency in their public buildings' stock, and reduce CO2 emissions. Soto et al. [15] describe the general auditing protocol devised by SHERPA and illustrate by carrying out an audit in one school building. They conclude that in the case of school buildings, in order to reach nZEB, energy efficiency is not always profitable (unless photo-voltaic energy is produced in situ). However, there are other benefits, such as improving comfort and preparing for the climate change.

In order to ease decision-making in future interventions, Jradi [16] identifies the impact of renovation measures on buildings, once enhancements in school buildings are made.

Before proposing different improvement energy measures, some authors establish a classification of school buildings according to different types based on energy factors, year of construction, building geometry, etc., which enable to find renovation solutions for each type. For example, Arambuela et al. [17] suggest a cluster analysis method that supports the definition of representative architectural types, and the identification of a small number of essential parameters, to assess energy consumption for air heating and the production of hot water in 60 schools in Treviso, Italy. Dimoudi et al. [18] also look into the development of school building types over the time within a Greece region, identify the most representative building types and propose seven improvement scenarios. Likewise, in order to classify the public school buildings in Rome, Santoli et al. [19] make use of data on schools, such as composition, (in terms of number, type and size of buildings), energy label of buildings in property of the municipality, which describe quality in terms of energy consumption for building's thermal envelopes and energy consumption, as heat is transferred from several thermal power plants to school buildings. Katafyogiotou et al. [20] and Castro [21] are authors that should be mentioned as an example of classification models. They propose improvement measures in representative school building types in Cyprus and northern Spain.

Through their analyses, Ferrara et al. [6] define two different methods used for the selection of measures in cost-optimal studies. The first one is a manual approach (selecting a defined number of sets of measures and calculating and comparing the global cost values), the other is an automated search (using computer-generated optimisation algorithms). They also establish two methods for energy performance calculation: one simplified (using simplified methods, for example, the quasi-steady state method defined by the UNI EN 13790 standard, and national implementations) and another dynamic (using dynamic simulation tools that allow detailed and precise energy results). According to this classification, this article uses a manual selection method and a simplified performance calculation method.

This article shows the results of adopting the cost-optimal methodology for housing developed by the IVE (Instituto Valenciano de la Edificación, the Valencia Institute of Building), for school building assessment. Specifically, this study applies this methodology to 3 schools within the city of Valencia, looking into energy performance and proposing a series of sets of improvement measures in thermal envelopes. Each school building is representative of a group in the city. As a result, a tool is obtained to identify the type-energy saving and CO_2 emissions through each set of measures, as well as the global cost over 30 years.

On the other hand, Spanish energy saving regulation, the CTE DB HE [22], sets out a number of requirements or demands for retrofitted buildings. This article also highlights these requirements, and shows to what extent they would be complied with each set of measures proposed.

Finally, the tool or system used and the analysis of the results provide a series of indicators on its usefulness as instruments and data for Public Administration to enable decision-making and planning energy renovation in similar school buildings.

2. Materials and Methods

2.1. Cost Optimisation Methodology

The methodology used in this study is that established for building energy renovation by the Directive 2010/31/EU on energy efficiency in buildings [23], and the Delegated Regulation 244/2012/EU complementing such directive, in particular the cost optimisation methodology [24]. According to Annex 1 of the regulation, this methodology is structured in the following sections:

1. Setting representative buildings.
2. Identification of energy efficiency measures, as well as improvement measures based on renewable energy sources and/or sets of variants of both types of measures applicable to each reference building.
3. Calculation of primary energy demand, resulting from implementing the measures and sets of measures defined for reference buildings.
4. Calculation of global cost as annual net value for each reference building.

Likewise, in the study for the energy analysis of representative buildings and improvement measures, Ce3X v2.3 is used, the Spanish computer software for energy calculation that also verifies the compliance with the CTE DB-HE.

For global cost calculation, a tool developed by the IVE was used, which had been previously applied to different studies on residential buildings.

2.1.1. Buildings under Study

In the city of Valencia there are approximately 90 public primary schools. For this research, general data regarding 135 school buildings was obtained, corresponding to 79 schools.

An analysis of construction and architectural features of these buildings has enabled to group them together into six different building types.

The study is focused on three building types (A, B and C). These were built before the entry of the first Spanish Regulation on thermal characteristics of buildings NBE CT 79 into force, so none of them has thermal envelope insulation, nor have they been recently renovated because of their relative age.

The main factors that differentiate these school building types are structure and date of construction. Type A building was built using brick load-bearing walls and metal joints, and types B and C were built using a concrete structure. In type A the roof was made of ceramic tile mounted on wood, whereas the roof in types B and C was built by using a slab flat or pitched under the layers that make up the roof. Type B and C buildings have wooden windows, whereas type C includes metal windows.

In addition to these construction differences, the design and interior spaces vary according to the three types. Type C buildings are elongated or L-shaped buildings, their facade has a historicist character, and were built between 1945 and 1955. Type B buildings were built in the 1960s and follow the designs of the Modern Movement. They have a greater number of floors and are elongated in shape. Type C buildings were built in the 1970s and they are X or XX-shaped (Figure 1).

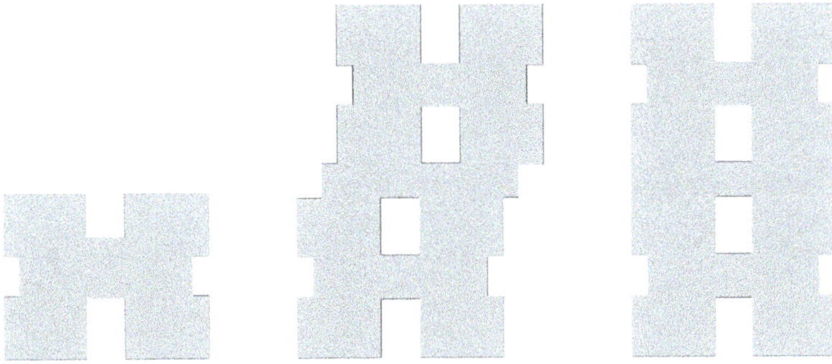

Figure 1. Examples of X or XX-shaped buildings.

These three building types (A, B and C), represent a total of 6, 11 and 29 schools respectively, that is to say, 46 school buildings.

For each type, a representative model school building is selected (Figure 2). They were built from school prototypes, adapting the architectural design to the site itself. This makes selected school buildings be representative in the city of Valencia, and also in other towns and municipalities.

Figure 2. Representative school buildings in the city of Valencia built before 1979. (a) Type A school; (b) type B school; (c) Type C School.

Type A and B buildings have a high historical value (listed or officially protected), or a unique architectural design. This means that any energy renovation cannot alter the building geometry and exterior design.

Although some of them have undergone minor renovations, such as window replacement, the initial condition of the buildings is taken for the energy study as a reference, and former potential improvements are ignored, so that they serve as a baseline to any building type. Table 1 includes the main characteristics of each building type, and Table 2 shows the thermal transmittance of construction elements in the thermal envelope for each building type in its initial state considered for energy study.

Table 1. Specifications and characteristics of building types.

Thermal Envelope Element	Type A	Type B	Type C
Construction date	1947	1957	1975
Constructed area	1634.77 m^2	1002.70 m^2	2595 m^2
Living area	785.89 m^2	525.42 m^2	1993.27 m^2
Number of floors	2	3/4	2
Climate zone	B3/IV	B3/IV	B3/IV
Use	8 h	8 h	8 h
Roofs	Curved ceramic roof tiles over ceramic brick board, wooden structure, highly ventilated air chamber, false ceiling made of cane, plaster.	Ceramic tiles, mortar, ceramic brick boards, waterproof protection, horizontal air chamber, reinforced with concrete slab floor, plaster.	Ceramic tiles, mortar, waterproof protection, reinforced with concrete slab floor, plaster.
Facades	Rendering mortar, brick wall (thickness: 40 cm/50 cm), plaster.	Rendering mortar, double perforated brick wall with a vertical air chamber (11 + 5 + 4 cm), plaster.	Double perforated brick wall with a vertical air chamber (11 + 7 + 7 cm), plaster.
Windows and doors	Wooden windows and metal single-glazed doors	Wooden windows and metal single-glazed doors	Metal single-glazed windows and doors. Rolling shutter
Heating System	Electric radiator/s in each room	Electric radiator/s in each room	Central heating through diesel oil boiler with multiple water circulation pumps and iron radiator/s in each room
Hot Water System	Electric hot water boiler for kitchen and pre-school toilets	Electric hot water boiler for kitchen and pre-school toilets	Electric hot water boiler for kitchen and pre-school toilets

Table 2. Thermal transmittance, U (W/m^2K), of construction elements in thermal envelope of representative school buildings in their initial state L0.

Thermal Envelope Element	Type A	Type B	Type C
Facades	1.1/0.92	2.94/1.33	1.41/1.29
Roofs	4.17	2.33/1.79	1.79
Windows (single glazed/structure)	5.7/2.2	5.7/2.2	5.7/5.7

2.1.2. Energy Efficiency Measures and Sets of Measures

This study is focused on proposing passive renovation measures, that is to say, measures to implement in the thermal envelope, so renewing air conditioning system, ventilation, lighting, renewable energy, etc., is not considered.

The buildings selected represent a group of school buildings according to specific architectural aspects, but many of them are part of a school complex, where other buildings are introduced, such as classrooms, canteens, gymnasiums, etc. This involves that requirements differ according to different typology, although the design and construction time of the main building are similar.

On the other hand, facility renovations are driven by the time when the existing ones break down or when a necessary renovation due to obsolescence is convenient and affordable. It would be unrealistic to plan the sudden replacement of all the facilities existing in school buildings. For all these reasons, the energy renovation measures proposed are aimed at reducing the energy demand in buildings.

The passive measures outlined are window and door replacement (W), including solar protection elements, and insulation upgrading in facades (F) and roofs (R). These measures can be implemented separately or combined with each other, resulting in the improvement of seven elements or combinations of elements of the thermal envelope.

In addition, three levels of energy demand are considered in relation to the transmittance of thermal envelope elements (L1, L2 and L3). These values are set according to specific regulatory requirements.

In particular, the intermediate level, L2, includes the minimum values set by current Spanish regulations on energy efficiency, the CTE DB-HE 2019 [22]. Level L3 corresponds to the guideline values of transmittance provided in Annex E of the aforementioned regulation, for pre-dimensioning construction solutions in private residential buildings. With these values, the requirements established for the global heat transmission ratio through the thermal envelope are fully met.

The strategic energy renovation in school buildings introduces a first level, L1, whose transmittance requirements are less restrictive than those set by current regulations. Thus, transmittance values in this level meet minimum requirements set by the same regulations in its initial version CTE DB-HE 2006 [25]. This is proposed with the aim of analysing whether requirements established by the former regulation reach cost-optimal results similar to those obtained with the conditions currently required.

Transmittance for each requirement level is shown in Table 3.

Table 3. Thermal transmittance, U (W/m^2K), of thermal envelope elements according to different energy demand levels.

Thermal Envelope Element	L1 [1]	L2 [2]	L3 [3]
Facades	0.82	0.56	0.38
Roofs	0.45	0.44	0.33
Windows	3.3	2.3	2

[1] CTE DBHE-2006. Values according to Table 2.2.-HE1 for climate zone B3. [2] CTE DBHE-2019. Values according to Table 3.1.1a-HE1 for climate zone B. [3] CTE DBHE- 2019. Values according to Table a-Annex and HE for climate zone B.

In total, 21 sets of improvement measures are proposed for the energy renovation strategy, according to each building type, combining three energy demand levels (L1, L2 and L3) and seven sets of improvement measures for thermal envelope elements (Table 4).

Table 4. Set of measures for thermal envelope according to a combination of energy demand levels and elements in thermal envelope to be renovated.

Thermal Envelope Element	L1	L2	L3
Windows (W)	L1 W	L2 W	L3 W
Facades (F)	L1 F	L2 F	L3 F
Roofs (R)	L1 R	L2 R	L3 R
Windows + Facades (WF)	L1 WF	L2 WF	L3 WF
Windows + Roofs (WR)	L1 WR	L2 WR	L3 WR
Facades + Roofs (FR)	L1 FR	L2 FR	L3 FR
Windows + Facades + Roofs (WFR)	L1 WFR	L2 WFR	L3 WFR

The improvement measures proposed for representative school buildings, type A and B, whose geometry and facade design cannot be altered, are found in facades, introducing inner insulation plasterboard lining with metal framing, and on the inside of sloping roofs in type A buildings, through a removable plaster false ceiling with thermal insulation. Both for facades and flat roofs in type C buildings, different exterior thermal insulation systems are proposed.

The type of insulation and the thickness used for facades and roofs, as well as the characteristics of glazing are different, depending on the transmittance to be obtained.

2.1.3. Calculation of Global Costs of Sets of Measures

The calculation of global costs of sets of measures is made for a period of 30 years, as established by the Delegated Regulation 244/2012/EU for public buildings.

The global costs include those related to the intervention itself, building consumption and maintenance during the calculation period. Following the regulation, for cost calculation of each set of measures, the initial investment is considered (C_I), as well as replacement costs, disposal costs, annual energy costs and annual rise in energy price, the annual maintenance cost of measures, as well as the residual value of elements added.

The Equation (1) used for global cost calculation is:

$$C_g(\tau) = C_I + \sum_j \left[\sum_{i=1(\tau)} (C_{a,i}(j) \times Rd(i)) - V_{f,\tau}(j)\right] \quad (1)$$

In which: τ indicates the calculation period; $C_g(\tau)$ indicates global cost (referred to starting year τ_0) over the calculation period; CI indicates initial investment costs for implementing measure or set of measures j; $C_{a,i}(j)$ indicates the cost per year, i for measure or set of measures, j; $V_{f,\tau}(j)$ indicates residual value of measure or set of measures j at the end of the calculation period (discounted of the starting year $\tau_{0)}$; and Rd (i) indicates discount factor for year i.

Consequently, the optimal cost of measures or sets of measures would be that with the lowest energy consumption at the lowest global cost per m^2 and year. As an example, to facilitate understanding and subsequent interpretation of global cost graphs, a graph is included (Figure 3). It shows the resulting cost-optimal curve (orange line), the initial state of the building (L0), the global cost in the initial state (red line) and the optimal-cost sets of measures (green line).

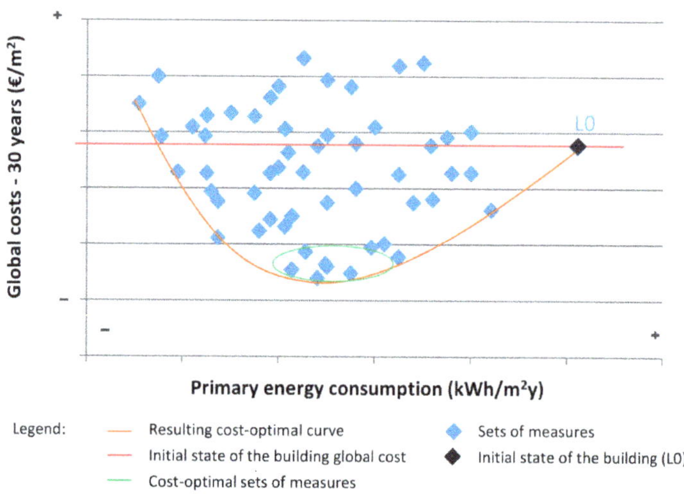

Figure 3. Example of cost-optimal graph. Correlation between the 30-year global costs of sets of measures proposed per m^2, and building's primary energy consumption per m^2 and year.

The sets of measures with global cost above the optimal cost in the initial state (L0) have a period to recover initial investment longer than the calculation period itself, in this case over 30 years.

2.2. Implementing Legislation

The Spanish regulation on energy efficiency, the Technical Building Code DB-Energy Saving (CTE DB-HE) [22] contains basic requirements on the matter, which have been modified in accordance with the directives and goals set by the EU.

Currently, regarding energy demand levels for energy saving, the CTE DB-HE establishes maximum transmittance (U_{lim}) for thermal envelope elements, both new and renovated, and a global ratio of heat transmission through thermal envelope (K_{lim}). It also limits consumption of non-renewable primary energy ($C_{ep,nren}$) and total primary energy consumption ($C_{ep,tot}$).

On the other hand, concerning renovation of specific buildings, the regulation allows greater flexibility in obtaining some values required, as in the case of buildings with significant architectural value. In the same way, it classifies interventions into small or large ones, and sets limits or requirements according to the type of performance.

This study also analyses requirements for each building type, and the level of compliance with regulation, based on the sets of measures to be implemented.

3. Discussion on Results

3.1. Results on Global Costs of Measures and Primary Energy Consumption

It should be emphasised that each school building is studied in the initial state at an energy demand level, L0, subsequently implementing sets of measures, as seen in Sections 2.1 and 2.1.2.

As a result of relating 30-year global costs per m^2 to each set of measures and the consumption of primary energy per m^2 and year, the graphs obtained show an optimal intervention cost for each school building type (Figure 4a–c).

Overall, it is noted that, in the three building types, for the same combination of measures, as in the WF thermal envelope elements, the energy demand levels (L1, L2 and L3) do not imply great differences in terms of global costs and resulting energy consumption. The graph clearly shows how different combinations of measures are grouped together according to construction elements enhanced.

Moreover, when comparing the graphs of representative building types it is clear what set of measures or thermal envelope elements obtains optimal costs.

For type A buildings, sets of measures introducing roof insulation obtain a greater reduction in energy consumption at the lowest global cost over 30 years. In this case, and depending on the energy demand level in terms of transmittance, consumption reduction regarding the initial state (L0) is between 24.5% and 25.3%, for interventions on the roof (R); between 28.1% and 30.2%, for improvements in facade and roof (FR); between 30.8% and 32.2%, for windows and roofs (WR); and between 34.8% and 38.1%, if the three thermal envelope elements (WFR) undergo improvements. The energy saving gained would enable to recover initial investment in such improvements within 4–5 years (R), 5–6 years (FR), 15 years (WR) and 14 years (WFR). The cost-optimal levels of interventions would be R and FR, thermal envelope elements, with a lower global cost. A greater saving is obtained through the OR and OFR, elements improved in the thermal envelope.

For type B and C buildings respectively, the sets of measures with the lowest global cost over 30 years include facade thermal insulation (F), and facade and roof thermal insulation (FR). These sets of measures represent a decrease in consumption between 15.8–18.6% (F) and 16.6–19.2% (FR) for type B buildings, and between 18.6–21.1% (F) and 20.5–22.9% (FR) for type C buildings. The repayment term is less than 7 years in type B buildings, and between 12–13 years (R) and 19 years (FR) in type C buildings.

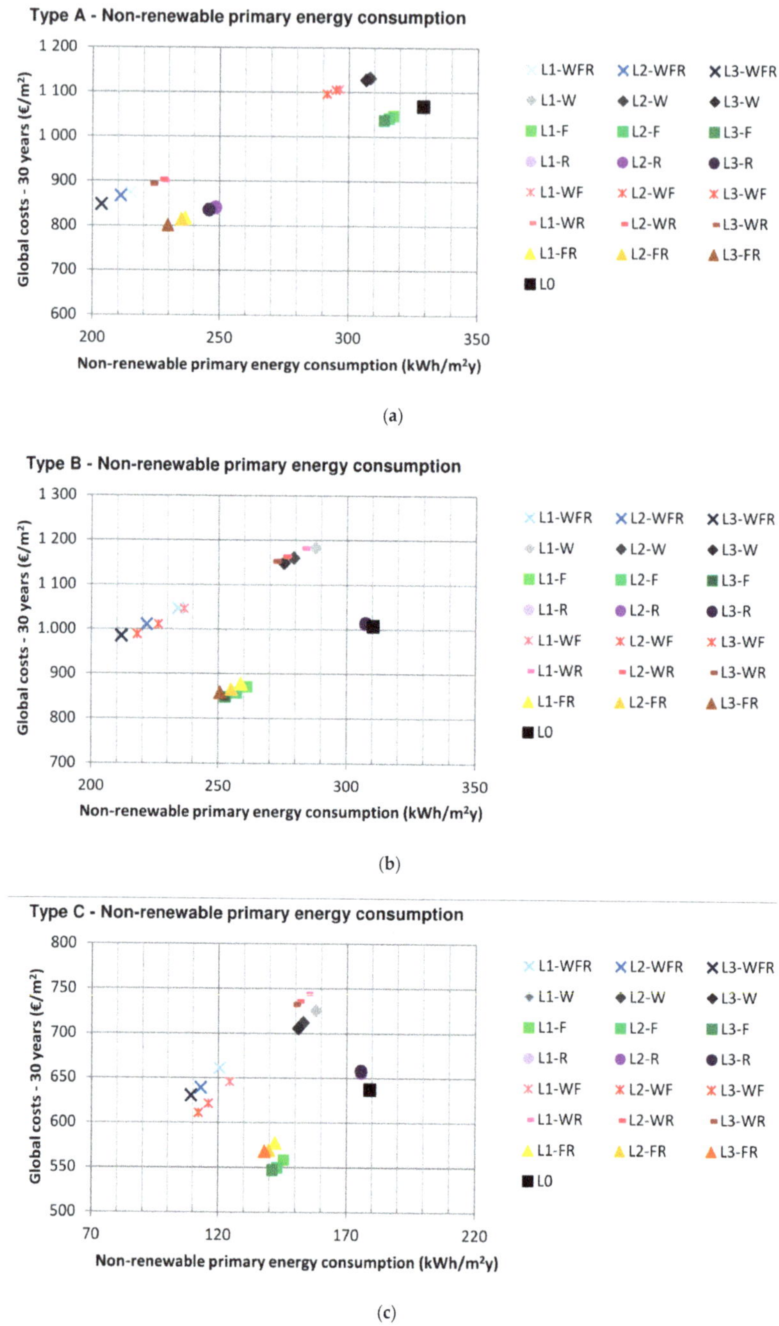

Figure 4. Cost-optimal graphs. Correlation between 30-year global cost per m^2 in the sets of measures proposed and building's primary energy consumption per m^2 and year, after the intervention, assuming a price increase rate of 1%. (**a**) Type A school building; (**b**) type B school building; (**c**) type C school building.

In building types B and C, a greater reduction in energy consumption is obtained after replacing facades and windows (WF), as well as renovating all thermal envelope elements (WFR). In both sets of measures, the global costs increase, which means a longer repayment term for the investment made. In type B buildings, the global cost would be very similar to 30-year global cost, without necessary performance in the building. That is, the initial state L0, even exceeding it in some cases, which implies that the 30-year investment recovery is greater.

Table 5 is drawn up in order to provide accurate figures and facilitate comparison with some results included in the previous graphs. It shows 30-year global costs per m^2 corresponding to the most relevant sets of measures which include the full amount to pay for energy consumption in 30 years for each m^2 (living area), the implementation of measures taken and maintenance. It also includes global cost per m^2 after 30 years of the initial state (L0), that is, if no action is taken, and identifies the full amount to pay for each m^2, especially energy consumption expenses.

Table 5. 30-year global cost of initial state L0 per m^2, according to type of school building, and global cost of sets of measures per m^2, including those with better energy performance.

Type of School Building	L0	Elements Improved	L1	L2	L3
Type A	€1069.29	R	€840	€842	€836
		FR	€818	€815	€802
		WR	€901	€904	€894
		WFR	€875	€868	€848
Type B	€1008.40	F	€871	€858	€849
		FR	€879	€866	€859
		WF	€1047	€1012	€990
		WFR	€1047	€1012	€986
Type C	€637.46	F	€559	€533	€547
		FR	€578	€570	€568
		WF	€646	€622	€611
		WFR	€661	€639	€631

The fact that the global cost of specific sets of measures is higher than that in the initial state entails that the starting investment is not amortised in 30 years. These sets of measures could be implemented together with an enhancement of building's facilities, namely, by taking measures to be implemented in thermal installations. In this way, global costs would be minimised through a reduction in energy consumption due to improvement in system performance.

The annual savings generated after implementing sets of measures concerning energy renovation in each building can be expressed from the viewpoint of emission reduction (Table 6), and the subsequent economic saving after minimizing annual energy consumption per year (Table 7). Table 8 shows the reduction percentage in energy consumption and recovery period of initial amortisation for sets of improvement measures with lower 30-year global cost and a greater consumption reduction, according to building type.

Table 6. Emission reduction compared to the initial state per year, according to type of school building and sets of measures. Expressed in $KgCO_2$ per year and according to building.

Type of School Building	Elements Improved	L1	L2	L3
Type A	R	10,766.4	10,766.4	11,080.8
	FR	12,338.2	12,573.9	13,281.2
	WR	13,517.0	13,595.6	14,145.7
	WFR	15,245.9	15,796.0	16,739.0
Type B	F	4413.5	4781.3	5149.1
	FR	4623.7	4938.9	5306.7
	WF	6410.1	7513.5	8249.1
	WFR	6830.5	7933.8	8827.1
Type C	F	17,740.1	19,135.4	20,132.0
	FR	19,534.0	20,730.0	21,726.6
	WF	28,503.8	32,291.0	34,483.6
	WFR	31,095.0	34,682.9	36,875.5

Table 7. Average annual saving according to type of school building due to reduction in energy consumption, according to sets of measures.

Type of School Building	Elements Improved	L1	L2	L3
Type A	R	€6868.81	€6868.81	€7078.37
	FR	€7858.39	€7991.44	€8458.78
	WR	€8630.09	€8693.29	€9022.59
	WFR	€9732.76	€10,057.07	€10,684.08
Type B	F	€2799.90	€3032.30	€3284.72
	FR	€2936.67	€3153.51	€3399.25
	WF	€4080.88	€4784.74	€5266.22
	WFR	€4343.30	€5053.84	€5624.27
Type C	F	€8297.11	€8954.29	€9410.83
	FR	€9133.15	€9741.87	€10,192.79
	WF	€13,510.74	€15,446.24	€16,439.34
	WFR	€14,599.98	€16,362.83	€17,343.28

Table 8. Reduction percentage in energy consumption and recovery period of initial amortisation in sets of measures with lower 30-year global cost and a greater consumption reduction, according to type of school building.

		% Reduction in Energy Consumption			Amortisation of Investment (Years)		
Type of School Building	Elements Improved	L1	L2	L3	L1	L2	L3
Type A	R	24.5%	24.5%	25.3%	4	5	5
	FR	28.1%	28.5%	30.2%	5	6	6
	WR	30.8%	31.0%	32.2%	15	15	15
	WFR	34.8%	35.9%	38.1%	14	14	14
Type B	F	15.8%	17.1%	18.6%	5	4	5
	FR	16.6%	17.8%	19.2%	7	7	7
	WF	23.1%	27.1%	29.8%	>30	>30	29
	WFR	24.6%	28.6%	31.8%	>30	>30	29
Type C	F	18.6%	20.1%	21.1%	13	12	13
	FR	20.5%	21.9%	22.9%	19	19	19
	WF	30.5%	35.1%	27.3%	>30	29	28
	WFR	32.8%	36.8%	38.9%	>30	>30	30

If the results obtained for each building type were implemented in all representative schools, the result obtained by applying the lowest overall cost measures N3 FR (Type

A) and N3 F (Types B and C) would be an average saving of 50,752.68 euros per year, 37,391.75 euros and 295,590.91 euros, respectively.

In addition, annual emission reductions would be 79,687.2 $KgCO_2$, 56,640.1 $KgCO_2$ and 583,828 $KgCO_2$, respectively.

3.2. Regulatory Requirements and Compliance

Regarding Spanish Regulations on energy saving and compliance when it comes to building renovation, the CTE DB-HE 2019 [22] sets a global energy consumption limitation (non-renewable consumption of primary energy ($C_{ep,nren}$) and primary energy total consumption ($C_{ep,tot}$)) when there are interventions in over 25% of thermal envelope, as well as in thermal installations.

Specifically, within the climate zone corresponding to the city of Valencia, the regulatory limit value established for the $C_{ep,nren}$ is 55 kWh/m2 per year, and 80 kWh/m^2 per year for the $C_{ep,tot}$.

In the case of the school buildings studied, the $C_{ep,nren}$ value in the initial state of type A buildings (L0) is 351.3 kWh/m^2 per year, 256.9 kWh/m^2 per year for type B buildings and 192.1 kWh/m^2 per year for type C buildings. The greatest reduction in energy consumption would be obtained through the set of measures L3-WFR. In particular, this set of measures would allow the $C_{ep,nren}$ values to be 207.6 kWh/m^2 per year for type A buildings, 183.1 kWh/m^2 per year for type B and 116.3 kWh/m^2 per year for type C.

Regarding $C_{ep,tot}$, its value in the initial state for (L0) type A buildings is 425.7 kWh/m^2 per year, 311.3 kWh/m^2 per year for type B and 205.3 kWh/m^2 per year for type C. The set of measures L3-WFR would allow $C_{ep,tot}$ values to be reduced in 251.6 kWh/m^2 per year for type A buildings, 221.9 kWh/m^2 per year for type B and 128.8 kWh/m^2 per year for type C.

As mentioned above, this study is focused on the renovation of thermal envelope elements, so this limitation is not mandatory. However, it should be considered in the case of performing in facilities.

Below, the study analyses the compliance with requirements on transmittance of thermal envelope elements.

In any intervention on thermal envelope, every single element renovated must meet requirements on transmittance set by regulations, for example, in the case of window replacement. These values partly correspond to those set for level L2 in Table 2. Level L1, as indicated in Section 2, does not meet the CTE DB-HE 2019 regulation.

Particularly, for major renovations over 25% of thermal envelope, the standard also sets a global heat transmission ratio (K). This depends on the climate zone where buildings are located and their compactness. According to the CTE DB HE, the K coefficient indicates the average value of heat transfer ratio for heat exchange surface of the thermal envelope (A_{int}).

As shown in Table 9, the K_{lim} level set in the regulations is reached only by implementing one set of measures (L3-WFR) in type C buildings. Consequently, to reduce global ratio of the sets of measures proposed, it is necessary to minimise transmittance, that is, by setting more restrictive energy demand levels in facades (F), roofs (R) and windows (W). Another way to minimise building's energy demand is by performing in other elements, such as interior partitions and floors in contact with unheated rooms. This is not always technically or economically feasible.

Table 9. Global heat transmission ratio K (W/m^2K). Regulatory requirements according to type of school building, building compactness and climate zone, Klim and ratios obtained for each set of measures and energy demand level.

			K (W/m^2K)		
Type of School Building	**Klim**	**Elements Improved**	**L1**	**L2**	**L3**
Type A	0.9	R	2.83	2.83	2.83
		FR	2.25	2.07	1.91
		WR	2.44	2.19	2.13
		WFR	1.86	1.43	1.21
Type B	0.85	F	2.27	2.18	2.09
		FR	1.94	1.84	1.73
		WF	1.90	1.62	1.46
		WFR	1.57	1.29	1.11
Type C	0.86	F	2.31	2.21	2.14
		FR	1.72	1.62	1.42
		WF	1.82	1.54	1.50
		WFR	1.23	0.95	0.78

In view of this situation, legislation provides for the fact that occasionally it is not possible to reach the level of benefit generally established. In these cases, some solutions may be adopted to achieve the highest level of adequacy. This is possible, as long as they are buildings with recognised historical or architectural value, when other solutions are not technically or economically feasible, or solutions involve substantial changes in elements of thermal envelope, or in thermal facilities without initial intervention.

For this study, type A and B school buildings could benefit from this flexibility criterion included in the standard, provided that measures implemented are close to limit values.

Regarding solar control, in the case of renovating over 25% of the total area of the final thermal envelope, the solar control parameter ($q_{sol;jul}$) should not exceed the limit value 4 kWh/m^2 per month, for uses other than private and residential.

According to the CTE DB HE, $q_{sol;jul}$ indicates the ratio between the solar gains of windows and doors on the thermal envelope during the month of July with mobile solar protection activated, and the useful floor area included in spaces within the thermal envelope (A_{useful}). Solar protection can be implemented in all the building or in part of it.

This solar control parameter is fulfilled in those sets of measures in which windows and doors (O) are replaced, since protection elements are included. In type A and B buildings, as it is not possible to alter facades, protection systems should be installed on the inside, for example by using shutters.

4. Conclusions

Through the implementation of optimum-cost methodology, the study aims at identifying the set of improvement measures with optimum intervention cost for each building type studied. It is found that for type A buildings, the most cost-effective intervention would be to renovate facades and roofs (L3-FR), whereas for types B and C, the improvement is only made in facades (L3-F). In all cases, the set of improvement measures that brings buildings closer to nZEB involves performing in all thermal envelope elements with the most restrictive values of thermal transmittance (L3-WFR).

All in all, the implementation of cost-optimal sets of improvement measures in 46 schools studied results in an average annual saving of 378,735.34 euros and an annual emission reduction of 720,155.3 $KgCO_2$.

These results show Public Administration in Valencia that the most cost-effective solution is not always the same for all schools, and depends on building typology. This suggests that it is not advisable to renovate the same thermal envelope element simultaneously in all buildings for instance, by renovating windows in all schools. In order to save costs when performing in more than one element or school at the same time, the typology classification proposed for the sample is suggested to be applied so that the same intervention is replicated for the same typology but not for all schools.

The results achieved allow school principals in Valencia to make decisions on carrying out priority renovations in buildings and potential strategies. For example, for those measures with a global cost similar or higher than that in the current state of the building, it is appropriate to perform a deep renovation, including enhancement of thermal envelope and heating and cooling systems, so that the 30-year global cost is reduced. The long-term cost reduction and the quick return of investment of some sets of measures may be a reason for public administrations in Valencia to invest in school renovation.

Regarding regulatory requirements, it is shown that an approach to building renovation with the aim of complying with current legislation for new buildings may not be the most cost-effective option.

In summary, the cost-optimal methodology applied to the renovation of school buildings studied provides quantitative data on costs and energy saving that can be obtained after implementing specific sets of measures in 46 school buildings in the city of Valencia. All this facilitates to identify, among other data, the initial cost of measures, consumption reduction gains and the return on investment periods. These data can provide public administrations in Valencia with criteria to design long-term intervention plans, which enable available resources to be efficiently invested.

Irrespective of the specific results reached, the adaptation of the methodology proposed to a wider scale (regional/national) can help to build support for deciding about the renovation of school buildings and designing long-term renovation strategies. The study can also be expanded to other buildings in the city, as well as other cities or regions.

Author Contributions: Conceptualization, M.E.L.-D.; Methodology, B.S.-L. and L.O.-M.; Writing—original draft, M.E.L.-D.; Writing—review & editing, M.E.L.-D., B.S.-L. and L.O.-M. All authors have read and agreed to the published version of the manuscript.

Funding: This research received no external funding.

Institutional Review Board Statement: Not applicable.

Informed Consent Statement: Not applicable.

Conflicts of Interest: The authors declare no conflict of interest.

References

1. European Commission Notice, 28 November 2018 (COM/2018/773 final). A Clean Planet for Everyone. Europe's Long-Term Strategic Vision for a Prosperous, Modern, Competitive and Climate-Neutral Economy. Available online: http://eur-lex.europa.eu (accessed on 1 April 2021).
2. Directive 2018/844/EU by the European Parliament and Council, 30 May 2018, Which Modifies Directives 2010/31/EU, on Energy Efficiency of Buildings, and 2012/27/EU, on Energy Efficiency. Available online: http://eur-lex.europa.eu (accessed on 1 April 2021).
3. European Commission Notice, 11 December 2019 (COM/2019/640 Final). European Green Pact. Available online: http://eur-lex.europa.eu (accessed on 1 April 2021).
4. European Commission Notice, 17 September 2020 (COM/2020/562 Final). Stepping up Europe's Climate Ambition by 2030: Investing in a Climate-Neutral Future for Citizens' Benefits. Available online: http://eur-lex.europa.eu (accessed on 1 April 2021).
5. EU Countries' 2018 Cost-Optimal Reports. Available online: https://ec.europa.eu/energy/topics/energy-efficiency (accessed on 1 April 2021).
6. Ferrara, M.; Monetti, V.; Fabrizio, E. Cost-Optimal Analysis for Nearly Zero Energy Buildings Design and Optimization: A Critical Review. *Energies* **2018**, *11*, 1478. [CrossRef]
7. Lou, S.; Tsang, E.K.; Li, D.H.; Lee, E.W.; Lam, J.C. Towards zero energy school building designs in Hong Kong. *Energy Proceia* **2017**, *105*, 182–187. [CrossRef]
8. Stocker, E.; Tschurtschenthaler, M.; Shrott, L. Cost-optimal renovation and energy performance: Evidence from existing school buildings in the Alps. *Energy Build.* **2015**, *100*, 20–26. [CrossRef]
9. Dalla Mora, T.; Righi, A.; Peron, F.; Romagnoni, P. Cost-Optimal measures for renovation of existing school buildings towards nZEB. *Energy Proceia* **2017**, *140*, 288–302. [CrossRef]
10. Gaitani, N.; Cases, L.; Mastrapostoli, E.; Eliopoulou, E. Paving the way to nearly zero energy schools in Mediterranean. *Energy Proceia* **2015**, *78*, 3348–3353. [CrossRef]

11. Ferrari, S.; Romeo, C. Retrofitting under protection constraints according to the nearly Zero Energy Building (nZEB) target: The case of an Italian cultural heritage's school building. *Energy Proceia* **2017**, *140*, 495–505. [CrossRef]
12. Marrone, P.; Gori, P.; Asdrubali, F.; Evangelisti, L.; Calcagnini, L.; Grazieschi, G. Energy Benchmarking in Educational Buildings through Cluster Analysis of Energy Retrofitting. *Energies* **2018**, *11*, 649. [CrossRef]
13. Mora, T.D.; Pinamonti, M.; Teso, L.; Boscato, G.; Peron, F.; Romagnoni, P. Renovation of a School Building: Energy Retrofit and Seismic Upgrade in a School Building in Motta Di Livenza. *Sustainability* **2018**, *10*, 969. [CrossRef]
14. *European Project "Sherpa" (Shared Knowledge for Energy Renovation in Buildings by Public Administrations)*; Technical Report, EU Project Financed by the Interreg Med Programme (2017–2019). Available online: https://sherpa.interreg-med.eu/ (accessed on 1 April 2021).
15. Soto, V.M.; Serrano, B.; Valero, V.; Navarro, M. Improving schools performance based on SHERPA project outcomes: Valencia case (Spain). *Energy Build.* **2020**, *225*, 110297. [CrossRef]
16. Jradi, M. Dynamic Energy Modeling as an Alternative Approach for Reducing Performance Gaps in Retrofitted Schools in Denmark. *Appl. Sci.* **2020**, *10*, 7862. [CrossRef]
17. Arambula Lara, R.; Pernigotto, G.; Cappelletti, F.; Romagnoni, P.; Gasparella, A. Energy audit of schools by means of cluster analysis. *Energy Build.* **2015**, *95*, 160–171. [CrossRef]
18. Dimoudi, A.; Kostarela, P. Energy monitoring and conservation potential in school buildings in the C ′Climatic zone of Greece. *Renew. Energy* **2009**, *34*, 289–296. [CrossRef]
19. Santoli, L.; Fraticelli, F.; Fornari, F.; Calice, C. Energy performance assessment and retrofit strategies in public school buildings in Rome. *Energy Build.* **2014**, *68*, 196–202. [CrossRef]
20. Katafygiotou, M.; Serghides, D. Analysis of structural elements and energy consumption of school building stock in Cyprus: Energy simulations and upgrade scenarios of a typical school. *Energy Build.* **2014**, *72*, 8–16. [CrossRef]
21. Vázquez, C.; Manuel, J. La Rehabilitación Energética en la Arquitectura Escolar. Hacia el Edificio Rehabilitado nZEB. (Energy Renovation in School Building Architecture. Towards the Renovated Building nZEB). Ph.D. Thesis, University of La Coruña, La Coruña, Spain, 2017.
22. Website of the Technical Building Code. CTE DB HE 2019_ Real Decreto 732/2019, de 20 de Diciembre, por el que se Modifica el Código Técnico de la Edificación, Aprobado por el Real Decreto 314/2006, de 17 de Marzo. (CTE DB HE 2019_ Royal Decree 732/2019, 20 December, Which Modifies the Technical Building Code, Approved by Royal Decree 314/17 March 2006). Available online: https://www.codigotecnico.org/pdf/Documentos/HE/DBHE.pdf (accessed on 1 April 2021).
23. Directive 2010/31/EU of the European Parliament and the Council, 19 May 2010 on Energy Efficiency of Buildings (Recast). Available online: http://eur-lex.europa.eu (accessed on 1 April 2021).
24. Commission Delegated Regulation (EU) No. 244/2012, 16 January 2012, Which Complements Directive 2010/31/EU of the European Parliament and the Council, on the Energy Efficiency of Buildings, Establishing a Comparative Methodological Framework to Calculate Optimal Levels of Profitability of the Minimum Requirements for Energy Efficiency of Buildings and Their Elements and Accompanying Guidelines. Available online: https://eur-lex.europa.eu (accessed on 1 April 2021).
25. Website of the Technical Building Code. CTE DB HE 2006_Royal Decree 314/2006, 17 March, Approving the Technical Building Code. Available online: https://www.codigotecnico.org/pdf/Documentos/HE/DBAnteriores/DBHE_200602.pdf (accessed on 1 April 2021).

applied sciences

MDPI

Article

Passive Solar Solutions for Buildings: Criteria and Guidelines for a Synergistic Design

Giacomo Cillari *, Fabio Fantozzi and Alessandro Franco

Department of Energy, Systems, Territory and Constructions Engineering, University of Pisa, Largo Lucio Lazzarino, 56126 Pisa, Italy; f.fantozzi@ing.unipi.it (F.F.); alessandro.franco@ing.unipi.it (A.F.)
* Correspondence: giacomo.cillari@phd.unipi.it

Featured Application: Optimization of passive solar strategies to minimize building energy demand.

Abstract: Passive solar system design is an essential asset in a zero-energy building perspective to reduce heating, cooling, lighting, and ventilation loads. The integration of passive systems in building leads to a reduction of plant operation with considerable environmental benefits. The design can be related to intrinsic and extrinsic factors that influence the final performance in a synergistic way. The aim of this paper is to provide a comprehensive view of the elements that influence passive solar systems by means of an analysis of the theoretical background and the synergistic design of various solutions available. The paper quantifies the potential impact of influencing factors on the final performance and then investigates a case study of an existing public building, analyzing the effects of the integration of different passive systems through energy simulations. General investigation has highlighted that latitude and orientation impact energy saving on average by 3–13 and 6–11 percentage points, respectively. The case study showed that almost 20% of the building energy demand can be saved by means of passive solar systems. A higher contribution is given by mixing direct and indirect solutions, as half of the heating and around 25% of the cooling energy demand can be cut off.

Keywords: solar energy; building energy performance; energy saving; passive solar design; synergistic design

Citation: Cillari, G.; Fantozzi, F.; Franco, A. Passive Solar Solutions for Buildings: Criteria and Guidelines for a Synergistic Design. *Appl. Sci.* **2021**, *11*, 376. https://doi.org/10.3390/app11010376

Received: 14 November 2020
Accepted: 29 December 2020
Published: 2 January 2021

1. Introduction

Energy use in the residential sector represents a great share of the global energy demand, attested between 20% and 40% of the total [1]. Each country has developed specific energy efficiency strategies for the civil sector. Firstly, the main attention concerned energy conservation, focusing on the building envelope insulation that can reduce energy demand by up to 28% [2]. Then the focus moved to increasing the efficiency of Heating Ventilation and Air Conditioning system (HVAC) systems, that are the main building energy consumption source and commonly do not operate efficiently [3]. In a nearly zero-energy building (nZEB) perspective, the integration of different strategies regarding envelope insulation, heat generation, shading, and control devices is fundamental to minimize building conditioning loads, so as to be able to maximize the share of energy demand covered by renewable energy sources. It must also be taken into account that residential space heating is responsible for 86% of building energy demand and can easily take advantage from renewable sources [4]. Among the various alternatives, solar energy represents the best option for buildings as one of the most accessible and easily exploitable, even if its current contribution to global energy supply is still imperceptible [5]: it can be directly exploited by heating up the building through solar irradiation, or indirectly by means of photovoltaic (PV) or solar thermal systems. As Table 1 data from International Energy Agency (IEA) [6] show, in a scenario where mandatory building energy codes will almost totally cover construction activity by 2030, with a nZEB share forecast of

50%, the sales of heat pumps, which can be coupled with PV as a renewable production system, are slowly but constantly increasing, between 0.3 and 0.5% on a yearly basis, with fossil fuel-based equipment that, after a peak in 2014, is gradually decreasing. These data highlight the core role that solar energy plays in building design in the near future: by properly combining photovoltaic and solar thermal systems, indeed, solar energy can provide at least 76% of the primary energy demand of a residential building, with a short payback period [7].

Table 1. (a) Heating technology sales, (b) global building construction area by type of building code.

Heating Technology	Fossil fuel-Based Equipment	Conventional Electric Equipment	Heat Pumps	District Heat	Other Renewables and Hydrogen-Based
2010	59.7%	21.1%	2.9%	8.3%	8%
2012	57.9%	19.9%	3.3.%	10.1%	8.8%
2014	60.1%	18.5%	3.4%	10.2%	7.8%
2016	57.9%	21.2%	3.9%	9.1%	7.9%
2018	57.1%	20.6%	4.3%	9.4%	8.6%

(a)

Year	Constructions without building energy code	Constructions with mandatory building energy code	nZEBs
2019	5150	3750	250
2025	3300	3100	3350
2030	250	5000	5250

(b)

Despite the main contribution in a modern building coming from active solar systems commonly integrated in a new building [8], passive solar strategies represent an energy reduction design feature with a high potential to dramatically reduce building energy demand [9] through energy promotion, buffering and prevention, influencing heating, cooling, ventilation loads, and building's system size. The main advantages of passive solutions include variety, versatility, simplicity, generally low initial and maintenance cost, and long lifetime [10] involving different building components, from walls to roofs. The effectiveness of passive solar solutions depends on a wide variety of design parameters [11]. Like all solar based devices, the performance is mainly related to the latitude, as it influences the average weather conditions and solar radiation on the collector surface. The maximum impact is expressed at medium latitudes, where cold winter temperature and long sunshine create the ideal conditions to exploit passive solutions. The benefit is also related to the interaction with the building and its users as heat delivery, shading, or ventilation must comply with the comfort requirements of the occupants: the magnitude of passive systems' gains becomes more relevant in highly energy consuming buildings like public or office ones.

The aim of the present paper is to provide a theoretical background of the parameters influencing the applicability and integrability of the various passive solar solutions in buildings and analyze how their synergy affects building energy performance. The scope of this analysis is to determine the impact of the different passive solutions while varying some controllable variables: the paper reports the first results of the investigation, where the most relevant influencing variables, such as latitude, orientation, building use, and shape factors, have been considered. Not much research has analyzed the topic from a general point of view in order to develop guidelines for passive solar application mainly providing a more qualitative outcome, describing the affecting parameters and defining rules of thumbs for sizing and related possible expected solar savings [12]. Other investigations focused on a specific topic, e.g., orientation, analyzing the effects by a deep theoretical background validated by measures and results from experimental activity [13]. The present paper follows a different approach as it intends to quantify the effects of passive solar strategies in terms of energy savings in different conditions: this will allow to determine the most suitable solutions in different cases and to define design guidelines according to the quantified impact of the variables analyzed.

The influencing parameters work together in defining the performance of a passive solar strategy: heating and cooling must be provided according to the living patterns that depend on building use, but also orientation that determines room distribution. Climate is linked to latitude and affects the energy balance and operativity by determining sun hours and lighting levels. Cost depends on the integrability in building design, thus, on the kind of structure and building shape: integration of massive systems in heavyweight buildings, for example, would be relatively easy, but could represent a high cost for lightweight ones. As all these elements work in synergy, a synergy of different strategies may achieve a better result in terms of energy saving. Recent investigations on passive solar systems and design integration in buildings have generally focused on spot solutions, either through a set of parametric simulation to verify and optimize the contribution of the asset to the building [14], or to characterize, evaluate, and maximize the performance of a specific device in individual applications [15]. The present paper intends to evaluate the efficacy of the synergy of different passive strategies and quantify the impact on building energy demand. Both direct, indirect, and isolated solutions can be designed to minimize each other's side effects. Direct systems that only provide heating, but faster than any other solution, are a perfect solution for a nonresidential building that needs to heat up quickly to reach indoor air comfort levels. On the other side, indirect systems can be exploited as solar chimneys to get ventilation and reduce discomfort due to overheating caused by wide glazed openings, or to provide cooling through roof ponds. In residential buildings, sunspaces can be used as a passive plenum of warmed air in winter, as a more manageable solution than high-inertia massive systems or direct systems with heat storage. By merging direct systems and sunspaces, more uniform heat delivery can be achieved, while a ventilated Trombe wall can be used in synergy with the sunspace to get higher indoor cross ventilation. A proper combination of direct, indirect, and shading systems can guarantee passive heating most of the winter season and minimize overheating in summer. Synergistic design must optimize the combination of different systems to minimize the need of HVAC contribution and promote the achievement of a zero-energy building. The paper analyzes the effect of synergistic, complementary strategies on a real office building, with different levels of intervention complexity, moving from simple, less expansive solutions, to composite and integrated ones, to quantify the reachable savings.

A brief description of the most common passive solutions is provided in Section 2, defining how extrinsic and intrinsic parameters work to determine the suitability of a specific passive system according to building kind. The influence of basic parameters affecting energy modelling of solar based devices is examined. Section 3 includes the methodology description and simulation settings of both the general investigation and the case study. Section 4 provides the results of the preliminary quantification of the impact of the different variables on building demand and then analyzes the outcomes of the case study of a public office building to evaluate the effect of possible combined passive strategies on a real building and determine the share of energy saving achievable through a synergistic design. Finally, after a brief discussion in Section 5 on the limit and contribution of the results achieved, the main conclusions are outlined, to show the potential of a structural, planned integration of passive solar systems in buildings in terms of energy demand reduction.

2. Passive Solar Systems

The action of a passive solar system for buildings can be summarized in a resistive and capacitive combined model, as the one described in Figure 1. According to this description, passive solar systems can be suitable both for increasing, during the summer period, or reducing, in the winter season, thermal resistance of direct solar radiation as well as to introduce appropriate capacitive systems able to store a part of solar radiation.

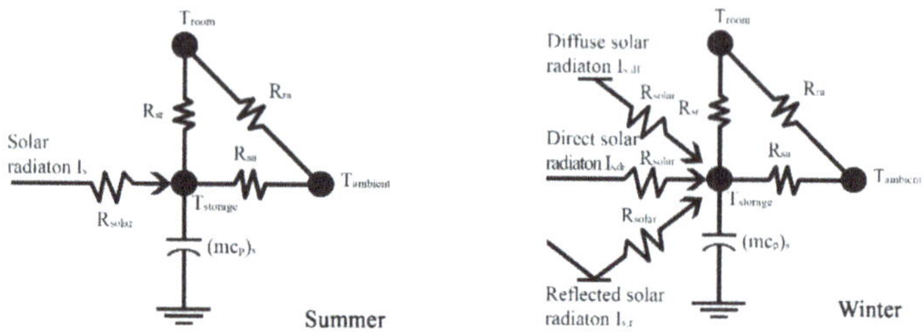

Figure 1. Conventional equivalent thermal circuit for passive solar heating building systems.

2.1. An Overview of the Common Solutions

Common categorization classifies passive solar systems into three classes. Direct gain systems collect solar energy through wide south-facing glazed envelopes: low construction cost, widely available and common technology represent the main advantages. No thermal storage is needed when these systems are sized to provide a small fraction of the heating load, but this could generate high indoor air temperature fluctuations. The most relevant disadvantage is related to glare discomfort or possible degradation from ultraviolet radiation [16]. The integration of insulation systems, thermal storage, external reflective surfaces, and shading systems (Figure 2) are the main possible improvements but can hugely impact the total cost. Indirect gain systems include a thermal mass placed between the solar collector and the indoor space that prevents indoor glare and ultraviolet degradation. A main disadvantage that can affect the system's lifespan is the maintenance, as proper access to the wall cavity for dust and condensation removal may be hard to achieve. In isolated gain systems, the collector and the storage are thermally isolated from the building: the main advantage is the independence of the system from the building.

(a) **(b)** **(c)** **(d)**

Figure 2. Example of a passive solar system, external adjustable blinds applied in the case study building: (**a**) external view, (**b**) interior view, (**c**) louvres detail, (**d**) connection detail.

Integration of shading systems and reflective surfaces are simple strategies for both new and existing buildings with a low economic impact. At low latitudes, blinds on the east and west windows reduce heat gain by around 20% more than on the south-faced windows, whose contribution increases with latitude [17]. Investigation carried out by Alhuwayil et al. [18] shows that self-shading envelope in a multistory building has a lower payback period, 2 years, than additional insulation and low-e windows, with an energy consumption decrease of 20%. In direct systems, U value and solar factors are the most relevant design parameters: for common glass, U values around 1–1.5 W/m^{2}K and a minimum solar factor of 0.3 is needed to have positive gain [19]. Low thermal resistance of the glass represents a critical weakness: use of triple glasses is a relevant energy saving measure in cold climates. Night insulation methods, such as roller-shade devices, are profitable but determine a not negligible extra cost.

In massive wall systems, the wall works as a storage mass buffering heat transfer. The greenhouse effect guaranteed by the external glazing improves the performance of the system. The thickness of the wall must be chosen according to the wall's material properties such as periodic and steady-state thermal transmittance, which define heat transfer rate, time lag, and decrement factor. Trombe wall adds convective heat exchange to the massive wall system by adding bottom and top openings to the wall: with the integration of outdoor vents, it can be used as a solar chimney [20]. This kind of configurations have good performance both in warm climates, cutting off 71.7% and 36.1% of heating and cooling demand, respectively, and in warm weather [14]. By exploiting fluid in the storage, water wall's heat transfer occurs by convection and the system is quite isothermal. This solution presents lower heat loss during night-time due to a lower surface temperature, and faster achievement of steady-state operating conditions, compared to other massive systems. Tests on a passive solar house with integrated water thermal storage walls showed an energy consumption reduction of 8.6% [21]. Roof pond systems, Figure 3, integrate water bags inside the roof structure: the large ceiling surfaces assure a more uniform heating distribution, but the effectiveness is limited to single-story buildings with an efficiency around 45%, as less than half of the collected heat is transferred downward. The high structural loads and the low efficiency at medium-high latitudes due to the low horizontal irradiance represent the main disadvantages. Results from a field study of a roof pond report a reduction of the indoor temperature swing around 1/4th [22]. The Barra–Costantini system is based on the collector loop configuration, but the warmed air flows inside a cavity in the ceiling and is finally released at the non-sun-facing rooms: this system guarantees a diffuse heat distribution and storage suitable for multistory buildings and is exploitable for building ventilation. The main disadvantage is the hard maintenance due to dust or condensation in the cavity. In temperate climates, this configuration can reach an annual heating energy saving percentage from 60 to 70%, but with a payback period of 25 years [23].

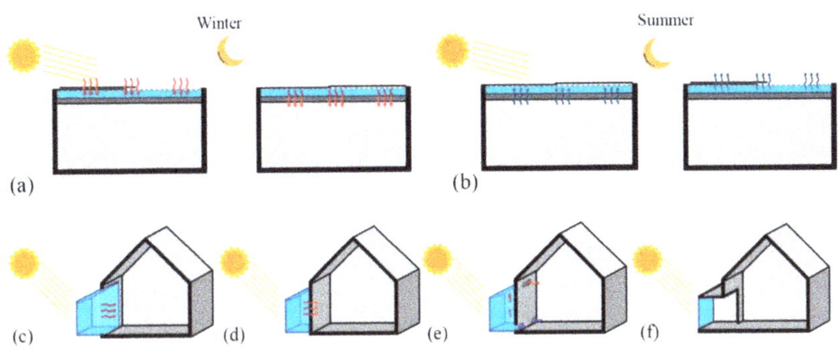

Figure 3. Water systems: (**a**) roof ponds winter mode (day and night), (**b**) roof ponds summer mode (day and night); sunspaces: (**c**) integrated, (**d**) attached, (**e**) attached ventilated, (**f**) sunporch.

Collector loops are based on convective heat exchange: these systems are perfectly suitable for lightweight insulated structures as easily integrable by adding an external glazing to the south façade. Convective loops also contribute to indoor ventilation. Likewise, thermosyphon systems consists of solar-glazed collectors with a black metal absorber. The most common application of these kinds of strategies in buildings is ventilated facades: in summer, the ventilated air gap and the reflective covering reduce the heat entering in the building, decreasing the cooling demand, while in cold weather conditions, the air gap reduces heat losses through the envelope, guaranteeing high air conditioning related savings. The system can also exploit the chimney effect to set up natural ventilation that helps in heat and moisture removal.

A sunspace layout is related to its use: as Figure 3 shows, integrated sunspaces act as large direct gain systems, while externally attached sunspaces work as indirect systems. A sunspace can be exploited to preheat external air [24], and works as a buffer zone. Although representing a possible additional living space is an advantage, the high construction cost and the compliance with local requirements limit sunspace implementation. The role of thermal storage is relevant: energy saving can be enhanced from 10–15% to 80% by adding a water-based storage system [15]. Achievable by installing windows or sheets in a south-facing balcony, sunporches represent a simple and cheaper solution for existing buildings, thus a valid energy retrofit strategy [25].

Building orientation and shape are basic passive strategies that highly affect both the base building energy load and the performance of other passive solutions by defining solar exposure. When designing a new building, difference in heating demand from a regular-squared to an irregular building can account for about 50%, while proper site orientation can save up to 40% [26].

2.2. Standard Modelling

As clearly evidenced in the models of Figure 1, the sizing of a distinct passive solar system with a defined objective appears to be a quite difficult task due to the synergistic effects of different parameters that act together in defining the effectiveness of a passive solar system. Therefore, energy performance can only be determined by considering this synergy. As Table 2 shows, parameters can be classified into extrinsic factors, external to the passive system, such as latitude or building shape, and intrinsic, such as collector inclination and orientation. These factors have a different impact on final performance: they act directly on the incident solar radiation or influence the operativity of the passive system and its integrability in building design. All these parameters contribute to the performance of passive solar systems at different stages of a project: while building orientation, shape, and position should be adapted to maximize the potential performance of passive solar systems, building use and latitude are usually given parameters, so they must be considered for a preliminary screening to limit the choice to the most suitable strategies. Then, another set of parameters allows to identify the most appropriate system for a specific building. The selection process is strictly country-dependent, as microclimatic conditions, architectural details, and related costs widely differ from place to place.

Table 2. Influence of design parameters.

Parameters	Kind	Affected Element	Influence on Design Process
Orientation	Intrinsic	Solar radiation	Design adaptation to maximize performance
Building shape	Extrinsic	Integrability	
Building positioning	Extrinsic	Solar radiation	
Latitude	Extrinsic	Solar radiation	Primary selection parameters
Building use	Extrinsic	Operativity	
Surface area to volume ratio	Extrinsic	Integrability	
N° of services involved	Intrinsic	Operativity	Secondary selection parameters
Overheating sensitivity	Intrinsic	Operativity	
Adaptability	Intrinsic	Integrability	
Range of the effect	Intrinsic	Operativity	
Cost of construction and maintenance	Intrinsic	Integrability	
Cost/performance ratio per unit area	Intrinsic	Integrability	Sizing parameters

The performance of each solar system is related to incident solar irradiance G (W/m^2), that depends on the different position of the site, latitude and longitude, on the different seasons, on the hour of the day, and on the position of the surface exposed to solar radiation. Solar irradiance can be evaluated with the following law [27]:

$$G_T = G_{on} \times \tau \times \cos\theta \tag{1}$$

with

$$G_{on} = G_{sc}\left(1 + 0.033\ \cos\left(\frac{360n}{365}\right)\right) \tag{2}$$

$$\begin{aligned} cosT = \ & sin\delta sin\phi cos\beta - sin\delta cos\phi sin\beta cos\gamma + cos\delta cos\phi cos\beta cos\omega + \\ & + cos\delta sin\phi sin\beta cos\gamma cos\omega + cos\delta sin\beta sin\gamma sin\omega \end{aligned} \tag{3}$$

$$\tau = e^{-b\lambda^{-\alpha} ma} \tag{4}$$

where G_T is global irradiance on a tilted surface, G_{sc} is the solar constant, θ is angle between surface normal and incident radiation (Figure 4a) and τ represents atmospheric attenuation of solar radiation. Figure 4b reports the maximum daily solar irradiance per month at different latitudes.

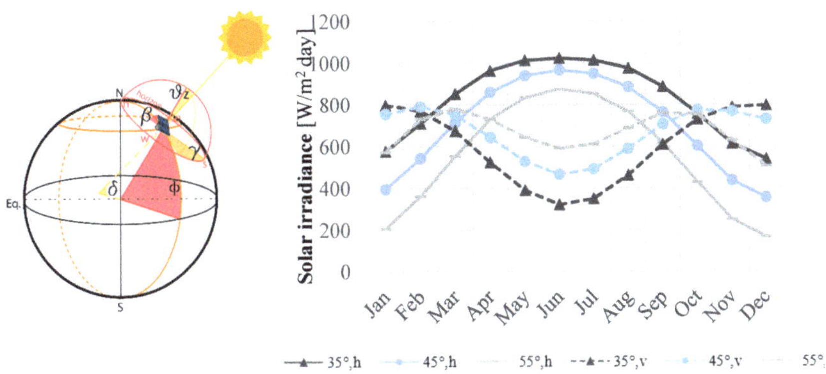

(**a**) (**b**)

Figure 4. (**a**) Angles related to sun position; (**b**) maximum daily solar irradiance per month at different latitudes on a horizontal, h, and a vertical, v, surface.

As Equation (3) states, latitude and surface inclination are the most impactful parameters to determine solar irradiance on a surface. Latitude affects the number of hours of sunlight on the surface. The best performance can be achieved at medium latitudes: at low ones, high solar radiation guarantees a high amount of collectable energy, but the short heating season makes the integration less cost-effective with a higher risk of overheating; at high latitudes, energy saving can be higher, but the lower solar radiation and shorter daytime limit the operativity, with a need of insulation, due to the lower average temperatures, that increases the total cost.

2.3. *Synergy of Influencing Parameters*

A generic model for passive solar systems, like the one in Figure 1, is internally ruled by three main resistances related to heat exchange among room, ambient, and storage, and characterized by a capacitive massive element. It is also externally stressed by the incident solar radiation on the collector. In summer, shading systems represent the most relevant resistance, while in winter, all kinds of solar radiation contribute with specific resistances. In building applications, materials and structures work in synergy with the elements previously analyzed to define passive system operativity. Glass properties are the first element that intervene. Once the incident solar irradiance has been determined with Equation (1), the energy transferred inside the building can be evaluated with the following balance:

$$H \times A = Q_t + Q_r + Q_{a,out} + Q_{a,in} \tag{5}$$

where H is global solar radiation on the window, A is windows surface, while subscript t represents the amount of energy transferred, r—the reflected share, a,out—the energy absorbed and retransferred outside, whereas a,in—the energy absorbed and transferred inside the building. The solar factor (g), the main parameter related to glass, can be defined as the percentage of solar energy incident on the glass that is transferred indoor, as showed in Equation (6).

$$g = \frac{Q_t + Q_{a,in}}{H \times A} \tag{6}$$

The solar factor depends on the composition and properties of the glass: from single to low emission glasses, the solar factor is almost halved, due to highly reduced transmittance.

The type of building structure and envelope materials affect the system's integrability [11] and the opportunity of exploiting building elements for heat storage with no extra cost. Integrated heat storage reduces indoor temperature fluctuation and extends the performance of a passive system. Density, specific heat capacity, and thermal diffusivity influence the rate of heat storage and the useful thickness involved in the process: high density and mass guarantee high heat capacity, while the higher the thermal conductivity, thus the diffusivity, the more readily the storage system absorbs heat and the deeper the material thickness involved is. The minimum thermal storage surface recommended by Balcomb [28] is six times the solar collector glazing area. Passive solar design solutions can be grouped according to the integrability, as shown in Figure 5, based on the kind of structure, light or heavyweight, and the suitability to new or existing buildings.

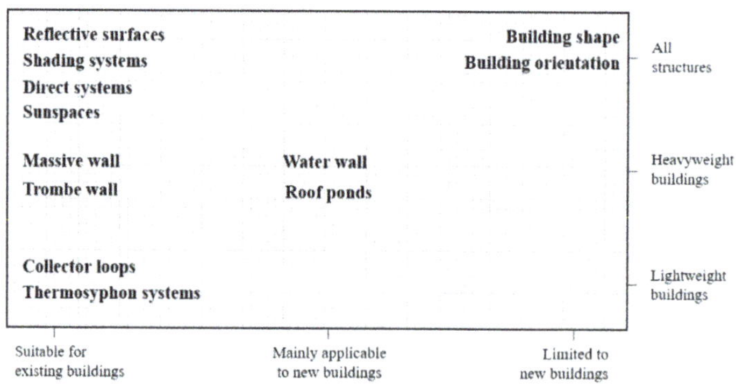

Figure 5. Classification of passive solar design strategies.

The building use and occupant's activity schedule are another set of parameters affecting the operation of a passive solar system: living pattern that varies according to the building use defines the period of comfort and that of possible setback. In office buildings, the main activity is focused on the middle of the day, from 7.00 a.m. to 5.00 p.m., while in residential buildings the time slot shifts to evening and morning. Educational buildings and offices, with a prevalent daytime use, are the most suitable for directly exploiting the heat provided by the sun. In residential buildings, the efficacy of passive solutions is related to heat storage effectiveness to deliver heat during night or cloudy days.

3. Performance Evaluation of Passive Solar Systems

3.1. General Investigation

As the first investigation, a preliminary quantification of the impact of different parameters on the three classes of passive solar systems has been carried out through a set of simulations. The energy analysis has been carried out with the EnergyPlus™ simulation tool developed by US DoE. In order to properly compare the outputs, the simulations considered a standard application of different strategies: the same construction and structures have been used, with no specific optimization for different solutions. The simulations have been run during the winter period to evaluate the impact of the solar strategies on the building's heating demand. Two kinds of buildings, an office complex and an apartment building, have been analyzed with different surface to volume ratio, 0.4 and 0.6, respectively, and different heating demand schedules: from 8.00 am to 7.00 pm on the weekdays and turned off in weekends for the office, while the opposite was set for the residential building. Another difference between residential and office building simulation occurred in schedule and magnitude of internal gains such as from equipment and appliances. Simulations have been set at Rome, 41° 50′, and Frankfurt, 50° 06′, to evaluate the incidence of latitude. The three strategies, a direct system, a ventilated Trombe wall, and a sunspace, have been simulated both with a north-south building axis and at 90° rotation (r), with east-west building axis. Both the models contained common building constructions with an external wall U-value of 0.40 W/m^{2}K. Base number and dimensions of the windows were set to comply with the minimum window-to-floor surface ratio of 1/8 required by Italian legislation: fenestration constructions have a U value of 1.76 W/m^{2}K and a g of 0.56. With these settings and no passive systems installed, a benchmark value of energy demand of both office and residential buildings has been evaluated as a reference for energy saving of the passive solar strategies. The reference values for Rome and Frankfurt houses are 160 and 290 kWh/m^2, respectively, while office building reference demand is 98 kWh/m^2 in Rome and 112 kWh/m^2 in Frankfurt.

Implementing the direct system, a wide glazed surface has been added on the south-faced wall or divided between the east and the west façade in the rotated case. Window extension was set at 15% of the floor, according to the common rule of thumb [29]: fenestration U-value and g are 1.06 W/m^{2}K and 0.6, respectively, while a 10 cm concrete slab provides the ground floor construction with a heat storage mass. This simulation is the only one with a specific low emission window kind, as direct systems installed are part of the thermal zone and directly affect building energy balance. Regarding the indirect and isolated systems, they belong to a different thermal zone, which are not considered in building energy balance. In the indirect system simulation, the Trombe wall area accounted for 0.2 m^2 per m^2 of floor area, with an air cavity thickness of 15 cm. The thickness of the external wall, 30 cm, already provided a good buffering of the heat wave. In this simulation, the same clear glass construction of other windows has been applied to the glazed face of the Trombe wall. Finally, the sunspace modelled in the isolated system was designed as a sunporch linked to the main building through a patio door: its envelope was glazed with the ordinary kind of window, except for the floor, which had the same construction as the indoor floor.

3.2. Analysis of a Case Study

After a general investigation to quantify the impact of the different variables on the performance of the three kinds of passive solar strategies, another simulation was run to define possible effects of the synergy of different solutions on a case study of a real building. The aim was to provide a quantification of the energy saving that could be obtained with the design of passive solutions in an existing building. The selection of the case study has been influenced by various elements, as the positive effects of passive solar systems can be evaluated referring to both specific civil buildings for residential or public use. Between residential and public buildings, the latter result to be more energy consuming: daily use and commonly wide glazed envelopes make them also the most suitable kind of buildings for the implementation of passive solar systems. The European Directive 2010/31/UE [30], in fact, firstly focused its attention on public buildings, anticipating the requirement for nearly zero-energy buildings (nZEB) in new construction. For these reasons, the case study is based on the office building of the Department of Energy, Systems, Territory and Constructions Engineering of the University of Pisa, shown in Figure 6. The building hosts offices and research laboratories on two floors: the technical room for the building system at the ground level and the adjacent building with mechanical laboratories were not included in the simulation. It develops along the south-north axis, and is characterized by a wide glazed envelope, mainly equally east- and west-exposed with a gross window-to-wall ratio of 57.65%.

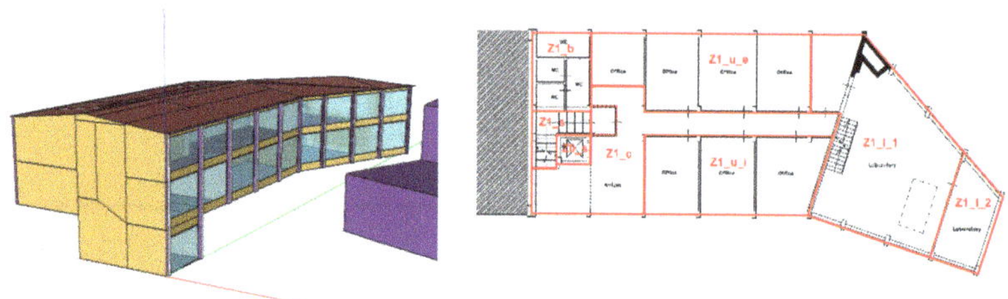

(**a**) (**b**)

Figure 6. Case study building: (**a**) energy model; (**b**) standard floor plan

For energy modeling, people, lights, and equipment loads were considered; no mechanical system has been modelled, only natural ventilation to simulate the passive behavior of the building and its energy demand. Table 3 lists the main construction involved and internal gains used.

Table 3. Model details.

Construction	Thermal Conductivity [W/m^{2}K]
External brick wall	1.05
External brickwork	0.94
External concrete wall	2.50
Office floor slab	1.07
Laboratories floor slab	1.47
Roof	0.428
Load	**Specific intensity [W/m^2]**
Lights	7.00
Electric equipment	5.00

Zone ideal loads have been evaluated with a heating setpoint at 20 °C and a cooling one at 26 °C. Natural ventilation through existing windows has been set according to the occupancy pattern and with an outdoor air temperature range between 15 and 30 °C.

The first solution investigated shading systems: given the high-glazed surface, this results as the easiest and most cost-efficient kind of passive solar solution. Fixed overhangs and external adjustable blinds, as the ones installed, were both simulated to compare their efficacy. The simulation has been set to activate the blind closing when 500 lux have been reached over the working plan or indoor air temperature rises above 25 °C, with a tilt angle configured to block direct beam solar radiation. A second passive strategy to reduce heat losses through the glazed envelope, focused on the integration of solar-control low-emissivity glasses. The windows chosen for this simulation had a solar control glass layer replacing common 4 mm clear glass, with a solar transmittance at normal incidence of 0.166: window U value moved from 2.7 to 1.65 W/m^{2}K while the solar heat gain coefficient decreased from 0.764 to 0.617. To fully exploit the potential of passive solar strategies, different solutions must be combined and balanced to take advantage of the energy savings coming from each system, as they can act on distinct services and in distinct times: for this reason, a combined solution, with external blinds and low-e glasses, has also been investigated. With reference to the models in Figure 1, the strategies that are discussed with this simulation focus on intercepting the incident solar radiation by acting on solar resistance. In terms of equivalent circuit, shading and glass element defined a rheostat with a variable resistance, as shown in Figure 7: both the integration of fixed overhangs and adjustable blinds or solar control glasses determined variable heat gain for the building.

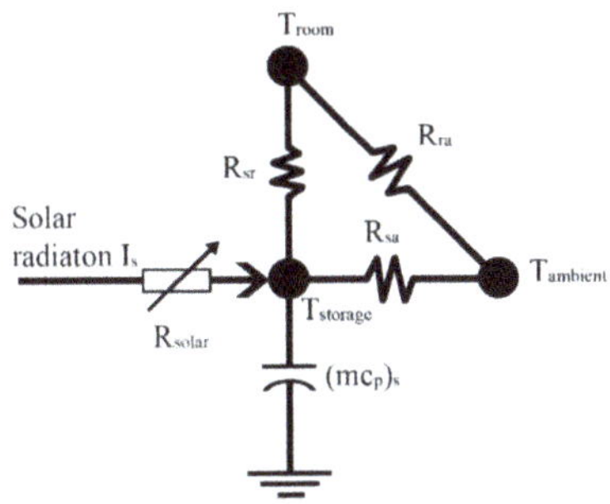

Figure 7. Conventional equivalent thermal circuit for passive strategies related to solar radiation resistance.

To get a deeper understanding of the combined effect of different solutions on the possible energy savings, a single module of the office building has been modelled, both in the basic case, in the combined case with a shading system activated, and in a Trombe wall configuration. In this last case, as shown in Figure 8, part of the external window has been replaced with the specified indirect system: while the direct gain system provides heating and lighting during the day, Trombe wall guarantees higher temperatures at night, leading to a lower demand in the morning when the HVAC system turns on. Moreover, the reduction of the glazed surface hugely reduces heat losses in winter or the unwanted heat gain in summer. The system has modeled both as an unventilated Trombe wall with a full conductive heat exchange, and as a naturally ventilated one. Due to the high window-to-wall ratio, 50% substitution rate has been chosen, as it does not highly impact the needed lighting level on the working plan.

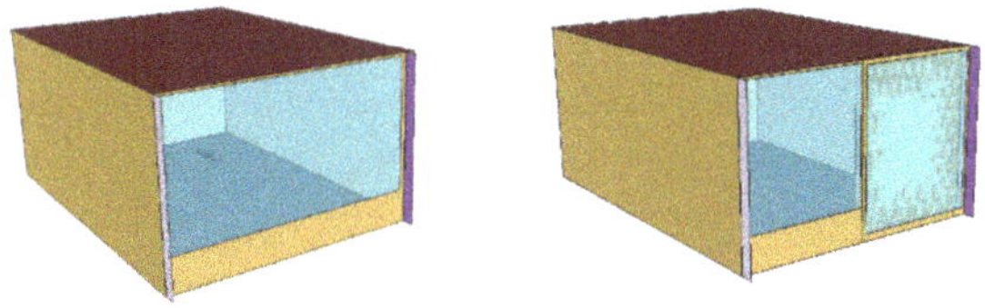

(a) (b)

Figure 8. Single office module: (**a**) standard configuration; (**b**) Trombe wall configuration.

4. Results

Results of the general investigation on the impact of variables over passive systems performance have been analyzed both in terms of absolute values and by comparing the percentages of energy saving achievable. Figure 9 shows the results of the simulation: in terms of energy demand, the difference between the two latitudes, as shown in Figure 9a,b, increased as the surface-to-volume ratio increases, moving from the office to the residential buildings. For the office building, direct systems proved to be the most profitable solution in terms of energy saving for both locations, as the highest savings, 28.2% and 38.5% for Rome and Frankfurt, respectively, occurred in the rotated direct system configuration, as described by Figure 9c. For the apartment building, Figure 9d, the Trombe wall rotated setting reached the highest saving in Frankfurt, 12.8%, and Rome, 22.3%. A very different trend can be defined for office and residential buildings in terms of passive system performance. Data collected confirmed that daily use office buildings prefer direct systems, as indirect and isolated systems showed much lower values. The trend highlighted an increase of the savings moving from Rome to Frankfurt, due to the higher demand of the Frankfurt facility that can better exploit solar radiation during its colder season, except for the sunspace simulation, and an increment from standard to rotated configuration certified the higher efficiency of south-faced passive solar systems. Quite the opposite trend can be detected in residential buildings, as the energy savings decreased from Rome to Frankfurt, and the variation among direct, indirect, and isolated systems percentages was lower. As in the previous analysis, however, the rotated configuration showed higher results for both locations, establishing again the higher performance of south-faced systems.

Figure 9. Heating energy demand in winter season: (**a**) office building; (**b**) apartment building; (**c**) percentage of energy saving in office building; (**d**) percentage of energy saving in residential building.

Even building destination and use affect the performance of passive solar systems. While at Rome latitudes the difference between the two kinds of buildings was quite constant, as it went to 7.6% in standard and 6.3% in rotated direct systems to 6.7% and 8.2% in indirect and 5.5% and 4.2% in isolated ones, it greatly varied for Frankfurt: direct systems efficacy changed around 28.6% in standard and 30.7% in rotated configuration due to the low performance in residential application, while the range of indirect and isolated systems was between 1–4%. Frankfurt values for residential buildings were generally lower than office ones, excluding the standard sunspace, while at Rome latitudes, except for the direct systems, residential values were higher than that of the office.

Latitude impact differs according to building destination: the office in Frankfurt showed higher energy saving potential than the Rome one, while the opposite occurred in the residential building analysis, where Rome percentages were higher than Frankfurt ones. The range of variation between the two latitudes was wider for direct systems, 12.5% and 10.3% in office or 8.5% and 14.1% in residential building, than in indirect and isolated solutions: for the north-south axis building, the gap was higher in the office application, around 12.5 percentage points in the direct system, 6 in indirect, and 1.2 in isolated, versus 8.5%, 2.8%, and 2.1% for residential building direct, indirect, and isolated solutions, respectively. In the rotated configuration, the opposite occurred, since the gap between Rome and Frankfurt in residential application was higher than the corresponding value for office buildings.

Looking at the orientation impact, south-facing systems showed the highest values, but they must deal with summer overheating risk. The sensibility decreases as the latitude increases, with Rome gaps between standard and rotated configurations higher than Frankfurt ones. The influence of different building axis depends on building use: for Frankfurt latitude, the impact of orientation moved from 7.7% of energy saving variation in direct systems, 7.9% in indirect, and 5.9% in isolated ones in the office study to lower values in the residential analysis, 5.6%, 6.1%, and 2.8% respectively. In Rome, the effect of orientation in residential application was higher than in office buildings, moving from 11.2% to 9.9% in direct systems and from 12.8% to 11.3% in indirect systems. Isolated sunspace solution represents an exception, as the residential gap of performance between standard and rotated configurations, 7.2%, was higher than the office one, 5.9%.

The lower residential building performance and the consequent lower sensitivity to orientation and latitude was attributable to the lack of optimization for the heat storage system. Analysing the results from the perspective of the systems, direct solutions showed the highest efficiency in reducing building energy demand, but they had higher sensitivity to orientation, 6 to 10 percentage points of variation, and latitude, 9–13%, than indirect or isolated solutions.

Moving to the case study of the office building, four different strategies have been studied and results are shown in Table 4. As a reference, the basic case, with no shading systems, determined an energy demand per total building area of 115.52 kWh/m^2, with an almost equal energy consumption for heating and cooling purposes, 47.59 kWh/m^2 and 45.66 kWh/m^2, respectively. Due to the east and west exposure, the effectiveness of fixed overhangs was negligible, while the impact of external blinds to reduce the summer undesired heat gain while guaranteeing proper lighting levels for office work was relevant. The results show that while fixed shadings led to a very low energy demand variation, with only 1.28% energy saving in cooling, blinds cut off one third of the cooling energy demand, with an impact of 12.58% on total energy consumption. This result highlights the hidden potential of passive solar solutions that can provide a reduction in embodied and operational energy requirements over a life cycle [31]. As shown in Figure 10, indoor air temperatures in the office zone in free running conditions rise compared to the basic case due to the higher thermal resistance of the windows: results show a reduction around 15% in heating demand, but the lower heat losses during the night generated a higher demand of energy in the cooling season. The analyzed synergy effect of the combined strategies individually, external blinds and low-e glasses, reduced by around a quarter the energy demand of the building, with a share for heating and cooling close to the highest values previously detected, 14.91% and 32.74%, respectively.

Table 4. Energy demand for run simulations.

Energy Demand		Basic Case [kWh/m^2]	Fixed Over-hang [kWh/m^2]	[%]	Adjustable Blinds [kWh/m^2]	[%]	Solar Control Glass [kWh/m^2]	[%]	Combined [kWh/m^2]	[%]
HVAC	Heating	47.59	47.81	0.46	48.20	1.29	39.75	−16.47	43.39	−14.91
	Cooling	45.66	45.07	−1.28	30.51	−33.17	46.26	1.32	33.05	−32.74
Energy Per Total Building Area		115.52	115.16	−0.32	100.99	−12.58	108.28	−6.27	93.48	−19.08

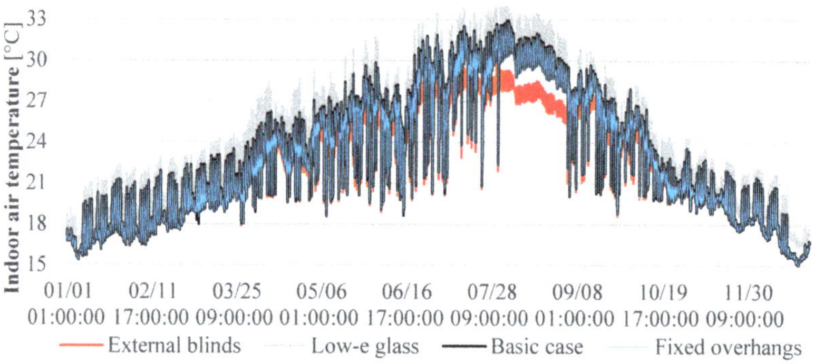

Figure 10. Indoor zone air temperature of the office zone in free running conditions.

Regarding the investigation of the office module, as in the previous analysis, the combined case showed a higher share of energy savings on the cooling side, with a reduction of 13.78% of energy demand, and zero effect on heating for the single office module. Replacing half of the wall-wide window with a Trombe wall drastically reduced the heating demand, more than half of the required energy in the basic case and cut off one third of the cooling demand. Figure 11a shows the results of the simulations. The lighting load moved from 2.28 kWh/m^2 in the full window configuration to just 2.65 kWh/m^2 in the half-window one. The naturally ventilated indirect system reached a higher share of energy savings for building heating, around 62%, but without a proper control system, its integration could be counterproductive: the cooling energy demand, indeed, rose from 3.21 to 3.91 kWh/m^2. This last configuration highlights the possible impact of an integrated passive solar design on building energy consumption, as the combined use of direct and indirect system could generate a reduction of 50% of a common office building's energy demand: a higher share could be achieved taking into account ventilation, by exploiting the Trombe wall cavity as a solar chimney during the cooling season. Assuming the office module as a functional unit, we compared a 10-units building, as a benchmark of a small office facility, with the department building previously analyzed, which contains 32 office zones. Results in Figure 11b confirm the impact on cooling demand of the combined case, whereas neglectable energy savings in terms of heating demand have been achieved, thus implying that the reduction of heating load, around 15%, achieved in the whole building analysis was mainly related to common spaces and wide laboratories.

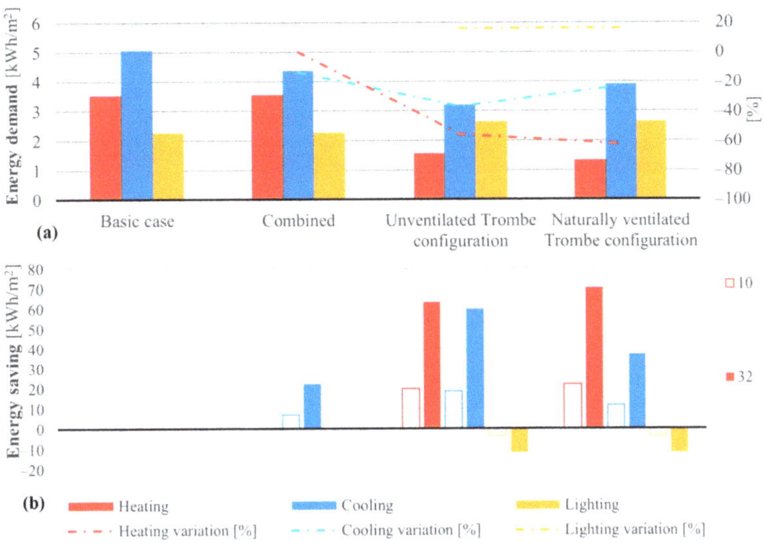

Figure 11. (**a**) energy demand of the office module and (**b**) energy saving of a 10- and 32-modules office building for different strategies.

The integration of the indirect system had a relevant impact on building energy saving: on heating demand, 65 to 70 kWh/m^2 can be saved yearly in unventilated and ventilated configuration, respectively, for the department building. On the cooling side, the unventilated solution reached the highest saving, around 20 kWh/m^2 for the small facility and 60 kWh/m^2 for the larger structure: the lower values of ventilated strategies are related to the lack of monitoring systems to prevent ventilation of warmed air from the Trombe wall cavity to cause overheating.

5. Discussion

As exposed in the paper, the interest on the promotion of passive solar system solutions is relevant to reduce the energy impact of civil/residential structures. Other studies investigated the impact of specific variables: Djordjević et al. [31] developed a mathematical model to estimate the indoor temperature trend in a building with a combined passive solar system made of a direct and an unventilated Trombe wall. Their results show that an 80° rotation of the room with a combined indirect passive system installed leads to indoor air temperatures about 40% lower than the temperatures in the south-oriented room. Results of Pathirana et al. [32] on shape factor influence indicate that the rectangular shape provides the higher thermal comfort, while window-to wall-ratio influence on thermal comfort is around 20–55%. Investigations on specific passive solar solutions report results close to the ones showed in this paper. As showed by Owrak et al. [15] the installation of a sunspace can lead to a 10–15% energy saving with no optimization: if a storage tank is added, the reduction can reach from at least 37% up to 87%. Other optimized studies show energy saving percentage even higher than the present paper's data: the integration of a south-faced sunporch in China led to a coal saving rate of 51.9% [33], while a Trombe wall-related saving on a residential building in Portugal was about 43% [20]. The difference in this case could both be due to the very low energy demand of the building, 14.95 kWh/m^2, and the lower indoor comfort temperature setpoint, 18 °C. Nevertheless, one of the major limits of the investigations on passive solar systems is the generalization of the results. Comparison of performance and savings from different research must take into account the boundary conditions of the passive strategies, such as latitude or kinds of building, which can hugely vary the impact on the total building energy demand. Due to the high number of variables defining the interaction of a building and a passive solar system, results from a specific case study or measurement from experimental activity cannot be generalized without proper assumptions. In addition, differently from common building systems, passive solutions deal with the building design at an architectural level to reach the high integration required, thus the sizing process and the optimization strategy of each solution must be tailored to the specific building. Thus, the main purpose of passive solar investigation should be to develop a general framework and guidelines for building design, or a protocol for the optimization process, as tools that can be adapted to different cases.

The sizing and design of a passive solar solution is not always simple. Moreover, the efficiency is related to multiple interconnected parameters, both on a technical and a climatic side, thus the evaluation and optimization of passive solar strategies needs to be carried out taking into accounts the different contribution of these factors to various elements: the energy balance, the cost of construction and maintenance, and the environmental impact. Another limit of passive solar analysis is proper technical and economic evaluation, usually based on a pure energy performance evaluation. As the synergy of intrinsic and extrinsic parameters defines the performance of passive solar solution, and the synergy of different systems can guarantee higher levels of energy savings as showed in the previous analysis, a synergy of evaluating factors determine the cost–benefit correlation to optimize this combination. Another element that can be considered in the analysis is the economic impact of the passive solar system, which is sometimes highly relevant. In general, the evaluation of the performance of a passive solar system can be carried out with different methods: an analysis exclusively based on energy balance or a life cycle analysis taking into account the environmental impact, with the related reduction of fuel consumption and greenhouses emissions, or economic analysis, to consider the payback time and the credit of original constructions elements replaced. A system design depends on the balance between the saving from energy consumption and the construction cost, during the lifespan. In these cases, as all the kinds of energy are equaled, the low available power and conversion efficiencies disadvantage the value of solar energy. In addition, this analysis is strictly country-dependent, as in some countries, the low price of energy, subsidized by the local government, results in even lower passive systems profitability [34]: in this perspective, the construction of passive solar systems strictly depends on the opportunity to exploit possible local benefits. As a result, a common economic analysis would not suggest the implementation of passive solar systems: to encourage the diffusion of this kind of strategies, the contribution to the reduction of greenhouse gas emissions or the energy degradation connected with the use of fossil fuels must constitute integral elements of the analysis. Possible alternatives may consist of two different approaches. A penalizing coefficient related to energy degradation could be introduced in the balance equation to promote the use of solar energy as a direct source: by counterbalancing the economic advantage of common energy vectors, this coefficient should enhance the use of natural energy. Otherwise, the evaluation must be based on a wider perspective through a multicriteria analysis that consider both technical, architectural, and energy stream-related issues.

6. Conclusions

Passive solar systems hide a huge potential as energy saving measures for buildings. The integration of different cooperating solutions in buildings is a relevant asset towards a nearly zero-energy building perspective. The design process is ruled by the synergy of various parameters, both intrinsic and extrinsic, that affect the performance and integrability of the passive solar system. In the paper, the authors have tried to define a general model for the definition of the performance of a passive solar system. The model is based on the combined consideration of direct and indirect thermal resistance and capacitive elements: it takes into account the massive capacitance of passive solar strategies in buildings and solar radiation variable resistances due to shading systems in summer and surrounding condition in winter, affecting the exploitation of diffuse and reflected solar radiation. As analyzed in the paper:

- passive solar systems' standard design depends both on intrinsic parameters such as latitude and orientation that influence the effectiveness of the solution, and on the correlation of extrinsic factors like building structure and destination, which affect their integration, heat storage, and cost, all operating in a synergistic way;
- as a result of the general analysis, direct systems showed the highest energy saving percentage for offices, 28.2% and 38.5% for Rome and Frankfurt, respectively, in the rotated configuration, while the indirect solution performed better in residential application, with the Trombe wall rotated setting that reached a saving of 12.8% in Frankfurt and 22.3% in Rome;
- building use also influenced the latitude impact on energy savings: while in offices the saving increased with the latitude, the opposite occurred in residential buildings. Direct systems showed higher sensitivity to latitude, around 10% variation of the energy saving, compared to the 2–6 percentage points in other cases, except for the indirect system in a residential building;
- the sensitivity to orientation variation decreased with increasing latitude, but generally south-oriented systems performed better than east-west ones: direct and indirect systems showed a higher range of variation, respectively, in Frankfurt for the office, 7.7%, and in Rome for the residential application, 12.8%;
- specific passive solar solutions on an existing public office building did not exceed 12%, while the synergy of different solutions showed a reduction of energy consumption of almost 20%. The application of synergistic passive solar systems can enhance the performance, saving more than 50% of the heating energy demand and up to 36% of the cooling load.

The integration of passive solar systems is hampered by performance evaluation: a simple energy balance disadvantages solar energy, characterized by low available power and low conversion efficiencies, while in an economic analysis, low energy costs, subsidized by some local government, makes them less cost-effective. Due to higher reliability and easier design, the integration of active devices has become a more common practice in the building market. In a long-term time frame, considering energy and carbon payback times, solar energy-based interventions, even for energy retrofit, proved to be profitable [7]. Future activity will continue both quantifying the impact of influencing parameters on passive solar system performance to develop a general framework useful to building design, analyzing combined solar systems. Research will focus on a proper way to evaluate the benefits of a passive solar design accounting on energy evaluation and energy degradation, by developing a multicriteria optimization function that considers each contribution of passive solar strategies.

Author Contributions: Conceptualization, A.F. and F.F.; methodology, A.F.; software, G.C.; writing—original draft preparation, G.C.; writing—review and editing, A.F. and F.F.; visualization and supervision, A.F. and F.F. All authors have read and agreed to the published version of the manuscript.

Funding: This research received no external funding.

Institutional Review Board Statement: Not applicable.

Informed Consent Statement: Not applicable.

Data Availability Statement: The data presented in this study are available in article.

Conflicts of Interest: The authors declare no conflict of interests.

Nomenclature

a	Single lumped wavelength exponent
β	Slope [°]
γ	Surface azimuth angle [°]
δ	Sun declination [°]
θ	Angle of incidence [°]
θz	Zenith angle [°]
l	Wavelength [mm]
t	Atmospheric attenuation
ϕ	Latitude [°]
ω	Hour angle [°]
A	Area [m^2]
b	Ångström turbidity coefficient
c_p	Specific heat capacity
g	Solar factor
G	Solar irradiance [W/m^2]
Gon	Extraterrestrial radiation incident on a normal surface [W/m^2]
Gsc	Solar constant [W/m^2]
G_T	Radiation incident on a tilted surface [W/m^2]
H	Global solar irradiation [kWh/m^2]
Is	Solar radiation
Is,df	Diffuse solar radiation
Is,dr	Direct solar radiation
Is,r	Reflected solar radiation
HVAC	Heat ventilation and air conditioning
m	Mass
m_a	Air mass
n	Day of the year
nZEB	Nearly zero-energy building
PV	Photovoltaic
Qa,in	Share of the absorbed energy transferred inside [kWh]
Qa,out	Share of the absorbed energy re-transferred outside [kWh]
Qr	Energy reflected [kWh]
Qt	Energy transferred [kWh]
Rra	Room-ambient resistance
Rsa	Storage-ambient resistance
Rsolar	Solar radiation resistance
Rsr	Storage-room resistance
Tambient	Ambient temperature
Troom	Room temperature
Tstorage	Storage temperature
U	Thermal transmittance [W/m^2K]
ZEB	Zero-energy building

References

1. IEA. World Energy Balances and Statistics. Available online: https://www.iea.org/subscribe-to-data-services/world-energy-balances-and-statistics (accessed on 28 November 2020).
2. Serghides, D.K.; Georgakis, C.G. The building envelope of Mediterranean houses: Optimization of mass and insulation. *J. Build. Phys.* **2012**, *36*, 83–98. [CrossRef]
3. Oldewurtel, F.; Parisio, A.; Jones, C.N.; Gyalistras, D.; Gwerder, M.; Stauch, V.; Lehmann, B.; Morari, M. Use of model predictive control and weather forecasts for energy efficient building climate control. *Energy Build.* **2012**, *45*, 15–27. [CrossRef]
4. IEA. *Tracking Buildings 2019*; IEA: Paris, France, 2019.
5. REN21. *Renewables 2019 Global Status Report*; REN21: Paris, France, 2019.
6. IEA (International Energy Agency). *Tracking Buildings 2020*; IEA: Paris, France, 2020.
7. Tsalikis, G.; Martinopoulos, G. Solar energy systems potential for nearly net zero energy residential buildings. *Sol. Energy* **2015**, *115*, 743–756. [CrossRef]
8. Franco, A.; Fantozzi, F. Experimental analysis of a self-consumption strategy for residential building: The integration of PV system and geothermal heat pump. *Renew. Energy* **2016**, *86*, 1075–1085. [CrossRef]

9. Fantozzi, F.; Filipeschi, S.; Mameli, M.; Nesi, S.; Cillari, G.; Mantelli, M.B.H.; Milanez, F.H. An Innovative Enhanced Wall to Reduce the Energy Demand in Buildings. *J. Phys. Conf. Ser.* **2017**, *796*, 012043. [CrossRef]
10. Balcomb, J.D.; Barley, D.; McFarland, R.; Perry, J., Jr.; Wray, W.; Noll, S. *Passive Solar Design Handbook—Volume Two of Two: Passive Solar Design Analysis*; U.S. Department of Energy: Washington, DC, USA, 1980.
11. Cillari, G.; Fantozzi, F.; Franco, A. Passive solar systems for buildings: Performance indicators analysis and guidelines for the design. *E3S Web Conf.* **2020**, *197*, 02008. [CrossRef]
12. Givoni, B. Building Design Guidelines for Solar Energy Technologies. Master's Thesis, University of California, Los Angles, CA, USA, 1989.
13. Olgyay, V. Bioclimatic orientation method for buildings. *Int. J. Biometeorol.* **1967**, *11*, 163–174. [CrossRef]
14. Bevilacqua, P.; Benevento, F.; Bruno, R.; Arcuri, N. Are Trombe walls suitable passive systems for the reduction of the yearly building energy requirements? *Energy* **2019**, *185*, 554–566. [CrossRef]
15. Owrak, M.; Aminy, M.; Jamal-Abad, M.T.; Dehghan, M. Experiments and simulations on the thermal performance of a sunspace attached to a room including heat-storing porous bed and water tanks. *Build. Environ.* **2015**, *92*, 142–151. [CrossRef]
16. Balcomb, J.D.; Barley, D.; McFarland, R.; Perry, J., Jr.; Wray, W.; Noll, S. *Passive Solar Design Handbook—Volume One of Two: Passive Solar Design Concepts*; U.S. Department of Energy: Washington, DC, USA, 1980.
17. Huang, Y.; Niu, J.; Chung, T. Comprehensive analysis on thermal and daylighting performance of glazing and shading designs on office building envelope in cooling-dominant climates. *Appl. Energy* **2014**, *134*, 215–228. [CrossRef]
18. Alhuwayil, W.K.; Abdul Mujeebu, M.; Algarny, A.M.M. Impact of external shading strategy on energy performance of multi-story hotel building in hot-humid climate. *Energy* **2019**, *169*, 1166–1174. [CrossRef]
19. Nielsen, T.R.; Duer, K.; Svendsen, S. Energy performance of glazings and windows. *Sol. Energy* **2001**, *69*, 137–143. [CrossRef]
20. Lohmann, V.; Santos, P. Trombe Wall Thermal Behavior and Energy Efficiency of a Light Steel Frame Compartment: Experimental and Numerical Assessments. *Energies* **2020**, *13*, 2744. [CrossRef]
21. Wang, W.; Tian, Z.; Ding, Y. Investigation on the influencing factors of energy consumption and thermal comfort for a passive solar house with water thermal storage wall. *Energy Build.* **2013**, *64*, 218–223. [CrossRef]
22. Krüger, E.; Fernandes, L.; Lange, S. Thermal performance of different configurations of a roof pond-based system for subtropical conditions. *Build. Environ.* **2016**, *107*, 90–98. [CrossRef]
23. Imessad, K.; Derradji, L.; Messaoudene, N.A.; Mokhtari, F.; Chenak, A.; Kharchi, R. Impact of passive cooling techniques on energy demand for residential buildings in a Mediterranean climate. *Renew. Energy* **2014**, *71*, 589–597. [CrossRef]
24. Suárez López, M.J.; Castro, S.S.; Manso, A.N.; Marigorta, E.B. Heat collection in an attached sunspace. *Renew. Energy* **2020**, *145*, 2144–2150. [CrossRef]
25. Allesina, G.; Ferrari, C.; Muscio, A.; Pedrazzi, S. Easy to implement ventilated sunspace for energy retrofit of condominium buildings with balconies. *Renew. Energy* **2019**, *141*, 541–548. [CrossRef]
26. Mokrzecka, M. Influence of building shape and orientation on heating demand: Simulations for student dormitories in temperate climate conditions. *E3S Web Conf.* **2018**, *44*, 00117. [CrossRef]
27. Duffie, J.A.; Beckman, W.A. *Solar Engineering of Thermal Processes*, 4th ed.; John Wiley & Sons Inc.: Hoboken, NJ, USA, 2012; ISBN 978-0470873663.
28. Balcomb, J.D. Heat storage and distribution inside passive solar buildings. In Proceedings of the 2nd International PLEA Conference on Passive and Low Energy Architecture, Crete, Greece, 28 June–1 July 1983.
29. Givoni, B. Characteristics, design implications, and applicability of passive solar heating systems for buildings. *Sol. Energy* **1991**, *47*, 425–435. [CrossRef]
30. EC. *Directive 2010/31/EU of the European Parliament and of the Council on the Energy Performance of Buildings*; European Commission: Belgium, Switzerland, 2010.
31. Djordjević, A.V.; Radosavljević, J.M.; Vukadinović, A.V.; Malenović Nikolić, J.R.; Bogdanović Protić, I.S. Estimation of Indoor Temperature for a Passive Solar Building with a Combined Passive Solar System. *J. Energy Eng.* **2017**, *143*, 04017008. [CrossRef]
32. Pathirana, S.; Rodrigo, A.; Halwatura, R. Effect of building shape, orientation, window to wall ratios and zones on energy efficiency and thermal comfort of naturally ventilated houses in tropical climate. *Int. J. Energy Environ. Eng.* **2019**, *10*, 107–120. [CrossRef]
33. Liu, Z.; Wu, D.; He, B.-J.; Wang, Q.; Yu, H.; Ma, W.; Jin, G. Evaluating potentials of passive solar heating renovation for the energy poverty alleviation of plateau areas in developing countries: A case study in rural Qinghai-Tibet Plateau, China. *Sol. Energy* **2019**, *187*, 95–107. [CrossRef]
34. Imessad, K.; Messaoudene, N.A.; Belhamel, M. Performances of the Barra–Costantini passive heating system under Algerian climate conditions. *Renew. Energy* **2004**, *29*, 357–367. [CrossRef]

applied sciences

MDPI

Article

Simulating and Comparing Different Vertical Greenery Systems Grouped into Categories Using EnergyPlus

Alberto Arenghi [1,*], Camilla Perra [2] and Marco Caffi [1,3]

1 Department of Civil, Architectural, Environmental Engineering and Mathematics (DICATAM), University of Brescia, 25123 Brescia, Italy; marco.caffi@unibs.it
2 Free Lance Architectural Engineer in Brescia, 25128 Brescia, Italy; camilla.perra1223@gmail.com
3 Green Building Council Italia, 38068 Rovereto, Italy
* Correspondence: alberto.arenghi@unibs.it; Tel.: +39-030-3711231

Abstract: The use of vegetation for the energy efficiency of buildings is an increasingly widespread practice; therefore, the possibility of representing these systems correctly with the use of simulation software is essential. VGS performances have been widely studied, but currently, the lack of a unique simulation method to assess the efficiency of different types of VGS and the absence of studies evaluating the performances of all the systems available, proposing simulation models for each of them, leads to an incomplete energy representation. The aim of this study is to achieve a consistent and complete simulation method, comparing the different systems' performances. The research is made up of five main steps. Firstly, a classification to group these systems into specific categories was proposed; secondly an in-depth analysis of existing literature was worked out to establish the methods used for different types of VGS. The study of plant physiology allowed the definition of an energy balance, which is valid for all vegetated surfaces; then, each category was associated to a mathematical formula and finally integrated into the EnergyPlus software. The results achieved for each model were compared evaluating two important parameters for the termohygrometric conditions control: outside walls face temperatures and operative temperatures.

Keywords: Vertical Greenery Systems (VGS); classification; comparison of different types of VGS; mathematical modeling and thermohygrometric analysis; EnergyPlus

Citation: Arenghi, A.; Perra, C.; Caffi, M. Simulating and Comparing Different Vertical Greenery Systems Grouped into Categories Using EnergyPlus. *Appl. Sci.* **2021**, *11*, 4802. https://doi.org/10.3390/app11114802

Academic Editors: Tiziana Poli, Luisa F. Cabeza, Andrea Giovanni Mainini, Gabriele Lobaccaro, Juan Diego Blanco Cadena and Mitja Košir

Received: 7 April 2021
Accepted: 20 May 2021
Published: 24 May 2021

1. Introduction

The search for technological systems and materials for urban and building regeneration, especially in recent years, has paid particular attention to sustainability. The integration between vegetation and buildings responds effectively to this request, providing benefits proven by several studies: the reduction of CO_2 emissions and the high temperatures that determine the "heat island" effect, modifying the urban microclimate [1–5], the increase in the quality of life [6,7] and the improvement of building hydrological [8] and thermal performances. Vertical Greenery Systems (VGS), especially, have a great potential, as buildings in urban areas develop mainly vertically, while Green Roofs (GR) only affect the higher floors in tall buildings [9,10]. Furthermore, GR do not include particularly diversified systems, a feature possessed by VGS instead, which in fact require different models.

Currently, the VGS are mainly included in projects carried out by internationally renowned architects, and have very high construction and maintenance costs. For example, Stefano Boeri designed a tower completely surrounded by trees in Milan [11] and Herzog & de Meuron integrated the Mur Vegetal, the system patented by the botanist Patrick Blanc [12], with the building envelope in their project at Caixa Forum in Madrid; the same system was chosen by Jean Nouvel for the Musée du Quai Branly in Paris. The design of systems with accessible costs, to allow a wider diffusion, can be encouraged through a more in-depth knowledge of VGS and their benefits on buildings. In fact, there are several technological solutions capable of providing different advantages, replacing traditional systems. The aim of the present study consists in identifying the contribution offered in the thermohygrometric field by the different types of VGS, translating the behavior of plants into a mathematical form that can be suitable for the EnergyPlus software. The validation of the methods proposed is obtained by a comparison between mathematical models and the experimental data, both as presented in the scientific literature.

2. Classification

The term VGS includes several systemic types that possess different characteristics and that are functional for specific plant varieties. Moreover, to build high-performance façades, it is necessary to act differently depending on the climate and the characteristics of the building [13], using the most suitable VGS. It is therefore necessary to provide a classification aimed at organizing plant verticalization into sets having common characteristics, which can be properly analyzed through accurate energy simulations and thus appreciate the differences. Some studies have proposed grouping, analyzing the scientific literature and systems characteristics, but without proposing a simulation method for each category [14–18]. On the contrary, in other research studies, such as the one conducted by Wong et al. [19], the field measurements for different VGS were analyzed but without proposing a classification.

In the present study, grouping into categories, based on the differences among the main technological solutions by considering both their geometrical and mathematical properties, is proposed. Each category has a different effect on the building envelope, based on its characteristics, through all or some of the following properties: shading, wind barrier, evaporation, transpiration and ventilated façade, as shown in Figure 1.

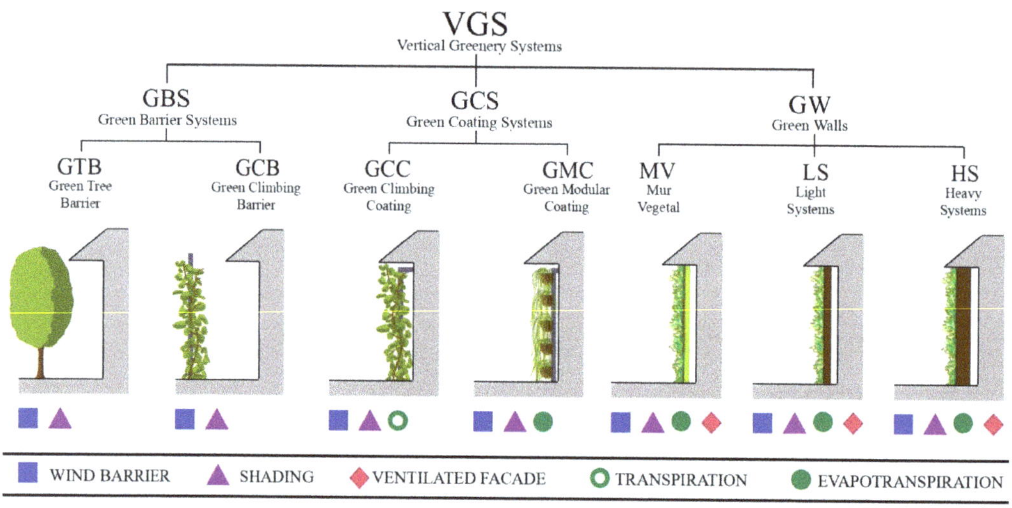

Figure 1. Proposed classification of VGS. Effects on buildings are specified for each category.

By re-elaborating the classification of VGS proposed by Bit [20], according to the energy benefits, three macro categories were defined: the Green Barrier Systems (GBS), the Green Coating Systems (GCS) and the Green Walls (GW). Each group interacts differently with the building envelope and therefore there are substantial differences in the approach to simulation. Further breaking down into sub-categories indicates differences in modeling or in the change of some parameters. GBS have the lowest level of integration with the building walls because there is no direct contact between them. The Green Barrier made up of trees (in the surroundings of the building or integrated with the structure) was defined as the Green Tree Barrier (GTB); systems composed by climbing species were defined as the Green Climbing Barrier (GCB).

The plant organisms that grow in adherence to a building closure or on a support system that forms a small gap with the wall were defined as the Green Coating Systems (GCS) divided into the Green Climbing Coating (GCC) and the Green Modular Coating (GMC). The main difference between these two sub-categories concerns the species of plants used: the first one is made up of climbing species and the second one consists in non-climbing species and requires particular modular systems in which a small amount of substrate is present. Finally, the Green Walls (GW) include systems which require greater technical specialization than the other categories. The technology package (which includes the plants and the substrate) is supported by a subsystem and forms a gap with the walls, thus behaving like a ventilated façade. In this case, three categories have been proposed: the Mur Vegetal (MV) that refers specifically to the system patented by the botanist Patrick Blanc [12] (in which the substrate is an inorganic fiber layer, and a PVC panel is present) and the Light Systems (LS) and the Heavy Systems (HS) that have a soil substrate and differ from each other in the soil thickness (less or greater than 15 cm, respectively).

3. Literature Review

In the literature, the topic of the Vertical Greenery Systems (VGS) has been faced with both experimental studies and through numerical modeling. Nevertheless, the authors have only considered one single type of VGS at a time in simulations, and therefore, a comparison among all of the different systems with a unique method is not available. The studies analyzed investigate different areas and types: the trees shading effect [21–27], the double skin façade with plants [28–32] and the Green Walls [3,8,33–36].

Table 1 shows the most significant studies for the model developed, which have the most compatible solutions with EnergyPlus, grouping the references according to the macro-categories identified.

Table 1. Studies about VGS divided into the three categories defined. The VGS effects on buildings considered in each study are specified.

Reference	Author/ Year	Considered Terms in Thermal Balance			Model/ Software Used	Category Analized/Method
		Shading	Convection	ETP		
				Green Barrier Systems (GBS)		
[32]	Stec et al. (2005)	YES	YES	YES	Simulink	**Green Climbing Barrier (GCB)** Radiative and convective thermal balances were mathematically represented through the definition of radiation absorption coefficient and convective heat transfer coefficient, respectively. The first one was found to be equal to 0.42, according to laboratory tests taken on the species *Hedera helix*. The second one was determined using the formula proposed by Stanghellini [37]: $Nu = 0.37\left(Gr + 6.417\,Re^2\right)^{0.25}$ Evapotranspiration (EPT) effect was represented through the FAO Penman Monteith formula [38]: $ET_0 = \frac{0.408\,\Delta*(R_n - G) + \gamma*\frac{900}{T_a+273.15}*U_2(e_s - e_a)}{\Delta+\gamma(1+0.34*U_2)}$
[23]	Hes et al. (2011)	YES	NOT	NOT	IES-VE	**Green Tree Barrier (GTB)** Two methods to simulate tree shading were proposed. The first one was modeling simplified objects, compatible with the shape of a tree. The changing density of the foliage was simulated through three models, with 0%, 50% and 100% of shading properties, a bare model, a perforated model and an opaque one, respectively. Instead, with the second method, an adjusted solar absorptance coefficient was introduced directly in the shaded wall properties. The adjustment considers a hypothetical shadowing, as shown in the following formula: *Adjusted Solar Absorptance* = 0.6 ∗ (1 − SC) Where SC is Shading Coefficient.

Table 1. *Cont.*

Reference	Author/Year	Considered Terms in Thermal Balance			Model/Software Used	Category Analized/Method
		Shading	Convection	ETP		
[31]	Larsen et al. (2014)	YES	YES	NOT	EnergyPlus	**Green Climbing Barrier (GCB)** Two different shading elements were used: Building Shading Object (BS) and Window Shading Device Object (WSD). To consider the wind barrier effect of the model, the convective heat exchange coefficient is calculated as proposed by [32]: $Nu = 0.37(Gr + 6.417\ Re^2)^{0.25}$ The modification of the view factor of the walls and windows is also proposed to take into account the diffused solar radiation reflected from the ground.
			Green Coating Systems (GCS)			
[28]	Yoshimi and Altan (2011)	YES	YES	NOT	ECOTECT	**Green Climbing Coating (GCC)** The evapotranspiration effect was greatly simplified, by modeling a layer of water vapor above the leaf layer that represents the water evaporated from the leaves. The thermal model of the species Hedera helix (Common Ivy) species is composed of 5 layers: water vapour, leaves, air gap, softwood (stem) and another air gap; the thermal properties are defined for each one. (ECOTECT Is not available any more as a stand alone tool)
			Green Walls (GW)			
[8]	Malys et al. (2014)	YES	YES	YES	Self-developed model in SOLENE-Microclimate	**Heavy System (HS)** The hydrothermal model, validated with field measurements, represents each layer as a node and each node is associated with a thermal or water balance. In addition, a parametric study was conducted: nine variable parameters were used to characterize the substrate and leaf layer. The quality of the combinations of parameters was determined by calculating the root of the mean square error.
[36]	Scarpa et al. (2014)	YES	YES	YES	Self-developed mathematical model	**Mur Vegetal (MV)** The Green Wall was divided into 11 thermal nodes; the behavior of each one was described with a thermal balance equation. The thickness of the cavity behind the system was also considered, evaluating its type (ventilated or not). The model shows a good agreement with the field measurement realized.
[34]	Dahana-yake e Chow (2017)	YES	YES	YES	Self-developed model integrated in EnergyPlus	**Light System (LS)** The heat balance equations of the model are based on the Green Roof module present in the EnergyPlus [39,40], on the FASST model proposed by Frankenstein and Koenig [41] and on the hydrothermal model validated by Malys et al. [8]. The correlation coefficients are close to unity, showing a good match between the simulation results and those of the experimental case.
[35]	Djedji et al. (2017)	YES	YES	YES	Self-developed model integrated in TRNSYS	**Heavy System (HS)** The mathematical model used was developed on the basis of a green roof model, analyzed by the authors in a previous study [42], analyzing solar and infrared radiation, convection and evapotranspiration. The validation of the analytical model of the green wall is based on the comparison of the external surface temperature of the substrate with that of a monitored green roof.

From Table 1, it is evident that all the mentioned studies consider just one type of VGS at time performing numerical analyses with diverse codes that evaluate different terms in thermal balance. Therefore, a consistent comparison among different VGS systems cannot be properly evaluated.

4. Methodology

The mathematical model used in this study was obtained by integrating some of the formulas analyzed in the literature review into EnergyPlus. The equations used have already been validated by previous research and are reported and specifically defined in Section 4.1. Then, according to the effects on buildings indicated in Figure 1, the components of the energy balance were integrated into the numerical code of the software, as presented in Section 4.2.

4.1. Study of Plant Physiology

The analysis of several studies and publications regarding plant physiology [37–51] allowed the determination of the energy balances that occur on the surface of plants.

Vegetated surfaces interact with the environment through energy exchange, basically involving four terms: net radiation R_n, sensible heat flux H, heat flux through soil G and latent heat flux L, that are linked together by the balance proposed by FAO [38]. In particular, G value is negligible on daily averages [34,38]. Moreover, the physical characteristics of the leaves influence the energy balance on vegetated surfaces [43], so the main parameters were analyzed.

4.1.1. Radiation Balance

Net radiation is the difference between incoming and outgoing radiation, considering both shortwave and longwave radiation [47]. It can be defined as the difference between net solar radiation and net longwave radiation:

$$R_n = R_{ns} - R_{nl} \tag{1}$$

and typically takes on a positive value during the day and a negative one at night. Net solar radiation R_{ns} is the percentage of radiation that is not reflected by the surface, and its value is therefore related to albedo by the following relation, as leaves are primarily absorptive [47]:

$$R_{ns} = (1 - \alpha) * R_s \tag{2}$$

where albedo (α) depends on different variables like density, thickness and the color of leaves [43].

The solar radiation transmitted to soil, beyond the foliage layer, was calculated according to the following equation:

$$R_{ns} = (1 - \alpha) * R_s * e^{-k_s * LAI} \tag{3}$$

that considers the geometrical properties of canopy.

4.1.2. Sensible Heat Flux

The energy exchange on a surface, due to sensible heat flux, is perceived as an increase or decrease in surface temperature. Heat transport occurs through convection, and after a comparison (using the EnergyPlus simulation program) with other studies [37,43,52], the mathematical formula proposed by Stec et al. [32] appears as the one that best fits what is stated by FAO [38]:

$$Nu = 0.37 \left(Gr + 6.417\ Re^2\right)^{0.25} \tag{4}$$

where Nu, Gr and Re are, respectively, the Nusselt, the Grashof and the Reynolds number. Thus, the convective heat transfer coefficient h can be expressed as follows:

$$h = \frac{k}{Nu \cdot L} \tag{5}$$

4.1.3. Latent Heat Flux

Latent heat flux allows a substance to affect a state change, by adding heat, without perceiving an increase in the temperature of the substance itself [47]. This type of heat exchange characterizes the vegetated surfaces: water vaporizes thanks to the transpiration from the leaves and evaporation from the ground. The whole process is known as evapotranspiration and results in a cooling effect of the surroundings [8,34,41]. Mass and energy flows are related by the following equation:

$$L = \lambda * ET \quad (6)$$

where L is the latent heat flux, λ is the latent heat of vaporization, and ET_0 is the evapotranspiration value, starting from the FAO Penman–Monteith equation [38], adapted to VGS by Davis and Hirmer [49]:

$$ET_0 = \frac{0.408\,\Delta * (R_n - G) + \gamma * \frac{900}{T_a + 273.15} * U_2(e_s - e_a)}{\Delta + \gamma(1 + 0.34 * U_2)} \quad (7)$$

To obtain the evapotranspiration value ET for a particular plant species, ET_0 value is multiplied by coefficient K_C, depending on the plant species [36,38].

4.1.4. Physical Parameters of Leaves

The physical characteristics of plants that influence the energy balance with the environment are essentially the LAI (Leaf Area Index) [53,54] and extinction coefficient k_s that depends on the spatial distribution of leaves and on the angle between the leaves and soil. Both these parameters contribute to the result of Equation (3).

4.2. Integration in the EnergyPlus

The use of a software that operates in dynamic regime is necessary to guarantee a simulation as close as possible to reality related to plant organisms, which are a living component of the building envelope and therefore respond to environmental conditions in a very complex way. The use of advanced programming language, the EnergyPlus runtime language (Erl) combined with the Energy Management System (EMS), and other functions integrated into the software make it possible to simulate every component of the energy balance on vegetated surfaces. EnergyPlus provides a model (EcoRoof Model [31]) to simulate greenery systems, but it works only for horizontal vegetated surfaces, and it does not consider the change of the emissivity values of foliage and soil and of the convective heat transfer coefficient, both influenced by the inclination of the analyzed surface.

4.2.1. The EMS Settings—The Solar Absorptance

The leaves can absorb long-wave radiation almost completely, so they are considered as perfect emitters [47]; consequently, the thermal absorptance value is equal to 1. Instead, the solar absorptance coefficient (ζ) depends on the specie's physical characteristics, according to [43]:

$$\zeta = 1 - e^{-k_s * LAI} - \alpha \quad (8)$$

and for each parameter a variation range was defined (Table 2).

Table 2. Variation range for the variable parameters that influence solar absorptance coefficient and related references.

Variable Parameters	Variation Range	Reference
k_s	0.16–1	[51]
LAI	0.30–5	[53]
α	0.20–0.25	[38]

The reference species, *Bergenia Crassifolia* and *Geranium Macrorrhizum* were associated with a LAI equal to 2.24 and 1.89, respectively. These values were obtained by the comparison with similar species analyzed by Candelari [55].

Equation (5) was integrated in the EMS, setting the solar absorptance coefficient as an actuator (i.e., value affected by programming) and the other variables in the formula as sensors (i.e., input values).

4.2.2. The EMS Settings—The Sensible Heat Flux

Sensible heat flux related to VGS was simulated using EMS, this time setting the convective heat transfer coefficient as an actuator. Analyzing different methods proposed in the literature [23,24,43,52], and comparing the daily sensible heat flux values obtained with the values of the other terms involved in the energy balance [38], the formula proposed by Stec et al. [32] resulted as the most suitable as reported above in Equation (4).

4.2.3. The Advanced Settings—The Latent Heat Flux

The EnergyPlus, among its advanced settings, provides the possibility to add another term to the traditional energy balance on the outer or inner surface of a wall. This function was to simulate the evapotranspiration effect. The hourly latent heat flux was calculated according to Equation (7). Different values of coefficient K_C were used to represent different plant species [34] to compare the related *ET* term.

4.2.4. The Advanced Settings—The Ventilated Façade

The Green Walls can be compared to a ventilated façade, due to the gap between the continuous surface of these systems and the building envelope. This effect was never considered in developing GW's models, despite having a significant effect in summer energy efficiency. Analyzing some studies about these systems [56–58] and their simulation methods in the EnergyPlus [39], the most suitable model was defined. The ventilated cavity was modeled as a separate thermal zone, and the setting *Zone Ventilation: Wind and Stack Open Area* was applied. This function allows the automatic calculation of the convective heat fluxes in the gap, considering wind and stack effects.

4.2.5. The Layers' Properties

The Green Walls were represented as a simplified model, composed of two layers: foliage and substrate, spaced from the wall by an air gap. The foliage properties are different for each plant species, so two reference species have been analyzed: *Geranium Machrorrhizum*, associated with Light Systems because of its superficial roots, and *Bergenia Crassifolia,* linked to Heavy Systems because of the roots developing in depth. Thermal and physical properties were attributed to these layers comparing the results obtained by Jayalakshmy and Philip [45] and Merzlyak et al. [46] for some plant species' leaves. Table 3 shows the properties obtained for the two reference species.

The study of the thermophysical properties of the soil, whose contribution is substantial due to its thermal mass, was carried out after a literature review [59–65]. Three types of soil have been analyzed in this study: sandy loam, loam and clay loam, suitable for cultivation, and for each one, a defined density was fixed, thermal conductivity and specific heat. For the two species selected, the composition of soil was chosen according to the plant considered, to ensure its proper growth. The properties of soil are related to its water content, that must be included in a specific range to allow the plants to fulfill their vital functions: between 5% and 10% for sandy soils and between 40% and 50% for clay soils, as stated by Pitts [65]. These values were used to find a linear correlation between sandy and clay soils, to identify the correct saturation percentage of the simulated soils as shown in Figure 2.

Table 3. Variation range for the variable parameters that influence solar absorptance coefficient and related references.

Properties	Units	*Bergenia Crassifolia*	*Geranium Macrorrhizum*
Single leaf thickness	$m \cdot 10^{-3}$	0.25	0.21
Canopy layer thickness	$m \cdot 10^{-3}$	5	5
Thermal conductivity	$W \cdot m^{-1} \cdot K^{-1}$	0.34	0.35
Density	$kg \cdot m^{-3}$	656	627
Specific heat	$J \cdot kg^{-1} \cdot K^{-1}$	2252	2232
Thermal effusivity	$W \cdot s^{\frac{1}{2}} \cdot m^{-2} \cdot K^{-1}$	714	702
Thermal diffusivity	$m^2 \cdot s^{-1} \cdot 10^{-6}$	0.23	0.25

Other properties related to each soil, were obtained analyzing the studies carried out by Clauser and Huenges [61], Abu-Hamdeh and Reeder [59] and Monteith and Unsworth [50]; the results are shown in Table 4.

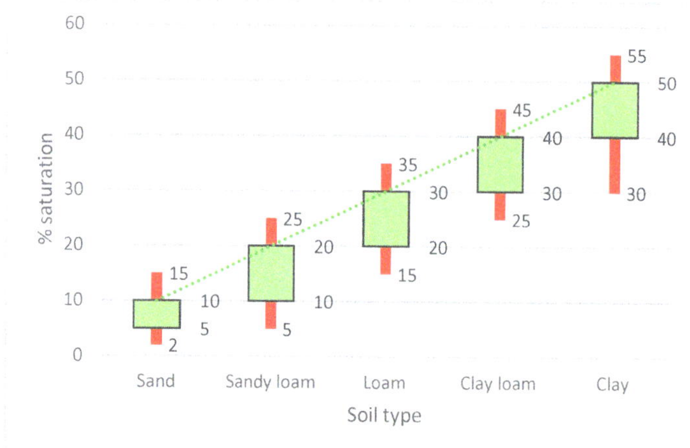

Figure 2. The optimal range of saturation percentage for each type of soil, to guarantee plant survival, is represented by the rectangular boxes; instead, the values indicated as a line generate stress in plants.

Table 4. Properties of cultivated soil (sandy loam, loam, clay loam) related to saturation percentage of pores.

Properties	Units	*Bergenia Crassifolia*				*Geranium Macrorrhizum*	
		Sandy loam		*Loam*		*Clay loam*	
% Saturation	-	min 10	max 20	min 20	max 30	min 30	max 40
Density	$\frac{kg}{m^3}$	1700	1800	1800	1900	1900	2000
Thermal conductivity	$\frac{W}{m \cdot K}$	1.26	2.03	1.02	1.22	1.34	1.45
Specific heat	$\frac{J}{kg \cdot K}$	1125	1328	1080	1209	1324	1465

4.3. Mathematical and Geometrical Models

Each category above defined was simulated in the EnergyPlus using simple models, by adopting small cubic samples (3 × 3 × 3 m) with 15 cm concrete walls. The concrete cubic sample was used as reference model in the comparisons with each category. The analysis of these models allowed the comparison between VGS categories and the execution of a large number of tests, avoiding the simulation of complex buildings. The dynamic simulations were performed in a free running mode, without plants. The meteorological data used refer to the Brescia–Ghedi weather file [66].

From the geometric point of view, GBS were modeled as shading objects: simplified trees for GTB and rectangular surfaces for GCB, both placed at a distance of 2 m from the walls of the sample. A transmittance schedule was associated with each shading object, to represent the different transmission of solar radiation in each season.

GCS were simulated adding two layers on the outer sample's surfaces: an anti-root membrane and a vegetated layer. The subcategory GMC also has a small layer of substrate, but its contribution can be negligible in terms of thermal insulation. In addition to the modification of the layers' properties, the behavior of the plants was reproduced through the modification of the solar absorptance, of the sensible heat flux and by adding the latent heat contribution. The latter, for the GCC only results from foliage transpiration, while for the GMC results from the whole process of evapotranspiration, due to the presence of soil.

The GW have the common feature of functioning as ventilated façades, and they were simulated as described in 4.2.4. The MV has a defined stratigraphy, and the outer layer was simulated respecting the characteristics of the system patented by Patrick Blanc [12]. The HS and LS can be composed of different types of plants and substrates, which determine a great variability in the results. Therefore, a parametric study was carried out to identify the most effective combination of variables in improving the thermohygrometric behavior of the sample. The parameters were analyzed one at a time, keeping the other characteristics of the system unchanged and applying only the setting concerning that specific variable. The variation of the extinction coefficient k_s, the thickness of the air gap and the type of soil appears to be almost irrelevant in influencing the external surface temperatures; while the variation of LAI, as shown in Figure 3, influences the external surface temperatures which is reduced by about 2 °C in summer for a LAI equal to 5 compared to a LAI equal to 1, comparable to the results obtained by [43].

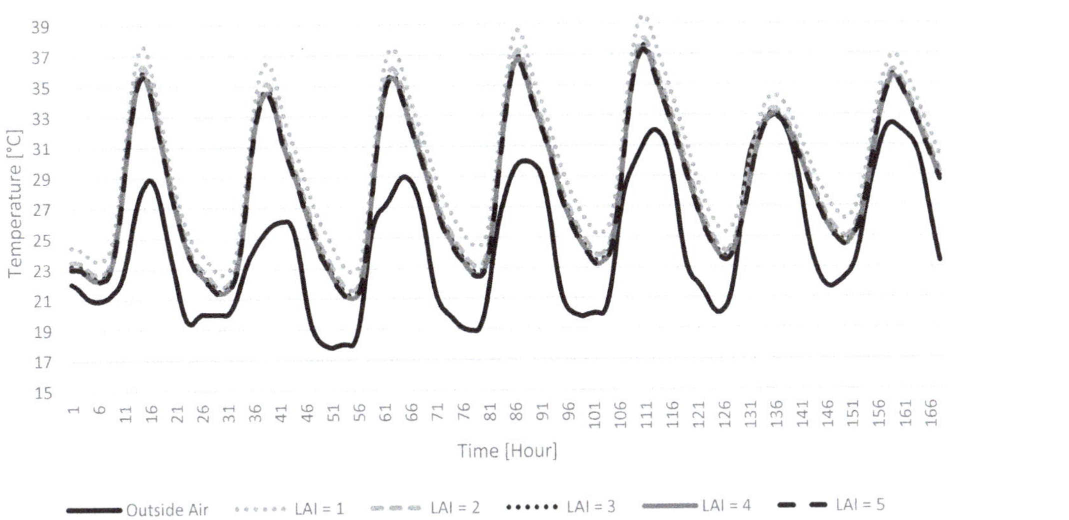

Figure 3. Effect of LAI on the surface external temperature of the walls. Model properties: Substrate in Loam (9 cm); k_s = 0.70; air gap thickness = 4 cm.

5. Results and Discussion

The overall assessment of the thermohygrometric benefits on the simulated model was made by comparing the results obtained for the VGS models and the reference concrete model, analyzing external surface temperatures and operative temperatures, both in summer and winter. All the graphs refer to the south-facing walls, and the physical parameters used are consistent with the species suitable for the different systems, as indicated above and in [67].

The trends of outside face temperature for each category and for the concrete model, compared with outside air temperature, are shown in Figures 4 and 5, in summer and winter conditions, respectively.

The summer chart shows the data obtained during the simulation performed in the last week of July (25-31/07), the hottest day of the summer according to the meteorological data analysis [66].

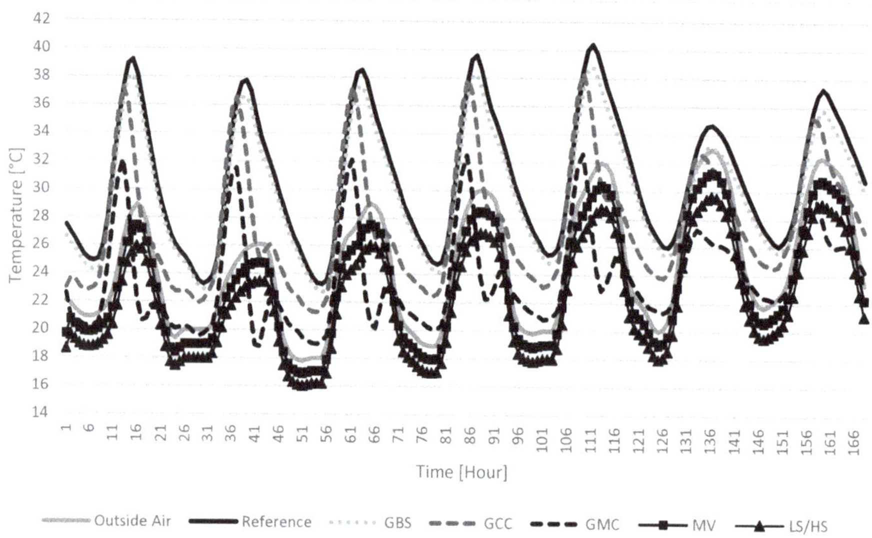

Figure 4. Summer (25-31/07). Comparison between the external surface temperatures of the south wall for each category, the concrete model and outside air.

The GBS reduce concrete surface outside face temperatures by about 1.5 °C, due to the shading effect of trees and climbing species. There are no significant differences in the trends of the GTB and GCB models as both act only through shading and do not influence the temperature of the wall with the transpiration process. Hes [23] and Stec et al. [32] instead reported up to a 25% reduction in the outside surface temperatures for the GBS. The Different results could be due to the different height and position of the trees or climbing species or to a greater window area of the analyzed model. The GCC reduce the temperature of the concrete model on average by about 5 °C, confirmed by the empirical results obtained by [68], and the GMC by about 6 °C, due to the presence of soil, which also adds the evaporation process. The GW systems remarkably reduce the surface temperatures, also lower than the ones of outside air. The most performing systems are the LS and HS that reduce the maximum temperature of the concrete model up to 13 °C, included in the range of values obtained by Mazzali et al. [69]. These systems record higher evaporation values from the soil, having thick and continuous layers.

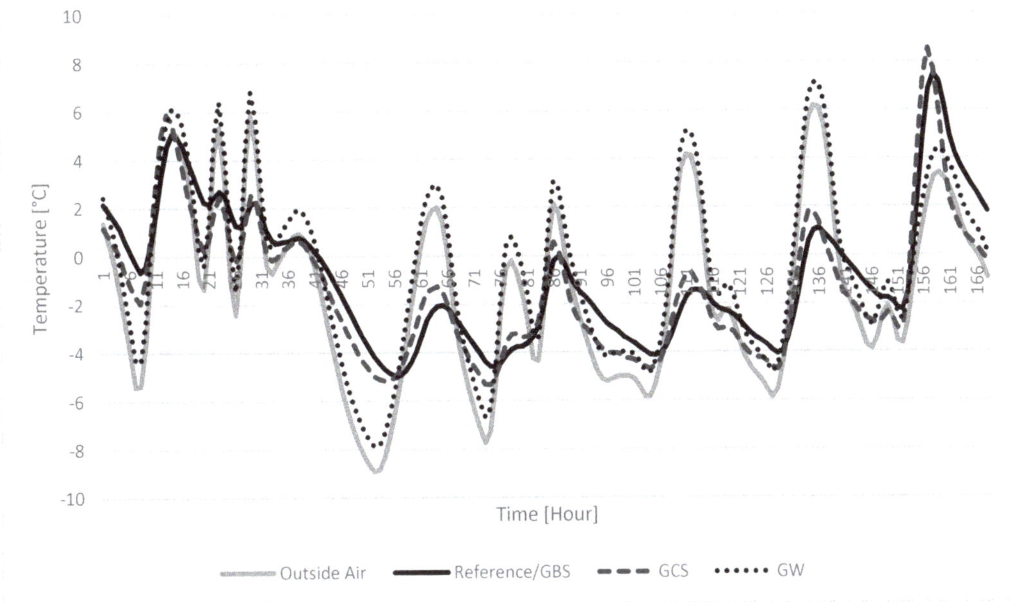

Figure 5. Winter (01-07/12). Comparison between the external surface temperatures of the south wall for each category, the concrete model and outside air.

The winter chart shows the data obtained during the simulation performed in the first week of December (01-07/12), the coldest day of the year. The trend of the external surface temperature of the GBS coincides with that of the bare wall, as in this season they only act as wind barriers; the latter effect could be more relevant with denser vegetation. The GCS increase the surface temperature by about 1 °C on average, during the day, similarly to the results obtained by [70], as they have a better behavior as wind barriers. The GW, instead, follow the trend of the outside air temperature, with negligible variations, increasing the external surface temperature up to 8 °C higher than the reference. No simulations were found in the winter period comparable with those carried out.

The influence of the different models on the summer and winter operative temperature is shown in Figures 6 and 7, respectively.

In summer (Table 5), the HS have the most significant impact in lowering the operative temperature: up to 9 °C less than the concrete model. These results derive from the higher thermal mass of the soil and from a higher LAI value than the other systems; in fact, the HS, thanks to the high thickness of the substrate, can host larger species and therefore greater foliar density. The LS, MV and GMC have similar trends, with a reduction in peak temperatures of between 4 and 8 °C, similarly to the field measurement on a living wall made by [71]. In these cases, the thermal mass of the soil, being less thick, has less effect on lowering temperatures. The GCC lower the temperatures between 3 and 6 °C, decreasing temperatures thanks to the transpiration of the foliage and shading. The GBS generally do not show a significant lowering in air temperature [72,73], even if in two of the days analyzed, there is a reduction in temperature of up to 4 °C, due to shading.

The best performing system in winter (Table 6) is the MV, which has 15 cm thick PVC panels in its stratigraphy, with a thermal conductivity lower than that of the soil, which allows a better insulation than the other systems; the operative temperature increases up to 4 °C compared to the bare wall. The LS and HS increase the operative temperature by about 1 and 2 °C, respectively, compared to the concrete model. In these systems, the soil provides thermal resistance but, being almost always wet, does not have a low enough thermal conductivity to be insulating. The GCS have a very similar trend to that of the concrete model and the GBS trend coincides with the latter, as they shelter from the cold winter wind but do not isolate the wall.

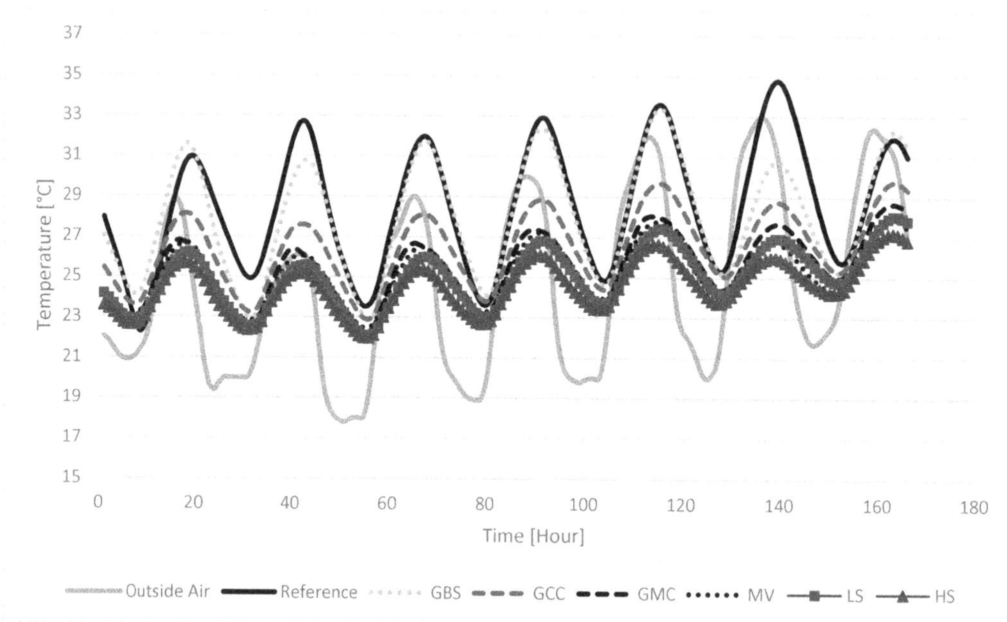

Figure 6. Summer. Comparison between the operative temperatures obtained for each category and outside air.

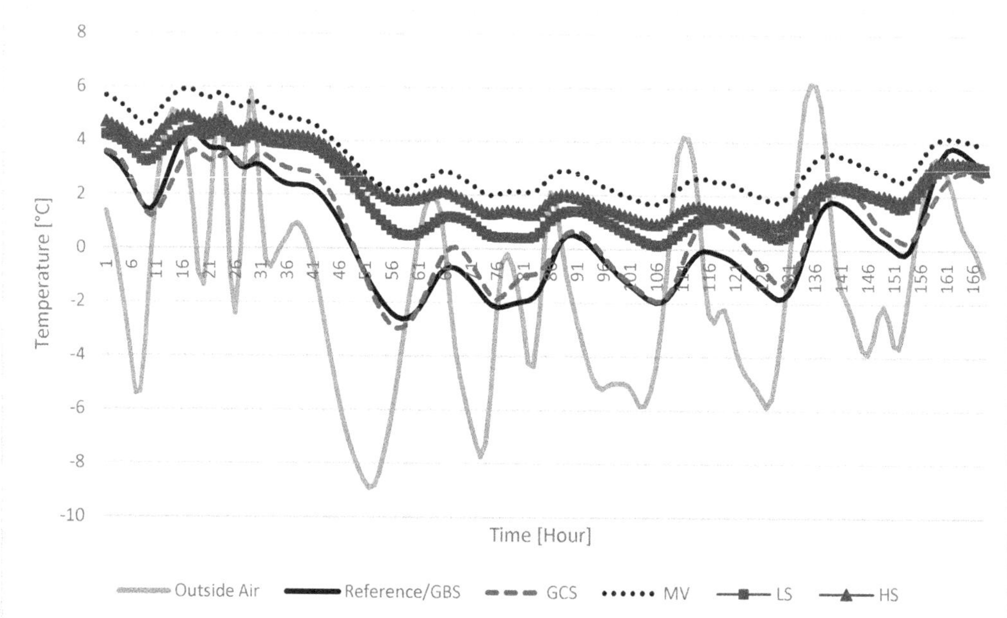

Figure 7. Winter. Comparison between the operative temperatures obtained for each category and outside air.

Table 5. Summer. Comparison of the results obtained on the south wall of the sample during the daytime and evaluation of the main variables involved for each category.

VGS	Maximum Reduction of the External Surface Temperature Compared to the Reference Sample	Maximum Reduction of the Operative Temperature Compared to the Reference Sample	Main Effects Involved
GTB	−1.5 °C (−4%)	−4 °C (−11%)	Shading
GCB	−1.5 °C (−4%)	−4 °C (−11%)	Shading
GCC	−5 °C (−12%)	−6 °C (−17%)	Shading/transpiration
GMC	−6 °C (−15%)	−7.5 °C (−21%)	Shading/evapotranspiration
MV	−10 °C (−26%)	−8 °C (−23%)	Evapotranspiration/Ventilated façade
LS	−13 °C (−32.5%)	−8 °C (−23%)	Evapotranspiration/Ventilated façade
HS	−13 °C (−32.5%)	−9 °C (−26%)	Evapotranspiration/Ventilated façade

Table 6. Winter. Comparison of the results obtained on the south wall of the sample during the daytime and evaluation of the variables involved for each category.

VGS	Maximum Increase of the External Surface Temperature Compared to the Reference Sample	Maximum Increase of the Operative Temperature Compared to the Reference Sample	Main Effects Involved
GTB	±0 °C (±0%)	±0 °C (±0%)	Wind barrier
GCB	±0 °C (±0%)	±0 °C (±0%)	Wind barrier
GCC	+1 °C (+13%)	+1 °C (+67%)	Wind barrier
GMC	+1 °C (+13%)	+1 °C (+67%)	Wind barrier
MV	+8 °C (+800%)	+4 °C (+200%)	PVC panel in the stratigraphy
LS	+8 °C (+800%)	+1 °C (+100%)	Soil thermal conductivity
HS	+8 °C (+800%)	+2 °C (+150%)	Soil thermal conductivity

6. Conclusions

The subdivision of the VGS into categories turned out to be a fundamental step in setting up the entire work, allowing to compare the advantages of using different VGS. The variations in modeling and in the attribution of diversified energy balances on the same sample building allowed comparison of results about trends of the external surface temperatures and the operative temperatures for each system. The in-depth literature review produced mathematical models and allowed comparison of the results with those from empirical studies, demonstrating a good match. The HS (with a consistent substrate thickness) proved to be the best performing in the summer season, while the Mur Vegetal results to be the best performing system in winter. Compared to the concrete model, generally all categories improve the thermohygrometric performance of the sample building. In particular, the benefits on the outside face temperatures of the walls, which remain below the temperature of the outside air on hot summer days, are significant.

The results obtained refer to the simulated sample and therefore may vary in relation to the characteristics of the building analyzed (stratigraphy, number and size of windows, location, etc.) The simulation of a single sample and the non-determination of the contribution of each term of the energy balance in energy efficiency can be considered limitations of the present research that can be developed in further research with a parametric study.

Author Contributions: Conceptualization, A.A., C.P. and M.C.; methodology, A.A., C.P. and M.C.; software, A.A., C.P. and M.C.; validation, A.A., C.P. and M.C.; formal analysis, A.A. and C.P.; investigation, A.A. and C.P.; data curation, A.A. and C.P.; writing—original draft preparation, C.P.; writing—review and editing, A.A.; visualization, A.A. and C.P.; supervision, A.A. and M.C.; project administration, A.A. All authors have read and agreed to the published version of the manuscript.

Funding: This research received no external funding.

Institutional Review Board Statement: Not applicable.

Informed Consent Statement: Not applicable.

Data Availability Statement: The data presented in this study are available in Perra, C.; Arenghi, A.; Caffi, M. Verde Verticale: Analisi Termoigrometriche in Regime Dinamico. *DICATAM Technical Report* **2020**, *7*, 1–199, [67].

Conflicts of Interest: The authors declare no conflict of interest.

References

1. Alexandri, E. Tackling the Heat Island Effect with Green Roofs and Green Walls: A Microclimatic Approach. In Proceedings of the WSA, 2nd Research Student Conference, Cardiff, Wales, May 2005.
2. Alexandri, E.; Jones, P. Temperature decreases in an urban canyon due to green walls and green roofs in diverse climates. *Build. Environ.* **2008**, *43*, 480–493. [CrossRef]
3. Djedjig, R.; Bozonnet, E.; Belarbi, R. Modeling green wall interactions with street canyons for building energy simulation in urban context. *Urban Climate* **2016**, *16*, 75–85. [CrossRef]
4. Li, J.; Zheng, B.; Shen, W.; Xiang, Y.; Chen, X.; Qi, Z. Cooling and Energy-Saving Performance of Different Green Wall Design: A Simulation Study of a Block. *Energies* **2019**, *12*, 2912. [CrossRef]
5. Rosenlund, H.; Kruuse, A.; Kronvall, J. Urban greenery in Sweden—Implications for microclimate and energy efficiency. In Proceedings of the World Green Roof Congress, London, UK, 3 September 2010. [CrossRef]
6. Farid, F.; Ahmad, S.; Raub, A.; Shaari, M. Green "Breathing Facades" for Occupants' Improved Quality of Life. *Procedia Soc. Behav. Sci.* **2016**, *234*, 173–184. [CrossRef]
7. Stav, Y. Transfunctional Living Walls-Designing Living Walls for Environmental and Social Benefits. Ph.D. Thesis, Queensland University of Technology, Brisbane City, Australia, 2016.
8. Malys, L.; Musy, M.; Inard, C. A hydrothermal model to assess the impact of green walls on urban microclimate and building energy consumption. *Build. Environ.* **2014**, *73*, 187–197. [CrossRef]
9. Dahanayake, K.C.; Chow, C.L. Comparing reduction of building cooling load through green roofs and green walls by EnergyPlus simulations. *Build. Simul.* **2018**, *11*, 421–434. [CrossRef]
10. Musy, M.; Malys, L.; Inard, C. Assessment of Direct and Indirect Impacts of Vegetation on Building Comfort: A Comparative Study of Lawns, Green Walls and Green Roofs. *Procedia Environ. Sci.* **2017**, *38*, 603–610. [CrossRef]
11. Boeri, S. *Un Bosco Verticale—Libretto di Istruzioni per il Prototipo di una Città Foresta*; Corraini Edizioni: Mantova, Italy, 2015.
12. Blanc, P. Vertical Gardens, the new Challenges. In *Green Cities in the World*, 2nd ed.; Briz, J., Köhler, M., de Felipe, I., Eds.; Editorial Agricola Espanola: Madrid, Spain, 2015; pp. 330–355.
13. Aksamija, A. High-Performance Building Envelopes: Design Methods for Energy-Efficient Facades. In Proceedings of the BEST4 Conference, Kansas City, MO, USA, 13–15 April 2015.
14. Safikhani, T.; Abdullah, A.; Ossen, D.; Baharvand, M. A review of energy characteristic of vertical greenery systems. *Renew. Sust. Energ. Rev.* **2014**, *40*, 450–462. [CrossRef]
15. Manso, M.; Castro-Gomes, J. Green wall systems: A review of their characteristics. *Renew. Sust. Energ. Rev.* **2015**, *41*, 863–871. [CrossRef]
16. Jim, C.Y. Greenwall classification and critical design-management assessments. *Ecol. Eng.* **2015**, *77*, 348–362. [CrossRef]
17. Pérez, G.; Coma, J.; Martorell, I.; Cabeza, L.F. Vertical Greenery Systems (VGS) for energy saving in buildings: A review. *Renew. Sust. Energ. Rev.* **2014**, *39*, 139–165. [CrossRef]
18. Jaafar, B.; Said, I.; Rasidi, M. Evaluating the Impact of Vertical Greenery System on Cooling Effect on High Rise Buildings and Surroundings: A Review. *Rev. Urban. Archit. Stud.* **2011**, *9*, 1–9. [CrossRef]
19. Wong, N.H.; Tan, A.; Chen, Y.; Sekar, K.; Tan, P.; Chan, D.; Chiang, K.; Wong, N. Thermal evaluation of vertical greenery systems for building walls. *Build. Environ.* **2010**, *45*, 663–672. [CrossRef]
20. Bit, E. *Il Nuovo Verde Verticale: Tecnologie, Progetti, Linee Guida*; Wolters Kluwer Italia S.r.l.: Assago (Milan), Italy, 2012.
21. Balogun, A.; Morakinyo, T.; Adegun, O. Effect of tree-shading on energy demand of two similar buildings. *Energy Build.* **2014**, *81*, 305–315. [CrossRef]

22. Giometto, M.; Christen, A.; Egli, P.; Schmid, M.F.; Tooke, R.T.; Coops, N.C.; Parlange, M. Effects of trees on mean wind, turbulence and momentum exchange within and above a real urban environment. *Adv. Water Resour.* **2017**, *106*, 154–168. [CrossRef]
23. Hes, D.; Dawkins, A.; Jensen, C.; Aye, L. A modelling method to assess the effect of tree shading for building performance simulation. In Proceedings of the Building Simulation 2011, 12th Conference of International Building Performance Simulation Association, Sydney, New South Wales, Australia, 14–16 November 2011.
24. Lin, B.-S.; Lin, Y.-J. Cooling effect of shade trees with different characteristics in a subtropical urban park. *HortScience* **2010**, *45*, 83–86. [CrossRef]
25. Pandit, R.; Laband, D. A hedonic analysis of the impact of tree shade on summertime residential energy consumption. *Arboric. Urban For.* **2010**, *36*, 73–80.
26. Simpson, J.R.; McPherson, E.G. Simulation of tree shade impacts on residential energy use for space conditioning in Sacramento. *Atmos. Environ.* **1998**, *32*, 69–74. [CrossRef]
27. Akbari, H.; Kurn, D.; Bretz, S.; Hanford, J. Peak power and cooling energy savings of shade trees. *Energy Build.* **1997**, *25*, 139–148. [CrossRef]
28. Yoshimi, J.; Altan, H. Thermal simulations on the effects of vegetated walls on indoor building environments. In Proceedings of the Building Simulation 2011, 12th Conference of International Building Performance Simulation Association, Sydney, New South Wales, Australia, 14–16 November 2011.
29. Coma, J.; Perez, G.; Solé, C.; Castell, A.; Cabeza, L.F. New Green Facades as Passive Systems for Energy Savings on Buildings. *Energy Procedia* **2014**, *57*, 1851–1859. [CrossRef]
30. Kontoleon, K.; Eumorfopoulou, E. The effect of the orientation and proportion of a plant-covered wall layer on the thermal performance of a building zone. *Build. Environ.* **2010**, *45*, 1287–1303. [CrossRef]
31. Flores Larsen, S.; Filippín, C.; Lesino, G. Thermal Simulation of a Double Skin Façade with Plants. *Energy Procedia* **2014**, *57*, 1763–1772. [CrossRef]
32. Stec, W.; Paassen, A.; Maziarz, A. Modelling the double skin façade with plants. *Energy Build.* **2005**, *37*, 419–427. [CrossRef]
33. Chen, Q.; Li, B.; Xioahu, L. An experimental evaluation of the living wall system in hot and humid climate. *Energy Build.* **2013**, *61*, 298–307. [CrossRef]
34. Dahanayake, K.; Chow, C. Studying the Potential of Energy Saving through Vertical Greenery Systems: Using EnergyPlus Simulation Program. *Energy Build.* **2016**, *138*. [CrossRef]
35. Djedjig, R.; El Ganaoui, M.; Belarbi, R.; Bennacer, R. Thermal effects of an innovative green wall on building energy performance. *Mech. Ind.* **2017**, *18*. [CrossRef]
36. Scarpa, M.; Mazzali, U.; Peron, F. Modeling the energy performance of living walls: Validation against field measurements in temperate climate. *Energy Build.* **2014**, *79*, 155–163. [CrossRef]
37. Stanghellini, C.; van Meurs, W.T.M. Environmental control of greenhouse crop transpiration. *J. Agric. Eng. Res.* **1992**, *51*, 297–311. [CrossRef]
38. Allen, R.; Pereira, L.; Smith, M. *Crop Evapotranspiration—Guidelines for Computing Crop Water Requirements*; FAO: Rome, Italy, 1998.
39. DOE. EnergyPlus Energy Simulation Software. 2015. Available online: https://energyplus.net/ (accessed on 31 March 2021).
40. Sailor, D.J. A green roof model for building energy simulation programs. *Energy Build.* **2008**, *40*, 1466–1478. [CrossRef]
41. Frankenstein, S.; Koenig, G. *Fast All-Season Soil Strength (FASST)*; Cold Regions Research and Engineering Laboratory: Hanover, NH, USA, 2004.
42. Anderson, R.; Zhang, X.; Skaggs, T. Measurement and Partitioning of Evapotranspiration for Application to Vadose Zone Studies. *Vadose Zone J.* **2017**, *16*, 1–9. [CrossRef]
43. Del Barrio, E.P. Analysis of the green roofs cooling potential in buildings. *Energy Build.* **1998**, *27*, 179–193. [CrossRef]
44. Gutschick, V.P. Leaf Energy Balance: Basics, and Modeling from Leaves to Canopies. In *Canopy Photosynthesis: From Basics to Applications*; Hikosaka, K., Niinemets, Ü., Anten, N., Eds.; Springer: Dordrecht, The Netherlands, 2016; pp. 23–58.
45. Jayalakshmy, M.S.; Philip, J. Thermophysical Properties of Plant Leaves and Their Influence on the Environment Temperature. *Int. J. Thermophys.* **2010**, *31*, 2295–2304. [CrossRef]
46. Merzlyak, M.N.; Chivkunova, O.B.; Melø, T.B.; Naqvi, K.R. Does a leaf absorb radiation in the near infrared (780–900 nm) region? A new approach to quantifying optical reflection, absorption and transmission of leaves. *Photosynth. Res.* **2002**, *72*, 263–270. [CrossRef] [PubMed]
47. Oke, T.R. *Boundary Layer Climates*, 2nd ed.; Routledge Taylor & Francis Group: London, UK, 1987; pp. 1–157.
48. Šuklje, T.; Medved, S.; Arkar, C. On detailed thermal response modeling of vertical greenery systems as cooling measure for buildings and cities in summer conditions. *Energy* **2016**, *115 Pt 1*, 1055–1068. [CrossRef]
49. Davis, M.M.; Hirmer, S. The Potential for Vertical Gardens as Evaporative Coolers: An Adaptation of the 'Penman Monteith Equation'. *Build. Environ.* **2015**, *92*. [CrossRef]
50. Monteith, J.L.; Unsworth, M.H. *Principles of Environmental Physics: Plants, Animals, and the Atmosphere*, 4th ed.; Elsevier: Oxford, UK, 2013.
51. Monteith, J.L. *Principles of Environmental Physics*; Edward Arnold Limited: London, UK, 1973.
52. Defraeye, T.; Blocken, B.; Carmeliet, J. Convective heat transfer coefficients for exterior building surfaces: Existing correlations and CFD modelling. *Energy Convers. Manag.* **2011**, *52*, 512–522. [CrossRef]

53. Asner, G.P.; Scurlock, J.M.O.; Hicke, J.A. Global synthesis of leaf area index observations: Implications for ecological and remote sensing studies. *Glob. Ecol. Biogeogr.* **2003**, *12*, 195–205. [CrossRef]
54. Goswami, S.; Gamon, J.; Vargas Zesati, S.; Tweedie, C. Relationships of NDVI, Biomass, and Leaf Area Index (LAI) for six key plant species in Barrow, Alaska. *PeerJ Prepr.* **2015**. [CrossRef]
55. Candelari, E. Caratterizzazione Sperimentale della Prestazione Termica e Acustica di un Living Wall. Ph.D. Thesis, Politecnico di Torino, Torino, Italy, 2015.
56. Gu, L. Airflow network modeling in energyplus. In Proceedings of the 10th International Building Performance Simulation Association, Conference and Exhibition, Beijing, China, 3–6 September 2007.
57. Peci-López, F.; Jensen, R.; Heiselberg, P.; Ruiz de Adana, M. Experimental analysis and model validation of an opaque ventilated facade. *Build. Environ.* **2012**, *56*, 265–275. [CrossRef]
58. Le, S.; Chen, Y.; Bi, Y.; Lu, X. Modeling and Simulation of Ventilated Double-Skin Facade Using EnergyPlus. In Proceedings of the 8th International Symposium on Heating, Ventilation and Air Conditioning, Sydney, New South Wales, Australia, 16–17 July 2014; Lecture Notes in Electrical Engineering. Li, A., Zhu, Y., Li, Y., Eds.; Springer: Berlin/Heidelberg, Germany, 2014; Volume 263, pp. 241–252. [CrossRef]
59. Abu-Hamdeh, N.H.; Reeder, R.C. Soil thermal conductivity: Effects of density, moisture, salt concentration and organic matter. *Soil Sci. Soc. Am. J.* **2000**, *64*, 1285–1290. [CrossRef]
60. Akpabio, G.T.; Ituen, E.E.; Ikot, A.N. A Comparative Study of the Thermal Properties of Different Soil Samples for Moulding Blocks for a Passively Cooled Building Design. *Int. J. Pure Appl. Phys.* **2010**, *6*, 517–521.
61. Clauser, C.; Huenges, E. Thermal conductivity of rocks and minerals. In *Rock Physics & Phase Relations: A Handbook of Physical Constants*; Ahrens, T.J., Ed.; American Geophysical Union: Washington, DC, USA, 1995; pp. 105–126.
62. Fuchs, M. Energy Balance. In *Encyclopedia of Soils in the Environment*, 1st ed.; Hillel, D., Ed.; Elsevier: Amsterdam, The Netherlands, 2005; pp. 438–441.
63. Han, S.; Chun, S.; Kim, K.; Lawrence, A.; Tia, M.; Said, Z. Evaluation of Soil Insulation Effect on Thermal Behavior of Drilled Shafts as Mass Concrete. *Cogent Eng.* **2018**, *5*. [CrossRef]
64. Nikoosokhan, S.; Nowamooz, H.; Chazallon, C. Effect of dry density, soil texture and time-spatial variable water content on the soil thermal conductivity. *Geomech. Geoeng.* **2015**, *11*, 1–10. [CrossRef]
65. Pitts, L. Monitoring Soil Moisture for Optimal Crop Growth. 2016. Available online: https://observant.zendesk.com/hc/en-us/articles/208067926-Monitoring-Soil-Moisture-for-Optimal-Crop-Growth (accessed on 31 March 2021).
66. EnergyPlus. Weather Data by Region. Available online: https://energyplus.net/weather-region/europe_wmo_region_6/ITA%20%20 (accessed on 31 March 2021).
67. Perra, C.; Arenghi, A.; Caffi, M. Verde Verticale: Analisi Termoigrometriche in Regime Dinamico. *DICATAM Tech. Rep.* **2020**, *7*, 1–199.
68. Perez, G.; Rincon, L.; Vila, A.; Gonzalez, J.M.; Cabeza, L.F. Green vertical systems for buildings as passive systems for energy savings. *Appl. Energy* **2011**, *88*, 4854–4859. [CrossRef]
69. Mazzali, U.; Fabio, P.; Romagnon, P.; Pulselli, R.M.; Bastianoni, S. Experimental investigation on the energy performance of Living Walls in a temperate climate. *Build. Environ.* **2013**, *64*, 57–66. [CrossRef]
70. Cameron, R.W.F.; Taylor, J.; Emmett, M. A Hedera green façade –energy performance and saving under different maritime-temperate, winter weather conditions. *Build. Environ.* **2015**, *92*, 111–121. [CrossRef]
71. Haggag, M.; Hassan, A.; Elmasry, S. Experimental study on reduced heat gain through green façades in a high heat load climate. *Energy Build.* **2014**, *82*, 668–674. [CrossRef]
72. Hoelscher, M.-T.; Nehls, T.; Jänicke, B.; Wessolek, G. Quantifying cooling effects of facade greening: Shading, transpiration and insulation. *Energy Build.* **2015**, *114*, 283–290. [CrossRef]
73. Perini, K.; Ottelé, M.; Fraaij, A.L.A.; Haas, E.; Raiteri, R. Vertical greening systems and the effect on air flow and temperature on the building envelope. *Build. Environ.* **2011**, *46*, 2287–2294. [CrossRef]

applied sciences

MDPI

Article

Phase Change Material Integration in Building Envelopes in Different Building Types and Climates: Modeling the Benefits of Active and Passive Strategies

Francesco Carlucci [1,*], Alessandro Cannavale [2,3], Angela Alessia Triggiano [1], Amalia Squicciarini [1] and Francesco Fiorito [1]

[1] Department of Civil, Environmental, Land, Building Engineering and Chemistry, Polytechnic University of Bari, 70125 Bari, Italy; a.triggiano2@studenti.poliba.it (A.A.T.); a.squicciarini2@studenti.poliba.it (A.S.); francesco.fiorito@poliba.it (F.F.)

[2] Department of Sciences in Civil Engineering and Architecture, Polytechnic University of Bari, 70125 Bari, Italy; alessandro.cannavale@poliba.it

[3] National Research Council, Institute of Nanotechnology (CNR-NANOTEC), Via Monteroni, 73100 Lecce, Italy

* Correspondence: francesco.carlucci@poliba.it

Featured Application: Reduction of energy consumption in residential and office buildings through the improvement of latent heat storage in active and passive strategies.

Citation: Carlucci, F.; Cannavale, A.; Triggiano, A.A.; Squicciarini, A.; Fiorito, F. Phase Change Material Integration in Building Envelopes in Different Building Types and Climates: Modeling the Benefits of Active and Passive Strategies. *Appl. Sci.* **2021**, *11*, 4680. https://doi.org/10.3390/app11104680

Academic Editors: Tiziana Poli, Andrea Giovanni Mainini, Mitja Košir, Juan Diego Blanco Cadena and Gabriele Lobaccaro

Received: 8 April 2021
Accepted: 17 May 2021
Published: 20 May 2021

Publisher's Note: MDPI stays neutral with regard to jurisdictional claims in published maps and institutional affiliations.

Abstract: Among the adaptive solutions, phase change material (PCM) technology is one of the most developed, thanks to its capability to mitigate the effects of air temperature fluctuations using thermal energy storage (TES). PCMs belong to the category of passive systems that operate on heat modulation, thanks to latent heat storage (LHS) that can lead to a reduction of heating ventilation air conditioning (HVAC) consumption in traditional buildings and to an improvement of indoor thermal comfort in buildings devoid of HVAC systems. The aim of this work is to numerically analyze and compare the benefits of the implementation of PCMs on the building envelope in both active and passive strategies. To generalize the results, two different EnergyPlus calibrated reference models—the small office and the midrise apartment—were considered, and 25 different European cities in different climatic zones were selected. For these analyses, a PCM plasterboard with a 23 °C melting point was considered in four different thicknesses—12.5, 25, 37.5, and 50 mm. The results obtained highlighted a strong logarithmic correlation between PCM thickness and energy reduction in all the climatic zones, with higher benefits in office buildings and in warmer climates for both strategies.

Keywords: phase change material; thermal energy storage; energy efficiency; passive strategies; active strategies; adaptive envelopes

1. Introduction

Thermal comfort and reduction of energy consumption are consolidated topics in scientific literature. Currently, an increasing interest in these fields is related to the benefits of adaptive technologies. These technologies, when applied to building envelopes, allow buildings to adjust their characteristics, in a reversible way, in response to external stimuli. As a result, an adaptation of their behavior to climate fluctuations is achieved, and, consequently, users' comfort requirements can be more efficiently met [1,2]. Currently, the most promising results of adaptive envelopes [3] are related to wall-integrated PCMs [4,5], switchable glazing [6–9], adaptive solar shadings [10–12], dynamic insulation [13,14], and multifunctional facades [15,16].

Among this wide range of technologies, PCMs have constantly grown their importance in recent years, thanks to broad experimentation and diffusion in different scientific fields [17,18]. These include mainly the aerospace industry [19], the design of low-energy buildings [20], the preservation of products [21], the electronic industry [22], and waste heat recovery systems [5]. The spread of PCMs in all these fields is related to the advantages of latent heat storage (LHS), which allows the storage and release of naturally available heat in low-volume elements, increasing, therefore, the energy storage density of the material [23].

In the building design domain, the importance of PCMs is related mainly to two different topics usually investigated in scientific literature: the reduction of HVAC energy consumption (i.e., active strategies) and the reduction of local and global thermal discomfort (i.e., passive strategies). Many studies have highlighted that PCMs can reduce the energy demand of HVAC systems by up to 30% if applied as a retrofit solution in residential buildings located in cold climates [24]. Moreover, thanks to the stabilization of the indoor radiant temperature, a significant reduction of thermal discomfort can be obtained [25]. Starting from these two important results, the aim of this paper is to merge and generalize these studies by considering: (i) a broader range of climates, (ii) both passive and active strategies, (iii) different building types, and (iv) different thicknesses of PCMs. Therefore, the benefits of PCM implementation are assessed through numerical analyses.

To design and model a PCM-integrated building element correctly and maximize its benefits, it is fundamental to understand the functioning of this technology and the different products available. The functioning is strictly related to the different ways in which materials can store or release heat: sensible heat storage (SHS), latent heat storage (LHS), and thermochemical heat storage. While SHS and LHS are applicable to buildings, thermochemical heat storage technologies are currently not applied in the civil field [23]. Regarding SHS, the heat absorbed/released is related to the increase/decrease of temperature in relation to the mass of the body (m), the specific heat (c), and the variation of temperature (dT), as described in Equation (1).

$$Q_S = \int_{T1}^{T2} m\, c\, dT \tag{1}$$

During LHS, the heat absorbed or released leads to a change of phase—for example, a melting from solid to liquid—without changing the temperature of the body. In this case, the heat stored depends on the mass, the fraction melted (f_m), and the variation of enthalpy of fusion per unit mass (Δh_m) (Equation (2)).

$$Q_L = m f_m \Delta h_m \tag{2}$$

Therefore, generalizing a phenomenon with SHS and LHS for a gypsum-PCM board, the heat exchanged corresponds to the total enthalpy variation ΔH_{tot} that depends on the enthalpy variation of each material contained in the board, according to Equations (3)–(5) [26]:

$$\Delta H_{tot} = m \cdot \left[(1 - f) \cdot \Delta h_{gypsum} + f \cdot \Delta h_{PCM}\right] \tag{3}$$

where, considering the complete melting of the PCM (f_m = 1), the partial enthalpies are:

$$\Delta h_{gypsum} = (1 - f) \int_{T1}^{T2} m c_S\, dT \tag{4}$$

$$\Delta h_{PCM} = f \left(\int_{T1}^{Tm} c_S\, dT + \Delta h_m + \int_{Tm}^{T2} c_L\, dT \right) \tag{5}$$

with f being the mass fraction of the PCMs in gypsum, c_S and c_L representing, respectively, the specific heat of the solid state and the liquid state, and T_m the melting temperature of the PCMs.

For this reason, in building applications, a PCM with a melting range within the thermal comfort range (20 °C–30 °C) can take advantage of LHS [27]. This is due to the storage of a good amount of heat in low-volume elements without increasing the surface temperature and therefore, without affecting thermal comfort. While, theoretically, PCMs work on four different possible changes of phase for building applications, namely, solid–solid, solid–liquid, gas–liquid, gas–solid [28], solid–liquid PCMs are usually considered.

PCMs can be classified into three main categories: organic, inorganic, and eutectic. Organic PCMs are composed of paraffins, fatty acids, fatty-acid esters, and sugar alcohols and, in general, can be classified as paraffin or non-paraffin. One of the main advantages of organic PCMs is that repeated melting–freezing does not lead to phase segregation; moreover, they are slightly affected by supercooling. Nevertheless, considering that paraffinic PCMs are derived from oil refining, organic PCMs have low ignition resistance, and, for this reason, envelope applications can be problematic. Organic PCMs can have a broad range of possible melting temperatures (−57 °C; +187 °C) with melting latent heat ranging from 85 to 300 J/g [29–31]; therefore, they should be accurately chosen to optimize their functioning. Inorganic PCMs are classified as salt-hydrates or metallic, and, despite having an enthalpy per mass similar to organic PCMs, they can reach higher melting latent heat per unit volume thanks to their higher density. Moreover, they have higher conductivity, can reach higher melting points, and are less expensive and less flammable than organic PCMs. However, inorganic PCMs present some limitations, such as supercooling, phase segregation, and corrosion. Lastly, eutectic PCMs are composed of at least two PCMs with the same melting point and can be classified as organic–organic, inorganic–inorganic, and organic–inorganic, depending on the types of PCMs used.

Often PCMs are encapsulated, primarily to hold both solid and liquid phases and to protect the PCMs from harmful interactions with the environment and other building materials. Encapsulation can also reduce phase segregation and corrosion, provide easier handling, and increase the heat transfer area [32,33]. Encapsulations can be classified, depending on capsule size, into macroencapsulation (d > 1 mm), microencapsulation (1 μm < d < 1 mm), or nanoencapsulation (d < 1 μm). Moreover, they can be made of different materials, e.g., aluminum, plastic, polyolefin, rubber, polymers, in different containers, such as balls, tubes, plates, and boxes [33].

Many studies have been performed to deepen the possible applications of PCMs to buildings, mainly classified into active storage systems and passive storage systems.

Active systems are characterized by heat exchangers and forced convection and can be, in turn, classified into direct and indirect systems. In direct systems, the heat transfer fluid is also the storage element of the system, while in indirect systems, the fluid serves as the transfer medium and another material is used as the storage element [34]. Active systems have been studied [35] and applied to suspended ceilings [36,37], ventilation systems [38,39], external double-skin façades [40], solar collectors [41,42], heat storage water tanks [43,44], integrated photovoltaics [45,46], and building cores enhanced with PCMs activated through the use of ducts or pipes [47].

When considering passive systems, there is no forced convection, and, in this case, these systems can be classified according to the way the PCMs are embedded in the building element: inside the material, as a new layer, and in windows or as sun protection [48]. Considering the PCMs embedded inside the material, encapsulated PCMs can be easily added to other construction materials such as concrete [49], plaster [50], cellulose, or glass fiber [51]. Another diffuse solution to applying PCMs in buildings is to add a new layer to increase the thermal inertia of lightweight constructions. The most common application of PCMs as a new layer is the PCM-enhanced gypsum plasterboard; many products, such as the Alba Balance (Rigips-Saint Gobain), are already commercially available. Other applications of PCMs as new layers are PCM sandwich panels [52] and macroencapsulated PCMs in plates or bags, such as the Delta-Cool24 (Dorken) or the Energain (Dupont). Lastly, PCMs can be used for both sun protection (for example, in internal blinds [53]) and inside windows (for example, with an extra air gap, behind the inner glass [54]).

Considering all these available PCMs, particular attention should be paid to their melting point because, due to the broad variability of this parameter, each application has its own more suitable range. Different studies concerning building energy performance [28,55–57] have identified that the most suitable melting points for cooling are up to 21 °C, while for heating, they are 22 °C or, in general, 2 °C higher than the heating setpoint temperature. Moreover, suitable melting point ranges for thermal comfort are between 22 and 28 °C; for hot water applications, they are between 29 and 60 °C. Finally, higher melting points—between 61 and 120 °C—are suitable for waste heat recovery applications. For the current study, a PCM layer was considered the inner layer of external walls, roofs, and floors, implementing a paraffinic microencapsulated organic PCM (Micronal) embedded in gypsum plasterboard with a melting point of 23 °C, the characteristics of which are described in the following sections. Considering that the models were run for active and passive strategies and for heating and cooling systems, the choice of this transition temperature allows us, as described before, to take advantage of heating, cooling, and thermal comfort applications.

2. Materials and Methods

Energy and comfort analyses were performed using EnergyPlus v.9.3. The main advantage of EnergyPlus [58] is that it can guarantee multidomain integration and physical interaction accounting for thermal, visual, mass-flow, and building service interactions [3,59]. A good starting point for energy-efficiency-oriented research in EnergyPlus [60] is offered by the U.S. Department of Energy (DOE), which developed 16 reference building models in different climatic zones. These models have been calibrated using several references such as the Commercial Buildings Energy Consumption Survey (CBECS), ASHRAE Standards 90.1, and many other academic sources [61].

Offices and apartments represent two of the most diffused building types in Europe, and they respectively account for 6% and 75% of total European building stock [62]. Therefore, two representative reference models, the small office and the midrise apartment, were chosen for this study. The geometrical data and main characteristics of the two reference buildings are summarized in the table and pictures below (Table 1, Figure 1).

Table 1. Main characteristics of reference buildings from the DOE.

Building Types	Number of Floors [–]	Gross Floor Area [m^2]	Floor-to-Ceiling Height [m]	Windows-to-Wall Ratio [%]	Number of Thermal Zones [–]
Midrise apartment	4	3134.59	3	20%	32
Small office	1	511.16	3	22%	6

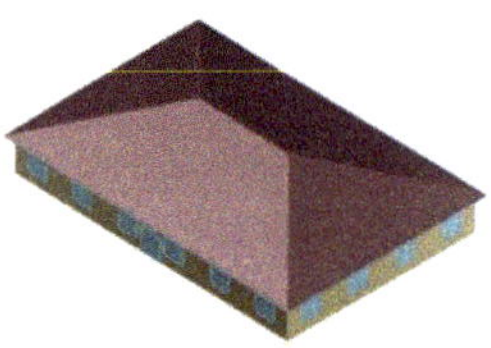

(a)

(b)

Figure 1. Reference buildings chosen from the DOE: small office (**a**) and mid-rise apartment (**b**).

The HVAC equipment of the reference buildings is modeled in accordance with ASHRAE Standard 90.1 and consider, for both the small office and the midrise apartment, a boiler for heating generation, a packaged air conditioning unit for cooling generation, and a single-zone constant-volume system for air distribution.

Particular attention was paid to the reliability of consumption and, considering that the analyses were run for different building types and different locations, to the independence of the results from the HVAC system sizing process. Therefore, the original HVAC systems were substituted with properly calibrated ideal-load air systems. These systems are similar to a traditional HVAC unit in EnergyPlus; the main difference is that it is not connected to a central system, and each ideal-load air system provides heating or cooling to satisfy the zone setpoint [63]. Starting from the ideal heating and cooling loads, the energy consumption is calculated by dividing the loads by an ideal energy efficiency ratio (EER) equal to 3 in cooling mode and by a coefficient of performance (CoP) equal to 3 in heating mode, which can be considered average values for packaged direct expansion air conditioning systems.

Existing HVAC systems are characterized by a series of controls, such as the management of outdoor air and a night-cycle availability manager. Therefore, to guarantee the proper behavior of the ideal-load system, an energy management system (EMS) program was written in order to account for these controllers and obtain results close to the ones of the original DOE reference building.

To confirm the reliability of the calibrated ideal-load model, a statistical analysis was performed and values of mean bias error (MBE) and cumulative variation of root mean squared error (CVRMSE) were obtained from hourly values of heating and cooling energy consumption of the reference and modified models. The analysis was performed for each thermal zone, starting from three different reference models, one for each of the chosen locations (Miami—ASHRAE Zone 1A, Chicago—ASHRAE Zone 5A, Fairbanks—ASHRAE Zone 8), for both small office and midrise apartment.

Table 2 summarizes the results of these analyses, considering the average of whole thermal zones and cities and excluding only the zones where the HVAC system is off for at least 95% of total hours since the thermal behavior of these zones is not considered representative of the behavior of the entire building.

Table 2. Average values of MBE and CVRMSE for the considered models, excluding the results of zones with a system functioning period of less than 5% of total hours.

Building Types	Heating		Cooling	
	Average ∣MBE∣ [%]	Average CVRMSE [%]	Average ∣MBE∣ [%]	Average CVRMSE [%]
Midrise apartment	4.9%	9.3%	5.7%	12.9%
Small office	1.2%	20.7%	4.5%	44.9% [1]

[1] Without considering Fairbanks' cooling consumption, the cooling average CVRMSE is 19.6%.

Values of MBE and CVRMSE were then compared with reference values from ASHRAE guideline 14 [64,65], which gives the hourly calibration criteria for real building modeling. Indeed, most of the thermal zones in all the configurations meet the ASHRAE hourly criteria, with $|MBE| \leq 10\%$ and $CVRMSE \leq 30\%$ [64,65]. All the results meet the criteria described before, except for the cooling CVRMSE of the small office, which was slightly over the threshold; in this case, the results were negatively influenced by the Fairbanks result, which can be considered nonrepresentative as it has a reduced functioning period (19% of total hours for small office, 17% for midrise apartment) and very low cooling energy consumption. Indeed, considering the Fairbanks yearly global HVAC consumption, the differences obtained by comparing reference and ideal-load models are very low: 1.1% and 1.5% for the office and apartment models, respectively. For these reasons, the model can be considered reliable.

The performance assessment of the integration of PCMs in buildings was carried out considering a whole year. In order to properly account for the transient behavior of PCMs, each simulation hour was divided into 20 analysis timesteps. Moreover, 25 different European cities were considered in order to assess the benefits of the integration of PCMs into buildings. To define a broad and nonredundant city sample, different European cities—most of them characterized by common EU directives—were studied, selecting cities with very different climates, as described in Figure 2 and Table 3.

Figure 2. Individuation of the sample of 25 cities used for the analyses.

Table 3. Characteristics of the 25 European cities considered (data from epw weather file).

City	Country	Köppen–Geiger Classification	$CDD_{18°}$	$HDD_{18°}$
Larnaca	Cyprus	BSh	1259	759
Seville	Spain	Csa	1063	916
Athens	Greece	Csa	1076	1112
Brindisi	Italy	Csa	834	1151
Santander	Spain	Cfb	209	1369
Rome	Italy	Csa	649	1444
Porto	Portugal	Csb	146	1491
Madrid	Spain	Csa	628	1965
Plovdiv	Bulgaria	Cfa	543	2471
Milan	Italy	Cfa	380	2639
Paris	France	Cfb	142	2644
London	England	Cfb	32	2866
Timisoara	Romania	Dfa	365	2896
Brussels	Belgium	Cfb	96	2912
Geneva	Switzerland	Dfb	193	2965
Ankara	Turkey	BSk	253	3307
Ljubljana	Slovenia	Dfc	168	3383
Copenhagen	Denmark	Dfb	29	3563
Prague	Czech Republic	Dfb	84	3703
Munich	Germany	Dfb	79	3738
Bergen	Norway	Cfb	21	3996
Moscow	Russia	Dfb	99	4655
Helsinki	Finland	Dfb	33	4712
Reykjavik	Iceland	Dfc	0	4917
Kiruna	Sweden	Dfc	0	6967

In order to adapt the models to the European climates, all thermal properties were adjusted to meet the different energy requirements [66–70]. The selected cities were then grouped in six different climatic zones—B, C, D, E, F, and G, from warmer to colder—in accordance with their values of heating degree days (HDDs). Hence, the thermal properties of roofs, walls, slabs, and windows were changed according to the climatic zones, as described in Table 4.

Table 4. Characteristics of the envelope properties for each climatic zone.

Climatic Zone	City	Envelope Component	Thermal Transmittance [W/m^{2}K]	Solar Heat Gain Coefficient [-]
B	Larnaca Seville	External wall Slab Roof Window	0.43 0.44 0.35 3	- - - 0.35
C	Athens Brindisi Santander	External wall Slab Roof Window	0.34 0.38 0.33 2.2	- - - 0.35
D	Rome Porto Madrid	External wall Slab Roof Window	0.29 0.29 0.26 1.8	- - - 0.35
E	Plovdiv Milan Paris London Timisoara Brussels Geneva	External wall Slab Roof Window	0.26 0.26 0.22 1.4	- - - 0.35
F	Ankara Ljubljana Copenhagen Prague Munich	External wall Slab Roof Window	0.24 0.20 0.21 1.1	- - - 0.35
G	Bergen Moscow Helsinki Reykjavik Kiruna	External wall Slab Roof Window	0.17 0.10 0.09 0.8	- - - 0.35

To understand energy demand variations, an ideal-load HVAC system was considered for active strategies, while, for passive strategies, no HVAC system was implemented and natural ventilation strategies were adopted. HVAC operational schedules were kept as standard, and setpoints of 21 °C in heating operational mode and 24 °C in cooling mode were considered.

With regard to passive strategies, a supplementary EMS program was developed in order to calculate the adaptive thermal comfort optimal range, in accordance with current European recommendations (EN 16798-1:2019 [71]). This EMS program calculates, starting from the temperature of the previous 7 days ($\theta_{ed\text{-}i}$), the running mean outdoor temperature (θ_{rm}, Equation (6)) and the optimal operative temperature (θ_c, Equation (7)). Then, considering the first comfort class, the program calculates the operative temperature (T_{OP}) comfort range (Equation (8)), as described in the following equations:

$$\theta_{rm} = \frac{(\theta_{ed-1} + 0.8\,\theta_{ed-2} + 0.6\,\theta_{ed-3} + 0.5\,\theta_{ed-4} + 0.4\,\theta_{ed-5} + 0.3\,\theta_{ed-6} + 0.2\,\theta_{ed-7})}{3.8}\ [^\circ\mathrm{C}], \quad (6)$$

$$\theta_c = 0.33\,\theta_{rm} + 18.8\ [^\circ\mathrm{C}], \quad (7)$$

$$\theta_c - 3\ \le\ T_{OP}\ \le\ \theta_c + 2\ [^\circ\mathrm{C}]. \quad (8)$$

Hence, the results of these calculations were used for both the definition of the number of discomfort hours—comparing the operative temperature with the optimal range in Equation (6)—and the definition of a natural ventilation schedule. Indeed, in the passive strategy case, natural ventilation is activated only when the operative temperature of the thermal zone oversteps the higher boundary of the optimal range and the outdoor dry-bulb air temperature is lower than the indoor dry-bulb air temperature. This choice is due to the behavior of the building users, who generally open windows only for warm thermal discomfort.

For both strategies, a linearly dimmable artificial lighting system was implemented, controlled by an illuminance sensor placed in the middle of each thermal zone, with a target illuminance of 300 lux in the apartments, 100 lux in the corridors, and 500 lux in the offices. Internal gains, occupancy, and indoor appliances were left in accordance with the original reference models. According to the ASHRAE Standards [72], indoor appliances consume a maximum of 10.76 W/m^2 in small offices and 5.38 W/m^2 in midrise apartments, with a radiant fraction equal to 0.5. An occupancy density of 0.054 persons/m^2 was considered for the offices, while a maximum number of 2.5 people was assumed for the apartments. In both cases, the metabolic emission rate was considered equal to 120 W/person.

Another fundamental step in energy analyses is to properly model the PCM in EnergyPlus. To that end, a commercial PCM-enhanced plaster panel—the Alba Balance by Saint-Gobain Rigips—was considered as a reference. Two phase-transition temperatures, 23 and 26 °C, are commercially available for these boards. In this study, a 23 °C phase transition with a latent heat of 300 kJ/m^2 is considered, as shown in Table 5, which reports the producer-declared technical data.

Table 5. Characteristics of PCM-enhanced plaster panel.

Technical Data		
Density	ρ	**1000 kg/m^3**
Areal density	ρ_A	25 kg/m^2
Latent heat	dH	300 kJ/m^2 = 83 Wh/m^2
Total storage capacity (10–30 °C)	-	866 kJ/m^2
Specific heat	c	28.3 kJ/m^{2}K
Thermal conductivity	λ	0.27 W/mK

In order to suitably model this material in EnergyPlus, an enthalpy table, ranging from −30 to +100 °C, was calculated, starting from the areal density, the specific heat, the latent heat, and the following enthalpy equation.

$$H = mc\,\Delta T \quad (9)$$

Once the material reaches the melting point (23 °C), it stores latent heat (300 kJ/m^2 = 12 kJ/kg). This heat storage is equally distributed as an increase of enthalpy at three temperature steps within the melting range (21, 22, and 23 °C). The plot of the enthalpy curve is included in Figure 3.

Finally, these calculated values were inputted in EnergyPlus, and, to model the behavior of the PCM properly, the heat balance algorithm was changed from the Conduction Transfer Function (CTF) to the Conduction Finite Difference (ConFD) algorithm, increasing the timesteps from 6 to 20 [73]. These settings significantly increase the running time but discretize the surfaces depending on the thermal diffusivity of the material and the time step, allowing us to model particular materials, such as PCMs or variable thermal conductivity materials [63]. This timestep setting has been largely validated and verified [63,73] and guarantees a good adherence of the numerical model with measured data. Further detailed hysteresis analysis would have led to errors of about 1% instead of a just-few-times-higher error [74], which can be considered acceptable for annual analyses.

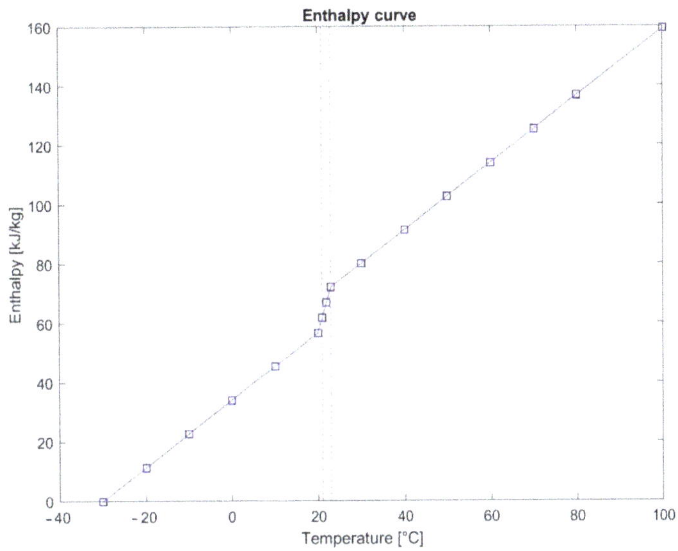

Figure 3. Enthalpy curve for the PCM-enhanced plasterboard.

The chosen plasterboards can be used in both walls and ceilings, and, for these analyses, they were considered the inner layer of all external walls and roofs. Despite Zwanzig et al. stating that in a multilayer wall, centrally located PCM better reduces heat fluxes [75], other studies have confirmed that PCMs located as an inner layer improve thermal comfort, thanks to the stabilization of radiant temperature [25]. Moreover, to understand how the thicknesses of the boards influence the results, four different boards, with thicknesses of 12.5, 25, 37.5, and 50 mm, were considered and modeled.

3. Results

The benefits of implementing PCMs were assessed by considering the reduction of energy demand in the active strategy and the reduction of total discomfort hours in the passive strategy. The following subparagraphs show the different results obtained for each strategy considered.

3.1. Active Strategies: Energy Demand Reduction

The reduction of energy demand is shown in the following figures (Figures 4 and 5), where the results are grouped with reference to the climatic zones, showing the changes of the control parameter as a change of PCM thickness. The contribution of PCMs in the reduction of heating and cooling energy consumption was split so as to differentiate the strategies. The percentage of energy reduction ($ER_{\%}$) refers to the difference between the total energy demand in the baseline case ($E_{T\ baseline}$)—without PCM—and the total energy demand in each PCM-implemented case ($E_{T\ PCM}$). This difference was then divided by the total energy demand of the baseline case, as described in the following equation (Equation (10)):

$$ER_{\%} = 100 * \frac{E_{T\ baseline} - E_{T\ PCM}}{E_{T\ baseline}} \quad (10)$$

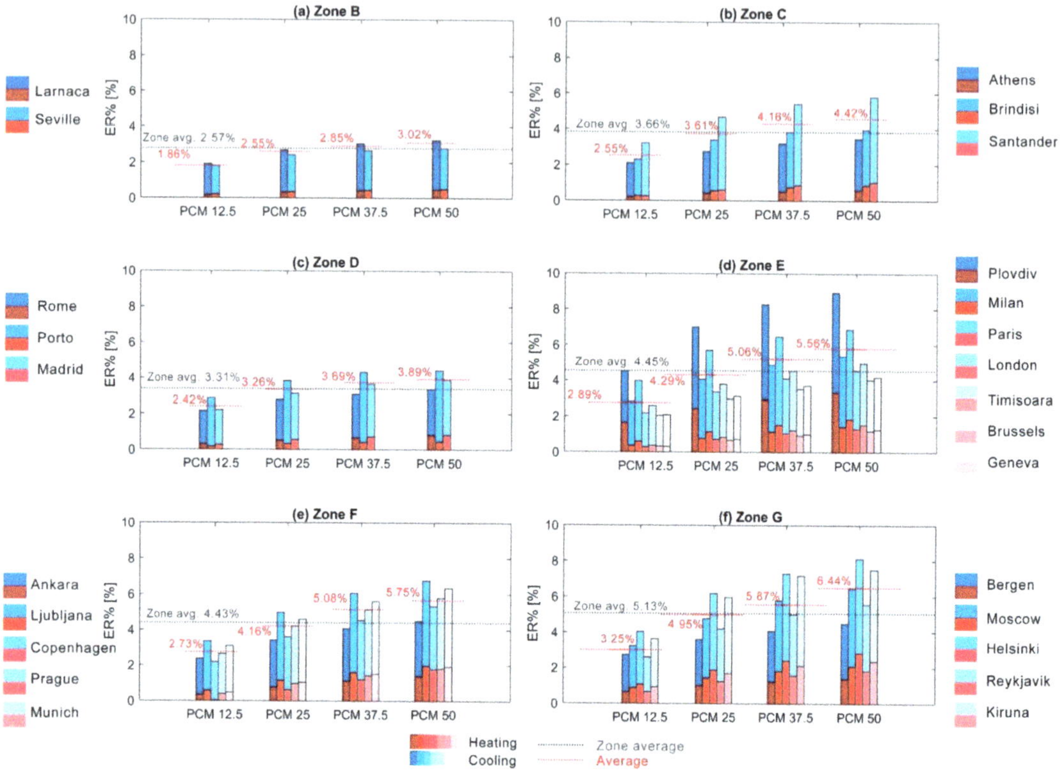

Figure 4. Small office, total energy demand reduction for the analyzed cities: (**a**) Climatic Zone B, (**b**) Climatic Zone C, (**c**) Climatic Zone D, (**d**) Climatic Zone E, (**e**) Climatic Zone F, and (**f**) Climatic Zone G.

Overall, it can be easily argued that the type and magnitude of energy reduction are strictly dependant on building typology, and outcomes show opposite trends when comparing the two models. On the one hand, in the small office model, the benefits of PCMs are clearer and increase in colder climates, with an average that rises from 2.57% for Zone B to 5.13% for Zone G. On the other hand, benefits for the midrise apartment model are lower and countertrend, considering that the main contribution is related to heating demand reduction and that the zone average decreases (with a less evident trend) in colder zones. However, both models confirm that an increase in PCM thickness corresponds to an increase in benefits, with a strong logarithmic correlation between thickness and energy reduction (Figure 6).

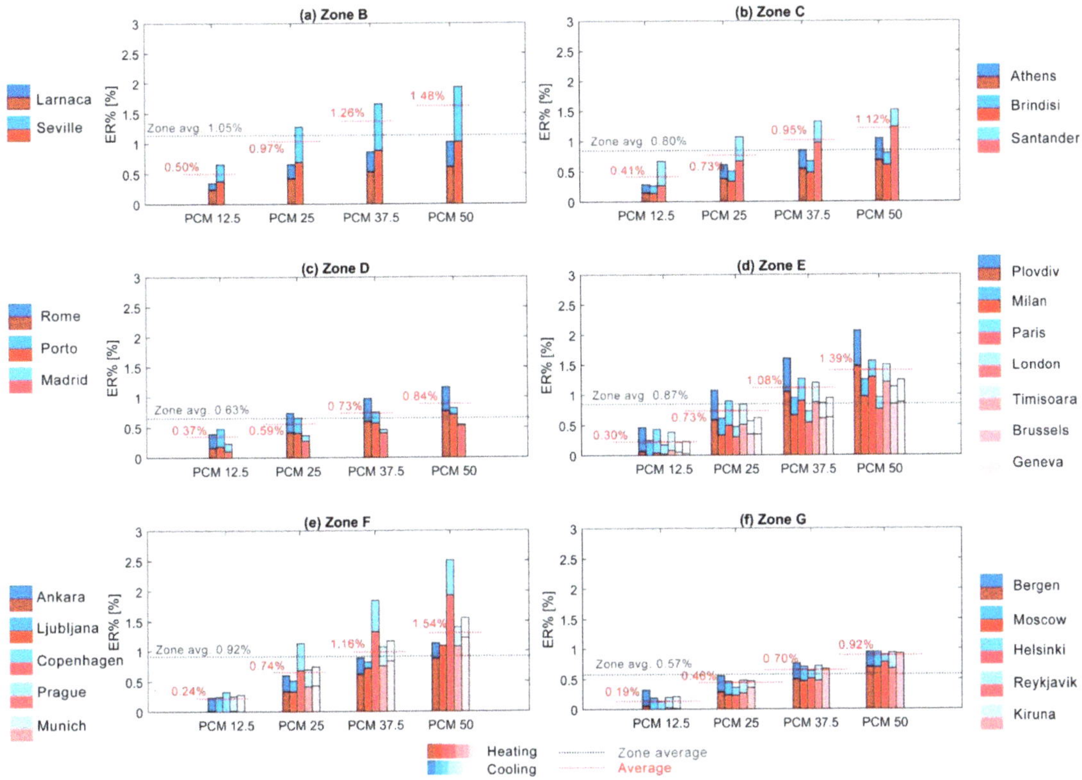

Figure 5. Midrise apartment, total energy demand reduction for the analyzed cities: (**a**) Climatic Zone B, (**b**) Climatic Zone C, (**c**) Climatic Zone D, (**d**) Climatic Zone E, (**e**) Climatic Zone F, and (**f**) Climatic Zone G.

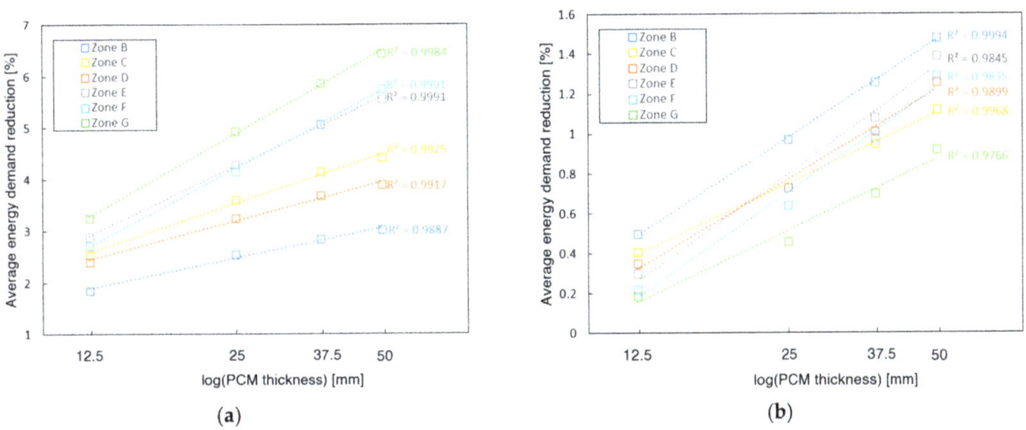

Figure 6. Logarithmic correlation between zone average and PCM thickness for (**a**) small office and (**b**) midrise apartment in the active strategies case.

3.2. Passive Strategies: Thermal Discomfort Reduction

In this second set of analyses, the index of the advantages related to the implementation of PCMs is the reduction of total discomfort hours (DH_T) in baseline ($DH_{T\ Baseline}$) and PCM ($DH_{T\ PCM}$) models, calculated by comparing the operative temperature of the thermal zones with the acceptance limits expressed in Equation (8). Analogously to the ER% described in the previous section, the percentage of discomfort hours reduction ($DHR_{\%}$) was calculated as described in Equation (11), and the results are reported in Figures 7 and 8.

$$DHR_{\%} = 100 * \frac{DH_{T\ Baseline} - DH_{T\ PCM}}{DH_{T\ Baseline}} \tag{11}$$

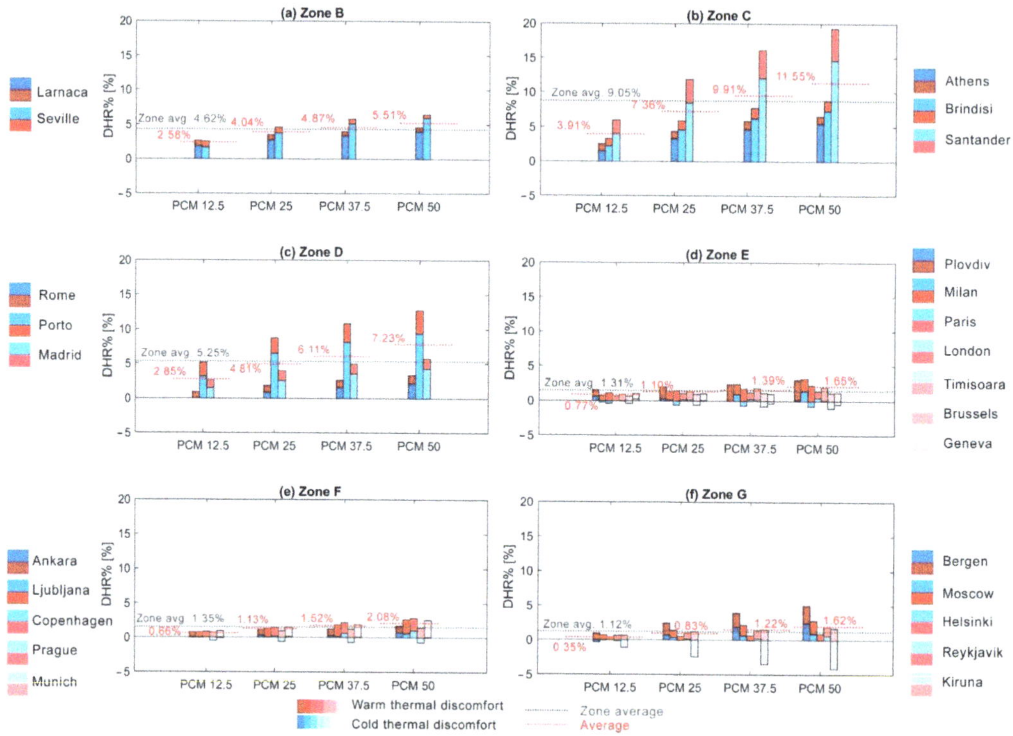

Figure 7. Small office, total discomfort hours reduction for the analyzed cities: (**a**) Climatic Zone B, (**b**) Climatic Zone C, (**c**) Climatic Zone D, (**d**) Climatic Zone E, (**e**) Climatic Zone F, and (**f**) Climatic Zone G.

Analogies and differences with results obtained for the active strategies can be easily noted from the graphs. Firstly, these simulations confirm that the implementation of PCMs is more effective in small offices than in midrise apartments.

With regards to the small office model, a clear different behavior can be identified when comparing the climatic zones: warmer climates (Zones B, C, and D) show higher benefits, mainly related to the reduction of cold thermal discomfort, while colder zones (E, F, and G) show lower benefits related to the reduction of warm thermal discomfort. This different behavior is also confirmed by the zone averages that fluctuate between 4.5% and 9% for the first three zones while oscillating around 1.2% for the other zones.

On the contrary, in midrise apartments, there is hardly ever a reduction of cold thermal discomfort; the results are mainly pushed by the reduction of warm thermal discomfort. In this case, the major advantages are registered in the intermediate climates—Zones C and D—with average values of 3.45% and 2.15%, respectively, while for other zones, the average benefits are lower and range from 0.73% to 1.57%.

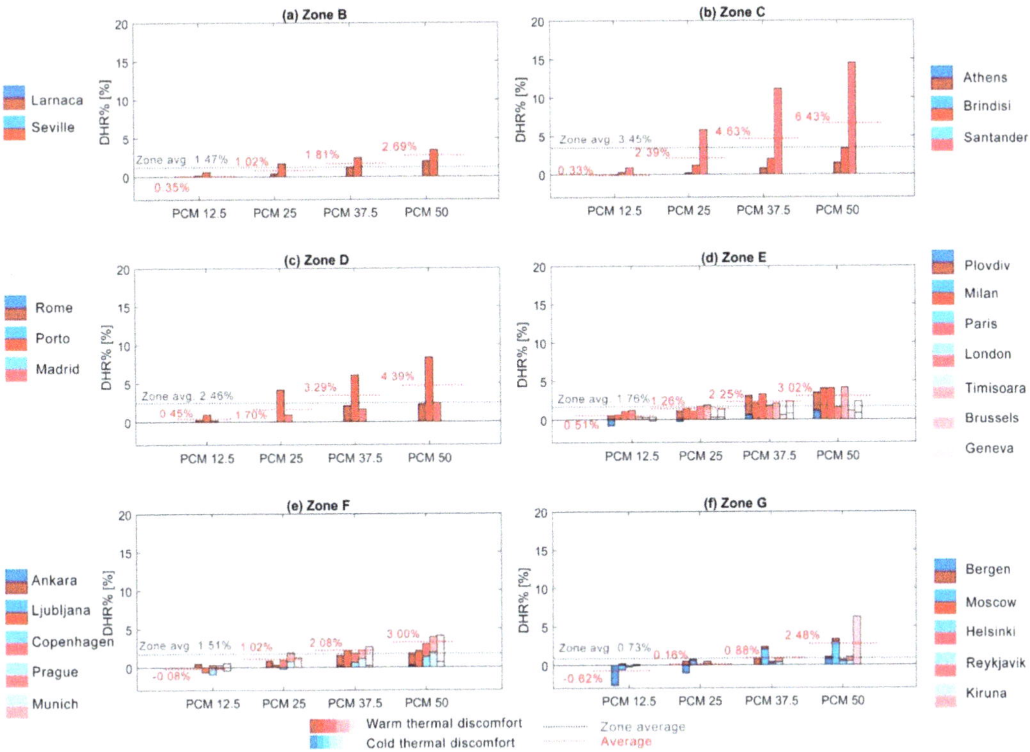

Figure 8. Midrise apartment, total discomfort hours reduction for the analyzed cities: (**a**) Climatic Zone B, (**b**) Climatic Zone C, (**c**) Climatic Zone D, (**d**) Climatic Zone E, (**e**) Climatic Zone F, and (**f**) Climatic Zone G.

Finally, comparing the effect of different PCM thicknesses, the comfort analysis confirms that thicker panels show higher advantages (Figure 9).

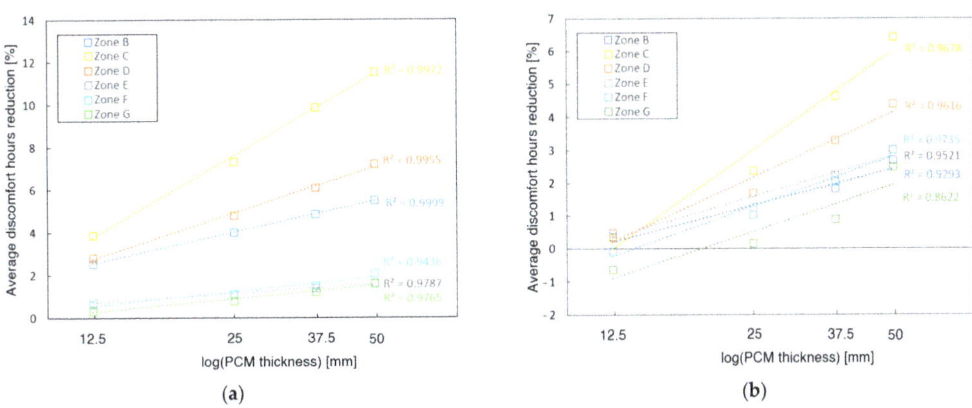

Figure 9. Logarithmic correlation between zone average and PCM thickness for (**a**) small office and (**b**) midrise apartment in the passive strategies case.

4. Discussion

The analyses performed point out a complex frame of results, where a few trends can be deduced, in general, for PCMs while others should be referred to each specifically analyzed case.

Firstly, one of the clearest conclusions that can be drawn is that the benefits of PCMs, integrated into the envelope's inner face, depend logarithmically on the thickness of the boards. The presented outcomes confirm that this trend is common to all analyses and does not depend on the climatic zones, building typology, or the comfort strategy adopted. Therefore, it can be stated that increasing board thickness is particularly advantageous, considering layers up to 50 mm; further increases lead to gradually lower improvements that may not be affordable. This behavior can be explained by considering the relatively low thermal conductivity of the boards (0.27 W/mK). In this case, increasing board thickness adds a further shift in the load peaks related to the thermal conduction inside the board that gradually reduces the effectiveness of the PCM.

Referring to different building typologies, the small office model shows higher benefits than the midrise apartment model, regardless of the climatic zone, PCM thickness, and comfort strategy adopted. The main reason for this trend can be identified in the different occupancy rates (35.3 m^2/person for the midrise apartment and 18.6 m^2/person for the small office) and the different internal loads of equipment and lights, which are sensibly higher in the office model. Considering the case of PCM boards located on the envelope's inner face, increasing the internal loads can increase the number of melting cycles, improving energy and comfort, especially in cooling-dominated climates.

Lastly, considering the climatic zones analyzed, it can be stated in general that the advantages of PCMs are higher in warm climates—Zones B, C, and D—with the only countertrend exception represented by the energy demand of the small office model in active strategies, where colder zones correspond to higher benefits, thanks to a constant increase in heating energy reduction.

Beyond these general considerations on the results obtained, referring to each analyzed case, on the one hand, the small office model has shown:

- significant overall benefits in the active approach, thanks to an important reduction of cooling demand; moreover, in colder zones, further reductions in heating consumption lead to higher benefits;
- a significant reduction of cold discomfort in hot climates in the passive approach and a slight reduction of warm discomfort in cold climates; nevertheless, few increases of warm discomfort can be found in extremely cold climates (Reykjavik).
- On the other hand, the midrise apartment model has shown:
- primarily reductions of heating demand in the active approach, with slight improvements of cooling consumption only for hot climates; considering the overall trend, the hot climates register slightly higher benefits;
- benefits in the passive approach are related mainly to warm discomfort, with higher effectiveness in intermediate climatic zones; moreover, in cold climates, reductions of cold discomfort have been registered, with a single case of increased cold discomfort in Bergen.

Comparing these results with those reported in similar studies, the main differences obtained can be explained by the positioning of the boards and the use of different building types, occupancy, and system availability schedules that can increase or decrease PCM effectiveness. Moreover, in this case, a PCM-embedded gypsum board was considered for adaptive implementation while, in other studies, pure PCM materials were applied to the envelope. Finally, as described by this study, the different locations can significantly change the effectiveness of the implementation of this technology.

5. Conclusions

The analyses presented in this paper are focused primarily on the effect of the PCM-enhanced boards studied, considering different boundary conditions, in order to define general trends and the specific behavior of this technology. To this end, the simulations considered different building typologies, climates, thicknesses, and active/passive strategies. Considering all the differences that can arise with this broad number of variables, the methodology adopted to create the energy building model tries to reduce all the fluctuations related to different national laws or the specific HVAC system in order to have comparable results.

Overall, this study highlights both general and specific trends for this technology, with interesting results. Undoubtedly, further studies could expand this research to other calibrated reference models or different climates in order to further generalize or specify the trends described in this paper. In this first work, our attention was focused on the European geographic area in order to analyze cities with similar energy containment directives. Future developments could expand this area—changing the envelope characteristics and also considering tropical and dry climates.

The economic aspects of PCM implementation are out of the boundaries of the current study as there are very different compositions of energy and energy prices in the considered countries. Future works could focus on smaller groups of locations and estimate a payback period on selected groups of cities with similar economic and energetic backgrounds.

Another interesting development of this first study could consider a simulation in future climatic scenarios. As shown in our research, office models are characterized by a strong cooling reduction in the active strategy and mainly by cold thermal discomfort in passive strategies. Considering future scenario simulations—characterized by warmer boundary conditions—the energy benefits should be higher than those obtained in our active strategy simulations while, on the contrary, in the passive strategy, the benefits could be lower. Finally, with regard to the midrise apartment model, the PCM benefits are related mainly to heating consumption reduction (active strategy) and the reduction of warm discomfort hours (passive strategy). Therefore, the benefits of PCM implementation could probably be reduced in future scenarios.

Author Contributions: Conceptualization, F.F. and A.C.; methodology, F.C., F.F. and A.C.; software, F.C.; validation, F.C. and F.F.; formal analysis, F.C.; investigation, A.S. and A.A.T.; resources, A.C.; data curation, F.C.; writing—original draft preparation, F.C.; writing—review and editing, F.F., A.C., A.A.T. and A.S.; visualization, F.C. and F.F.; supervision, F.F. and A.C. All authors have read and agreed to the published version of the manuscript.

Funding: This research received no external funding.

Institutional Review Board Statement: Not applicable.

Informed Consent Statement: Not applicable.

Data Availability Statement: Data is contained within the article.

Conflicts of Interest: The authors declare no conflict of interest.

References

1. Aelenei, D.; Aelenei, L.; Vieira, C.P. Adaptive Façade: Concept, Applications, Research Questions. *Energy Procedia* **2016**, *91*, 269–275. [CrossRef]
2. Fiorito, F.; Cannavale, A.; Santamouris, M. Development, testing and evaluation of energy savings potentials of photovoltachromic windows in office buildings. A perspective study for Australian climates. *Sol. Energy* **2020**, *205*, 358–371. [CrossRef]
3. Loonen, R.C.G.M.; Favoino, F.; Hensen, J.L.M.; Overend, M. Review of current status, requirements and opportunities for building performance simulation of adaptive facades. *J. Build. Perform. Simul.* **2017**, *10*, 205–223. [CrossRef]
4. Lee, K.O.; Medina, M.A.; Raith, E.; Sun, X. Assessing the integration of a thin phase change material (PCM) layer in a residential building wall for heat transfer reduction and management. *Appl. Energy* **2015**, *137*, 699–706. [CrossRef]
5. Akeiber, H.; Nejat, P.; Zaimi, M.; Majid, A.; Wahid, M.A.; Jomehzadeh, F.; Zeynali, I.; Kaiser, J.; Richard, B.; Ahmad, S. A review on phase change material (PCM) for sustainable passive cooling in building envelopes. *Renew. Sustain. Energy Rev.* **2016**, *60*, 1470–1497. [CrossRef]
6. Casini, M. Active dynamic windows for buildings: A review. *Renew. Energy* **2018**, *119*, 923–934. [CrossRef]
7. Baetens, R.; Jelle, B.P.; Gustavsen, A. Properties, requirements and possibilities of smart windows for dynamic daylight and solar energy control in buildings: A state-of-the-art review. *Sol. Energy Mater. Sol. Cells* **2010**, *94*, 87–105. [CrossRef]

8. Tällberg, R.; Jelle, B.P.; Loonen, R.; Gao, T.; Hamdy, M. Comparison of the energy saving potential of adaptive and controllable smart windows: A state-of-the-art review and simulation studies of thermochromic, photochromic and electrochromic technologies. *Sol. Energy Mater. Sol. Cells* **2019**, *200*, 109828. [CrossRef]
9. Cannavale, A.; Martellotta, F.; Cossari, P.; Gigli, G.; Ayr, U. Energy savings due to building integration of innovative solid-state electrochromic devices. *Appl. Energy* **2018**, *225*, 975–985. [CrossRef]
10. Fiorito, F.; Sauchelli, M.; Arroyo, D.; Pesenti, M.; Imperadori, M.; Masera, G.; Ranzi, G. Shape morphing solar shadings: A review. *Renew. Sustain. Energy Rev.* **2016**, *55*, 863–884. [CrossRef]
11. Barozzi, M.; Lienhard, J.; Zanelli, A.; Monticelli, C. The Sustainability of Adaptive Envelopes: Developments of Kinetic Architecture. *Procedia Eng.* **2016**, *155*, 275–284. [CrossRef]
12. Yi, H.; Kim, D.; Kim, Y.; Kim, D.; Koh, J.S.; Kim, M.J. 3D-printed attachable kinetic shading device with alternate actuation: Use of shape-memory alloy (SMA) for climate-adaptive responsive architecture. *Autom. Constr.* **2020**, *114*, 103151. [CrossRef]
13. Favoino, F.; Jin, Q.; Overend, M. Design and control optimisation of adaptive insulation systems for office buildings. Part 1: Adaptive technologies and simulation framework. *Energy* **2017**, *127*, 301–309. [CrossRef]
14. Jin, Q.; Favoino, F.; Overend, M. Design and control optimisation of adaptive insulation systems for office buildings. Part 2: A parametric study for a temperate climate. *Energy* **2017**, *127*, 634–649. [CrossRef]
15. Favoino, F.; Goia, F.; Perino, M.; Serra, V. Experimental assessment of the energy performance of an advanced responsive multifunctional façade module. *Energy Build.* **2014**, *68*, 647–659. [CrossRef]
16. Favoino, F.; Goia, F.; Perino, M.; Serra, V. Experimental analysis of the energy performance of an ACTive, RESponsive and Solar (ACTRESS) façade module. *Sol. Energy* **2016**, *133*, 226–248. [CrossRef]
17. Kuznik, F.; David, D.; Johannes, K.; Roux, J.J. A review on phase change materials integrated in building walls. *Renew. Sustain. Energy Rev.* **2011**, *15*, 379–391. [CrossRef]
18. Elias, C.N.; Stathopoulos, V.N. A comprehensive review of recent advances in materials aspects of phase change materials in thermal energy storage. *Energy Procedia* **2019**, *161*, 385–394. [CrossRef]
19. Neri, G.; Koehler, A.; De Parolis, M.N.; Zolesi, V. ESA conditioned container: A system for passive temperature controlled transportation of experiments for the international space station. In Proceedings of the International Astronautical Congress, Naples, Italy, 1–5 October 2012.
20. Soares, N.; Costa, J.J.; Gaspar, A.R.; Santos, P. Review of passive PCM latent heat thermal energy storage systems towards buildings' energy efficiency. *Energy Build.* **2013**, *59*, 82–103. [CrossRef]
21. Manivel, R.; Muthukumaran, V.; Nekilesh, S.; Kandharooban, S. Design of thermal storage using phase change material (PCM) for agro products preservation. *Int. J. Recent Technol. Eng.* **2019**, *8*, 1669–1671. [CrossRef]
22. Farzanehnia, A.; Khatibi, M.; Sardarabadi, M.; Passandideh-Fard, M. Experimental investigation of multiwall carbon nanotube/paraffin based heat sink for electronic device thermal management. *Energy Convers. Manag.* **2019**, *179*, 314–325. [CrossRef]
23. De Gracia, A.; Cabeza, L.F. Phase change materials and thermal energy storage for buildings. *Energy Build.* **2015**, *103*, 414–419. [CrossRef]
24. Berardi, U.; Soudian, S. Benefits of latent thermal energy storage in the retrofit of Canadian high-rise residential buildings. *Build. Simul.* **2018**, *11*, 709–723. [CrossRef]
25. Fiorito, F. Phase-change materials for indoor comfort improvement in lightweight buildings. A parametric analysis for Australian climates. *Energy Procedia* **2014**, *57*, 2014–2022. [CrossRef]
26. De Matteis, V.; Cannavale, A.; Martellotta, F.; Rinaldi, R.; Calcagnile, P.; Ferrari, F.; Ayr, U.; Fiorito, F. Nano-encapsulation of phase change materials: From design to thermal performance, simulations and toxicological assessment. *Energy Build.* **2019**, *188–189*, 1–11. [CrossRef]
27. Kośny, J. *PCM-Enhanced Building Components—An Application of Phase Change Materials in Building Envelopes and Internal Structures*; Springer: Berlin/Heidelberg, Germany, 2015; ISBN 9783319142852. [CrossRef]
28. Cabeza, L.F.; Castell, A.; Barreneche, C.; De Gracia, A.; Fernández, A.I. Materials used as PCM in thermal energy storage in buildings: A review. *Renew. Sustain. Energy Rev.* **2011**, *15*, 1675–1695. [CrossRef]
29. Kenisarin, M.M. Thermophysical properties of some organic phase change materials for latent heat storage. A review. *Sol. Energy* **2014**, *107*, 553–575. [CrossRef]
30. Yuan, Y.; Zhang, N.; Tao, W.; Cao, X.; He, Y. Fatty acids as phase change materials: A review. *Renew. Sustain. Energy Rev.* **2014**, *29*, 482–498. [CrossRef]
31. Su, W.; Darkwa, J.; Kokogiannakis, G. Review of solid-liquid phase change materials and their encapsulation technologies. *Renew. Sustain. Energy Rev.* **2015**, *48*, 373–391. [CrossRef]
32. Milián, Y.E.; Gutiérrez, A.; Grágeda, M.; Ushak, S. A review on encapsulation techniques for inorganic phase change materials and the influence on their thermophysical properties. *Renew. Sustain. Energy Rev.* **2017**, *73*, 983–999. [CrossRef]
33. Salunkhe, P.B.; Shembekar, P.S. A review on effect of phase change material encapsulation on the thermal performance of a system. *Renew. Sustain. Energy Rev.* **2012**, *16*, 5603–5616. [CrossRef]
34. Gil, A.; Medrano, M.; Martorell, I.; Lázaro, A.; Dolado, P.; Zalba, B.; Cabeza, L.F. State of the art on high temperature thermal energy storage for power generation. Part 1-Concepts, materials and modellization. *Renew. Sustain. Energy Rev.* **2010**, *14*, 31–55. [CrossRef]

35. Navarro, L.; de Gracia, A.; Colclough, S.; Browne, M.; McCormack, S.J.; Griffiths, P.; Cabeza, L.F. Thermal energy storage in building integrated thermal systems: A review. Part 1. active storage systems. *Renew. Energy* **2016**, *88*, 526–547. [CrossRef]
36. Weinläder, H.; Körner, W.; Strieder, B. A ventilated cooling ceiling with integrated latent heat storage—Monitoring results. *Energy Build.* **2014**, *82*, 65–72. [CrossRef]
37. Lizana, J.; de-Borja-Torrejon, M.; Barrios-Padura, A.; Auer, T.; Chacartegui, R. Passive cooling through phase change materials in buildings. A critical study of implementation alternatives. *Appl. Energy* **2019**, *254*. [CrossRef]
38. Fiorentini, M.; Cooper, P.; Ma, Z. Development and optimization of an innovative HVAC system with integrated PVT and PCM thermal storage for a net-zero energy retrofitted house. *Energy Build.* **2015**, *94*, 21–32. [CrossRef]
39. Real, A.; García, V.; Domenech, L.; Renau, J.; Montés, N.; Sánchez, F. Improvement of a heat pump based HVAC system with PCM thermal storage for cold accumulation and heat dissipation. *Energy Build.* **2014**, *83*, 108–116. [CrossRef]
40. de Gracia, A.; Navarro, L.; Castell, A.; Cabeza, L.F. Energy performance of a ventilated double skin facade with PCM under different climates. *Energy Build.* **2015**, *91*, 37–42. [CrossRef]
41. Sharif, M.K.A.; Al-Abidi, A.A.; Mat, S.; Sopian, K.; Ruslan, M.H.; Sulaiman, M.Y.; Rosli, M.A.M. Review of the application of phase change material for heating and domestic hot water systems. *Renew. Sustain. Energy Rev.* **2015**, *42*, 557–568. [CrossRef]
42. Khan, M.M.A.; Ibrahim, N.I.; Mahbubul, I.M.; Ali, H.M.; Saidur, R.; Al-Sulaiman, F.A. Evaluation of solar collector designs with integrated latent heat thermal energy storage: A review. *Sol. Energy* **2018**, *166*, 334–350. [CrossRef]
43. Navarro, L.; Barreneche, C.; Castell, A.; Redpath, D.A.G.; Griffiths, P.W.; Cabeza, L.F. High density polyethylene spheres with PCM for domestic hot water applications: Water tank and laboratory scale study. *J. Energy Storage* **2017**, *13*, 262–267. [CrossRef]
44. Najafian, A.; Haghighat, F.; Moreau, A. Integration of PCM in domestic hot water tanks: Optimization for shifting peak demand. *Energy Build.* **2015**, *106*, 59–64. [CrossRef]
45. Hasan, A.; McCormack, S.J.; Huang, M.J.; Norton, B. Energy and cost saving of a photovoltaic-phase change materials (PV-PCM) System through temperature regulation and performance enhancement of photovoltaics. *Energies* **2014**, *7*, 1318–1331. [CrossRef]
46. Lu, W.; Liu, Z.; Flor, J.F.; Wu, Y.; Yang, M. Investigation on designed fins-enhanced phase change materials system for thermal management of a novel building integrated concentrating PV. *Appl. Energy* **2018**, *225*, 696–709. [CrossRef]
47. Faheem, A.; Ranzi, G.; Fiorito, F.; Lei, C. A numerical study on the thermal performance of night ventilated hollow core slabs cast with micro-encapsulated PCM concrete. *Energy Build.* **2016**, *127*, 892–906. [CrossRef]
48. Navarro, L.; de Gracia, A.; Niall, D.; Castell, A.; Browne, M.; McCormack, S.J.; Griffiths, P.; Cabeza, L.F. Thermal energy storage in building integrated thermal systems: A review. Part 2. Integration as passive system. *Renew. Energy* **2016**, *85*, 1334–1356. [CrossRef]
49. Cabeza, L.F.; Castellón, C.; Nogués, M.; Medrano, M.; Leppers, R.; Zubillaga, O. Use of microencapsulated PCM in concrete walls for energy savings. *Energy Build.* **2007**, *39*, 113–119. [CrossRef]
50. Ramakrishnan, S.; Wang, X.; Sanjayan, J.; Wilson, J. Thermal performance assessment of phase change material integrated cementitious composites in buildings: Experimental and numerical approach. *Appl. Energy* **2017**, *207*, 654–664. [CrossRef]
51. Kosny, J.; Kossecka, E.; Brzezinski, A.; Tleoubaev, A.; Yarbrough, D. Dynamic thermal performance analysis of fiber insulations containing bio-based phase change materials (PCMs). *Energy Build.* **2012**, *52*, 122–131. [CrossRef]
52. Castellón, C.; Medrano, M.; Roca, J.; Cabeza, L.F.; Navarro, M.E.; Fernández, A.I.; Lázaro, A.; Zalba, B. Effect of microencapsulated phase change material in sandwich panels. *Renew. Energy* **2010**, *35*, 2370–2374. [CrossRef]
53. Weinlaeder, H.; Koerner, W.; Heidenfelder, M. Monitoring results of an interior sun protection system with integrated latent heat storage. *Energy Build.* **2011**, *43*, 2468–2475. [CrossRef]
54. Weinläder, H.; Beck, A.; Fricke, J. PCM-facade-panel for daylighting and room heating. *Solar Energy* **2005**, *78*, 177–186. [CrossRef]
55. Barrenechea, C.; Navarro, H.; Serrano, S.; Cabeza, L.F.; Fernández, A.I. New database on phase change materials for thermal energy storage in buildings to help PCM selection. *Energy Procedia* **2014**, *57*, 2408–2415. [CrossRef]
56. Heim, D.; Clarke, J.A. Numerical modelling and thermal simulation of PCM-gypsum composites with ESP-r. *Energy Build.* **2004**, *36*, 795–805. [CrossRef]
57. Zhang, Y.; Zhou, G.; Lin, K.; Zhang, Q.; Di, H. Application of latent heat thermal energy storage in buildings: State-of-the-art and outlook. *Build. Environ.* **2007**, *42*, 2197–2209. [CrossRef]
58. US Department of Energy EnergyPlus Version 9.5 Documentation: Getting Started. *Energyplus^{TM}*. 2021. Available online: https://energyplus.net/sites/all/modules/custom/nrel_custom/pdfs/pdfs_v9.5.0/GettingStarted.pdf (accessed on 17 May 2021).
59. Favoino, F.; Giovannini, L.; Loonen, R. Smart glazing in Intelligent Buildings: What can we simulate? In Proceedings of the All eyes on glass: Conference proceedings of Glass Performance Days 2017, Tampere, Finland, 28–30 June 2017; pp. 8–15.
60. Field, K.; Deru, M.; Studer, D. Using Doe Commercial Reference Buildings for Simulation Studies. In Proceedings of the SimBuild 2010 Fourth National Conference of IBPSA-USA, New York, NY, USA, 11–13 August 2010.
61. Deru, M.; Field, K.; Studer, D.; Benne, K.; Griffith, B.; Torcellini, P.; Liu, B.; Halverson, M.; Winiarski, D.; Rosenberg, M.; et al. *U.S. Department of Energy Commercial Reference Building Models of the National Building Stock*; National Renewable Energy Laboratory: Golden, CO, USA, 2011.
62. Economidou, M. *Europe's Buildings under the Microscope. A Country-by-Country Review of the Energy Performance of Buildings*; Buildings Performance Institute Europe (BPIE): Brussel, Belgium, 2011; ISBN 9789491143014.
63. US Department of Energy EnergyPlus Engineering Reference: The Reference to EnergyPlus Calculations. *US Dep. Energy* **2021**. Available online: https://energyplus.net/documentation (accessed on 17 May 2021).

64. *ANSI/ASHRAE ASHRAE Guideline 14-2002 Measurement of Energy and Demand Savings*; ASHRAE: Atlanta, GA, USA, 2002.
65. ASHRAE. *Guideline 14-2014: Measurement of Energy, Demand, and Water Savings*; ASHRAE: Atlanta, GA, USA, 2014.
66. Decreto Ministeriale del 26 giugno 2015—Applicazione delle metodologie di calcolo delle prestazioni energetiche e definizione delle prescrizioni e dei requisiti minimi degli edifici. *Gazz. Uff. Della Repubb. Ital.* 2015. Attachment 1—Appendix A. Available online: https://www.gazzettaufficiale.it/eli/gu/2015/07/15/162/so/39/sg/pdf (accessed on 17 May 2021).
67. Gagarin, V.G. Thermal performance as the main factor of energy saving of buildings in Russia. *Procedia Eng.* **2016**, *146*, 112–119. [CrossRef]
68. Schettler-Kohler, H.-P.; Ahlke, I. EPBD implementation in Germany. *Concert. Action Energy Perform. Build.* **2019**, 211–229. Available online: https://www.epbd-ca.eu/wp-content/uploads/2018/08/CA-EPBD-IV-Germany-2018.pdf (accessed on 17 May 2021).
69. Danish Knolwledge Centre for Energy Savings in Buildings. *Energy Requirements of BR18: A Quick Guide for the Construction Industry on the Danish Building Regulations 2018*; Danish Knowledge Centre for Energy Savings in Buildings: Taastrup, 2018; Available online: https://byggeriogenergi.dk (accessed on 17 May 2021).
70. Brekke, T.; Isachsen, O.K.; Strand, M. EPBD Implementation in Norway. *Concert. Action Energy Perform. Build.* **2019**, 387–398. Available online: www.epbd-ca.eu (accessed on 17 May 2021).
71. CEN EN 16798-1:2019—Energy Performance of Buildings—Ventilation for Buildings—Part 1: Indoor Environmental Input Parameters for Design and Assessment of Energy Performance of Buildings Addressing Indoor Air Quality, Thermal Environment, Lighting and Acoustics. 2019. Available online: https://www.cen.eu/news/brief-news/pages/en-2019-022.aspx (accessed on 17 May 2021).
72. ANSI/ASHRAE/IES. *Standard 90.1-2019—Energy Standard for Buildings Except Low-Rise Residential Buildings*; American Society of Heating, Refrigerating and Air-Conditioning Engineers, Inc.: Atlanta, GA, USA, 2019.
73. Tabares-Velasco, P.C.; Christensen, C.; Bianchi, M. Verification and validation of EnergyPlus phase change material model for opaque wall assemblies. *Build. Environ.* **2012**, *54*, 186–196. [CrossRef]
74. Zastawna-Rumin, A.; Kisilewicz, T.; Berardi, U. Novel Simulation Algorithm for Modeling the Hysteresis of Phase Change Materials. *Energies* **2020**, *13*, 1200. [CrossRef]
75. Zwanzig, S.D.; Lian, Y.; Brehob, E.G. Numerical simulation of phase change material composite wallboard in a multi-layered building envelope. *Energy Convers. Manag.* **2013**, *69*, 27–40. [CrossRef]

applied sciences

MDPI

Article

Influence of Traditional and Solar Reflective Coatings on the Heat Transfer of Building Roofs in Mexico

Iván Hernández-Pérez

División Académica de Ingeniería y Arquitectura (DAIA-UJAT), Universidad Juárez Autónoma de Tabasco, Carretera Cunduacán-Jalpa de Méndez km. 1, Cunduacán, Tabasco CP 86690, Mexico; ivan.hernandezp@ujat.mx; Tel.: +52-777-227-4111

Abstract: Building roofs are sources of unwanted heat for buildings situated in zones with a warm climate. Thus, reflective coatings have emerged as an alternative to reject a fraction of the solar energy received by roofs. In this research, the thermal behavior of concrete slab roofs with traditional and solar reflective coatings was simulated using a computational tool. The studied slab configurations belong to two groups, non-insulated and insulated roofs. In the second group, the thermal insulation thickness complies with the value recommended by a national building energy standard. Weather data from four cities in Mexico with a warm climate were used as boundary conditions for the exterior surface of the roofs. The computational tool consisted of a numerical model based on the finite volume method, which was validated with experimental data. A series of comparative simulations was developed, taking a gray roof as the control case. The results demonstrated that white roofs without insulation had an exterior surface temperature between 11 and 16 °C lower than the gray roof without insulation. Thus, the daily heat gain of these white roofs was reduced by a factor ranging between 41 and 54%. On the other hand, white roofs with insulation reduced the exterior surface temperature between 17 and 21 °C compared to the gray roof with insulation. This temperature reduction caused insulated white roofs to have a daily heat gain between 37 and 56% smaller than the control case. Another contribution of this research is the assessment of two retrofitting techniques when they are applied at once. In other words, a comparison between a non-insulated gray roof and an insulated white roof revealed that the latter roof had a daily heat gain up to 6.4-times smaller than the first.

Keywords: solar reflective coatings; heat transfer; daily heat gains; cool roofs

Citation: Hernández-Pérez, I. Influence of Traditional and Solar Reflective Coatings on the Heat Transfer of Building Roofs in Mexico. *Appl. Sci.* **2021**, *11*, 3263. https://doi.org/10.3390/app11073263

Academic Editor: Tiziana Poli

Received: 28 February 2021
Accepted: 2 April 2021
Published: 6 April 2021

1. Introduction

The buildings sector used 36% of the total final energy around the world and had 39% of the energy-related CO_2 emissions in 2018 [1]. Because the population is expected to increase by 2.5 billion people by 2050, the energy use in the building sector is set to rise sharply. In warm locations, the energy consumption from air conditioners is high due to the heat flow received by buildings situated in these zones. The building envelope plays a vital role in the thermal interaction between the outdoor and indoor environments. Thus, it is important to minimize the energy gain from the building envelope to avoid the excessive use of electricity for comfort purposes.

Today, several technologies are available for building energy retrofitting. There are advanced facades [2], highly insulated windows [3], high insulation levels for roofs and walls [4], reflective coatings [5], phase change materials [6], and well-sealed structures [7], to mention a few. In particular, reflective surfaces are becoming popular for two main reasons. First, the most direct way to reduce the incident solar energy is to reflect it. When new, solar reflective coatings can reflect to the sky up to 90% of the solar energy received by a surface. The second reason is that applying reflective coatings to opaque building components is probably the most simple passive measure because most of these coatings can be installed in the same way as ordinary paint [8]. These coatings are usually used on building roofs because these components are subject to solar radiation for more hours than other building envelope components [9]. Therefore, when a roof is retrofitted with a reflective coating, it is known as a cool roof.

Several researchers around the world have studied the benefits that cool roofs can bring to buildings in different types of climates using a number of approaches. Several experimental studies are available in the literature. For instance, Pisello and Cotana [10] performed a two-year monitoring campaign to test cool roofs in a residential building located in Italy. The results showed that the cool roof solution reduced the peak temperature of the air in the attic by around 5 °C. Further, the cool roof provided an overall year-round energy savings that corresponded to 14 kWh/m^2 per year. In China, Quin et al. [11] used a building cell to test different samples of concrete tiles with different solar reflectance. The authors performed a series of experiments and found that the tiles with a reflective coating reached lower interior surface temperatures than the control tile. Moreover, an additional 2–6 °C cooler temperature around the noon time was achieved by installing at the bottom of the tile a low-emissivity sheathing. Hernández-Pérez et al. [12] evaluated several building roofs with different coatings using two outdoor test cells located in Cuernavaca, Mexico. They showed that in summer, the white roof was 29 °C cooler than the black roof and just 1.5 °C warmer than the ambient air. Further, the average daily energy gain of the white roof was 73% less than the black roof.

Other studies aiming to investigate the year-round energy savings from cool roofs by using building energy simulation tools are available. Algarni [13] studied the influence of solar reflective roofs on the energy consumption of residential buildings of Saudi Arabia. The author used eQuest building simulation software to perform simulations of a building prototype using weather data of 13 major cities of this country. The researcher found that a reflective roof reduced the annual energy consumption required for building cooling by between 110.3 and 181.9 kWh/m^2. The maximum increase in annual energy consumption due to winter heating was only 4.4 kWh/m^2. Piselli et al. [14] assessed the effectiveness of cool roofs with optimal insulation levels in different weather conditions worldwide. The authors coupled a dynamic energy simulation tool with an optimization technique to find the best-combined building roof thermal insulation and solar reflectance and minimize the annual energy consumption. The results showed that a high value of solar reflectance reduced the annual energy consumption for most of the analyzed climate zones. Thus, they concluded that the optimal roof configuration must have high solar reflectance and no/a low insulation level. Dominguez-Delgado et al. [15] developed an energy and economic life cycle assessment of cool roofs applied to residential buildings in Southern Spain. The simulations demonstrated that the cool roof decreased the annual energy consumption. The maximum decrease found by the authors was 32%, and it was obtained when a roof with a solar reflectance of 0.1 was retrofitted with a reflective coating with a solar reflectance of 0.9. The LCA analysis reported that savings were 18.33 €/m^2, and the payback period was around thee years.

Some researchers have used validated models to predict the benefits of cool roofs. For instance, Tong et al. [16] studied the thermal behavior of ventilated and non-ventilated roofs during a typical day in Singapore. The authors conducted experiments to validate the concrete roof models. Compared with the roofs with a solar reflectance of 0.1, each 0.1 increase in reflectance reduced the daytime heat gain by 11% on the ventilated roofs and the non-ventilated roofs. The reflective coatings application reduced the daily heat gain by 234 and 135 Wh/m^2 in the ventilated and non-ventilated roofs, respectively. Further, they indicated that compared to non-ventilated roofs (both reflective and non-reflective), the individual use of ventilation and 2.5 cm of expanded polystyrene (EPS) reduced the heat gain by 42 and 68%, respectively. Zingre et al. [17] developed a model to study heat transfer in roofs. They used a spectral approximation method to solve the unsteady one-dimensional heat conduction equation. Furthermore, the model was validated using experimental data obtained from measurements made in two identical apartments with concrete roofs located in Singapore. Model predictions showed that on a sunny day, the reflective coating reduced the maximum roof temperature, indoor air temperature, and daytime heat gain by 14.1 °C, 2.4 °C, and 0.66 kWh/m^2 (54%), respectively.

As shown in the literature review, reflective or cool roofs have brought a series of benefits to the buildings in which they are installed. Thus, it is essential to understand how a cool roof behaves in a particular area. For this purpose, it is necessary to have tools capable of predicting the thermal behavior of a cool roof before its installation to determine if it is feasible from an energy point of view. This work presents the development of a computational tool for modeling the heat transfer of concrete slab-type roofs with traditional and solar reflective coatings. This tool is then used to estimate the temperature reductions and, consequently, the ability of reflective roofs to modulate the heat gains by implementing this technology in buildings situated in four different warm climates in Mexico.

2. Physical Model

Figure 1 shows the physical model of the roof, which is a concrete slab with traditional or solar reflective coating. Two configurations were considered: (a) a single roof with a thickness of 10 cm and (b) an insulated roof with a thickness of 13 cm. The insulated roof was made of concrete and polystyrene, and a thin plaster layer of concrete protected the polystyrene because it should not be exposed to solar radiation. The polystyrene had a thickness equal to 2.5 cm, and the plaster layer had a thickness of 0.5 cm. The width of the roof (W) was considered 1 m. Both configurations were studied with traditional and solar reflective coatings. The thickness of the coatings was considered negligible. The materials of both roof configurations were considered homogeneous. The solar absorptance and thermal emissivity of the roofs were also regarded as constant. Further, it was considered that the side surfaces of the two roof configurations were adiabatic. Finally, it was supposed that solar radiation had a normal incidence on the roof and that this was a gray body that only absorbed the incident solar radiation.

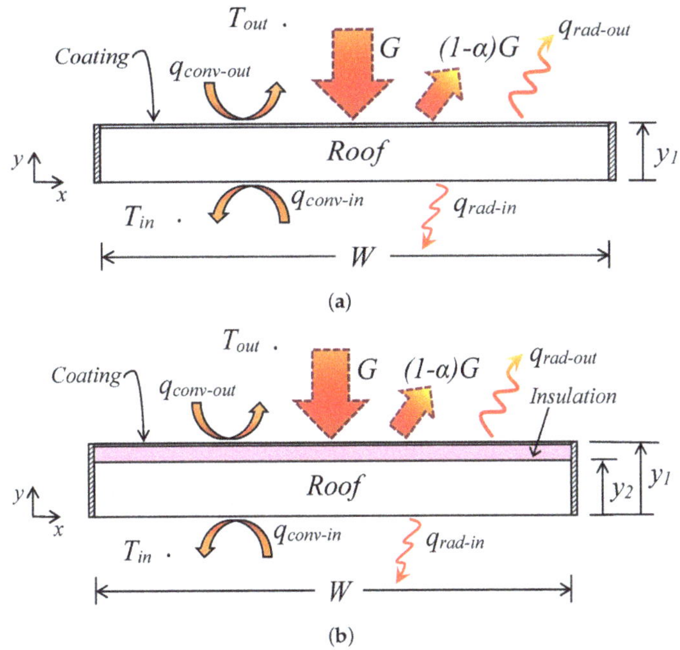

Figure 1. Physical model. (**a**) Single roof; (**b**) insulated roof.

3. Mathematical Model

The partial differential equation for transient heat conduction of a two-dimensional solid is [18]:

$$\frac{\partial(\rho c_p T)}{\partial t} = \frac{\partial}{\partial x}\left(\lambda\frac{\partial T}{\partial x}\right) + \frac{\partial}{\partial y}\left(\lambda\frac{\partial T}{\partial y}\right) \tag{1}$$

where λ is the thermal conductivity, ρ is the density, and c_p is the specific heat. The boundary conditions used to solve in the physical model were as follows: The exterior surface ($y = y_1$) is in contact with the outdoor environment or environmental conditions of different cities of Mexico. Therefore, the roof has convective and radiative exchange with the surroundings. The lateral surfaces are thermally insulated ($x = 0$ and $x = W$), and the interior surface ($y = 0$) also exchanges heat by convection and radiation with the indoor environment. Mathematically, the boundary conditions can be expressed as:

$$-\lambda\frac{\partial T}{\partial y} = \alpha G + h_{out}(T - T_{out}) + \sigma\varepsilon(T^4 - T_{sky}^4) \quad \text{for} \quad y = y_1, \quad 0 \le x \le W \tag{2}$$

$$-\lambda\frac{\partial T}{\partial y} = h_{in}(T - T_{in}) \quad \text{for} \quad y = 0, \quad 0 \le x \le W \tag{3}$$

$$\frac{\partial T}{\partial x} = 0 \quad \text{for} \quad x = 0, \quad 0 < y < y_1 \tag{4}$$

$$\frac{\partial T}{\partial x} = 0 \quad \text{for} \quad x = W, \quad 0 < y < y_1 \tag{5}$$

In Equation (2), G is the solar radiation received by the roof, α is the solar absorptance of the coating, ε is the thermal emittance, and σ is the Stefan–Boltzmann constant. The beam-solar radiation received by roofs has different angles during the day. The solar radiation data used for the simulations were obtained from measurements performed using weather stations, and these stations used pyranometers for measuring total (beam plus diffuse) radiation over a horizontal surface. The detectors of the pyranometers had a response that did not depend on radiation wavelength over the solar spectrum. Further, these devices had a response independent of the angle of incidence of solar radiation. Thus, it can be supposed that the roofs received total radiation in a perpendicular direction. To calculate the outdoor convective heat transfer coefficient (h_{out}) in Equation (2), the following empirical correlation was used [19]:

$$h_{out} = 2.8 + 3.0v \tag{6}$$

where v is the wind speed in m/s and h_{out} is the outdoor convective heat transfer coefficient in W/m^2K. The value of T_{sky} is calculated with the following expression [19]:

$$T_{sky} = 0.0552 T_{out}^{1.5} \tag{7}$$

In Equation (7), T_{out} is the outdoor air temperature, and both temperatures T_{sky} and T_{out} are expressed in K. The heat transfer coefficient in the interior environment h_{in} in Equation (3), which considers both convection and thermal radiation, is equal to 6.13 W/m^2 when the heat flux heat goes to the indoor air and 9.26 W/m^2K when the heat flux goes from the indoor air to the interior surface [20]. For the roof configuration (b) of Figure 1, due to the different materials involved and the thermophysical properties' changes across the interface between concrete and thermal insulation, the harmonic mean was used to calculate these properties in the interface. Furthermore, perfect contact between both materials was considered, such that heat flux through the interface was the same for the materials involved. The properties of the concrete used for the simulations were ρ = 2400 kg/m^3, c_p = 1080 J/(kg·K), and λ = 1.8 W/(m·K). On the other hand, the properties of the insulation were ρ = 28 kg/m^3, c_p = 1800 J/(kg·K), and λ = 0.033 W/(m·K).

4. Solution Methodology for the Roof Model

This section present the methodology followed to numerically solve the heat conduction Equation (1). The following steps were developed:

- Generation of the computational mesh.
- Discretization of the mathematical model.
- Solution of the system of algebraic equations.

The discretization technique used in this work was the finite volume method (FVM), and the following subsections describe each of these steps.

4.1. Generation of the Computational Mesh

This step consisted of dividing the domain into small control volumes where the nodes were situated, and the value of the temperature $T(x,y)$ was determined. In this analysis, the main node P was located in the center of the control volume (Figure 2), so that the its interface (point w, e, n, and s) was at the middle from one node to another.

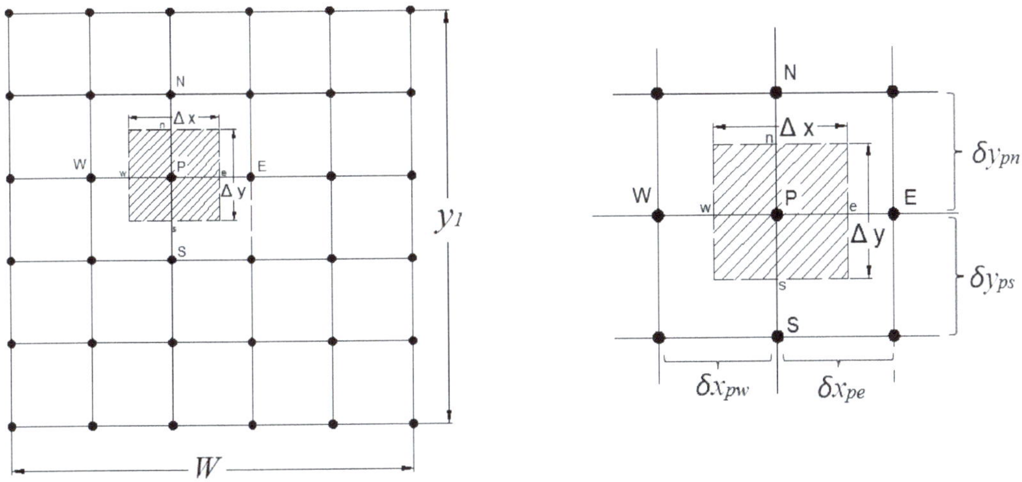

Figure 2. A control volume in the computational mesh.

Equation (8) describes the distribution of the nodes in the mesh in the direction of x:

$$x(i) = \frac{W}{N_x - 1}(i-1) \quad \text{for} \quad i = 1,2,3,\ldots,N_x \tag{8}$$

where N_x is the number of nodes in the x direction and W is the width of the building roof (Figure 1). This equation provides the direction coordinate of the nodes x. For the nodes in direction y, the coordinate is given by:

$$y(j) = \frac{y_1}{N_y - 1}(j-1) \quad \text{for} \quad j = 1,2,3,\ldots,N_y \tag{9}$$

where N_y is the number of nodes in the y direction and y_1 is the height of the building roof (Figure 1). It is also necessary to know the thickness of the control volume in both directions, and Equation (10) is used to calculate such thicknesses:

$$\Delta x = \frac{W}{N_x - 1} \qquad\qquad \Delta y = \frac{y_1}{N_y - 1} \tag{10}$$

4.2. Discretization of the Mathematical Model

Discretization consists of applying a technique to transform the partial differential equation of the mathematical model into a set of algebraic expressions to facilitate its solution. For the internal nodes, the Equation (1) governs their behavior, and this equation is known as the general differential equation for diffusion processes and can be represented by:

$$\frac{\partial \rho\phi}{\partial t} = \frac{\partial}{\partial x}\left(\Gamma\frac{\partial \phi}{\partial x}\right) + \frac{\partial}{\partial y}\left(\Gamma\frac{\partial \phi}{\partial y}\right) \quad \text{for} \quad 0 < x < W; \quad 0 < y < y_1$$

where: $\phi = T, \Gamma = \lambda/c_p$. Taking the domain of interest, for each term of the previous equation, between the points w and e in the x direction of the control volume in Figure 2 and between the points n and s in the y direction and, furthermore, considering the time step, a domain of interest is taken from an earlier time (t_0) to a later time ($t = t_0 + \Delta t$):

$$\begin{aligned}\int_{t_0}^{t}\int_{s}^{n}\int_{w}^{e}\frac{\partial \rho\phi}{\partial t}\,dx\,dy\,dt &= \int_{s}^{n}\int_{w}^{e}\frac{\partial}{\partial x}\left(\Gamma\frac{\partial \phi}{\partial x}\right)dx\,dy + \int_{s}^{n}\int_{w}^{e}\frac{\partial}{\partial y}\left(\Gamma\frac{\partial \phi}{\partial y}\right)dx\,dy \\ &+ \int_{s}^{n}\int_{w}^{e} S\,dx\end{aligned}$$

Integrating in the space the previous equation over the control volume, the following is obtained:

$$\left[\frac{\partial \overline{\rho\phi}}{\partial t}\right]\Delta x\Delta y = \left[\left(\Gamma\frac{\partial \phi}{\partial x}\right)_e - \left(\Gamma\frac{\partial \phi}{\partial x}\right)_w\right]\Delta y + \left[\left(\Gamma\frac{\partial \phi}{\partial y}\right)_n - \left(\Gamma\frac{\partial \phi}{\partial y}\right)_s\right]\Delta x + \overline{S}\Delta x\Delta y$$

Now, using an implicit interpolation scheme for time:

$$\begin{aligned}\left[\frac{(\rho\phi)_P^t - (\rho\phi)_P^{t_0}}{\Delta t}\right]\Delta x\Delta y &= \left[\left(\Gamma\frac{\partial \phi}{\partial x}\right)_e^t - \left(\Gamma\frac{\partial \phi}{\partial x}\right)_w^t\right]\Delta y + \left[\left(\Gamma\frac{\partial \phi}{\partial y}\right)_n^t - \left(\Gamma\frac{\partial \phi}{\partial y}\right)_s^t\right]\Delta x \\ &+ \overline{S}\Delta x\Delta y\end{aligned}$$

Because the conditions at the interfaces were unknown and the centered scheme interpolation was used, using the known values of the nodes adjacent to that control volume interface (nodes E, W, N, and S), the following was obtained:

$$\begin{aligned}\left[\frac{(\rho\phi)_P^t - (\rho\phi)_P^{t_0}}{\Delta t}\right]\Delta x\Delta y &= \Gamma_e\left(\frac{\phi_E - \phi_P}{\delta x_{PE}}\right)\Delta y - \left(\frac{\phi_P - \phi_W}{\delta x_{PW}}\right)\Delta y + \Gamma_n\left(\frac{\phi_N - \phi_P}{\delta y_{PN}}\right)\Delta x \\ &- \Gamma_s\left(\frac{\phi_P - \phi_S}{\delta y_{PS}}\right)\Delta x + \overline{S}\Delta x\Delta y\end{aligned}$$

It is convenient to group the terms of the equation into coefficients as follows:

$$\begin{aligned}&\underbrace{\phi_P\left[\left(\frac{\rho\Delta x\Delta y}{\Delta t}\right) + \left(\frac{\Gamma_e}{\delta x_{PE}} + \frac{\Gamma_w}{\delta x_{PW}}\right)\Delta y + \left(\frac{\Gamma_n}{\delta y_{PN}} + \frac{\Gamma_s}{\delta y_{PS}}\right)\Delta x\right]}_{a_P} = \phi_E\underbrace{\left(\frac{\Gamma_e\Delta y}{\delta x_{PE}}\right)}_{a_E} \\ &+\phi_W\underbrace{\left(\frac{\Gamma_w\Delta y}{\delta x_{PW}}\right)}_{a_W} + \phi_N\underbrace{\left(\frac{\Gamma_n\Delta x}{\delta y_{PN}}\right)}_{a_N} + \phi_S\underbrace{\left(\frac{\Gamma_s\Delta x}{\delta x_{PS}}\right)}_{a_S} \underbrace{+\phi_P^{t_0}\underbrace{\left(\frac{\rho\Delta x\Delta y}{\Delta t}\right)^{t_0}}_{a_P^0} + \overline{S}\Delta x\Delta y}_{b}\end{aligned}$$

Therefore:

$$a_E = \frac{\Gamma_e \Delta y}{\delta x_{PE}} \qquad a_W = \frac{\Gamma_w \Delta y}{\delta x_{PW}} \qquad a_N = \frac{\Gamma_n \Delta x}{\delta y_{PN}} \qquad a_S = \frac{\Gamma_s \Delta x}{\delta x_{PS}}$$

$$a_P^0 = \frac{\rho \Delta x \Delta y}{\Delta t}$$

$$b = S\Delta x \Delta y + a_P^0 \phi_P^0$$

$$a_P = a_E + a_W + a_N + a_S + a_P^0 - \overline{S}\Delta x \Delta y$$

We obtain here Equation (11), which is the generative equation of the system of algebraic equations in the notation of grouped coefficients:

$$a_P \phi_P = a_E \phi_E + a_W \phi_W + a_N \phi_N + a_S \phi_S + b \tag{11}$$

It is also necessary to discretize the boundary nodes. According to the physical model, there is a third class condition (interior surface of the roof), two second class conditions at the left and right ends, and another third class with radiative loss or gain in the exterior surface.

4.3. Solution of the System of Algebraic Equations

If the system of algebraic equations resulting from the discretization of a two-dimensional model is adjusted in a matrix way, a matrix of pentagonal and diagonally dominant coefficients is obtained. For the one-dimensional case, the Thomas algorithm or tridiagonal matrix algorithm (TDMA) is applied as a direct method by the dominant tridiagonal matrix of coefficients; however, Thomas's algorithm can be used for the two-dimensional case by combining it with iterative equation solving methods. In this work, the system of algebraic equations was solved using the line-by-line Gauss–Seidel method with alternating directions (LGS-ADI).

5. Verification and Validation of the Mathematical Model

The mathematical model as mentioned above was solved using the finite volume method; thus, it was necessary to develop a computer simulation tool that could follow the solution methodology described in Section 4. The numerical code was developed in the Fortran programming language. First, the mathematical model was used to solve a reference exercise available in the literature to verify that it was able to provide reliable results. It was verified by comparing its results against the analytical solution for a composite solid reported by Chen and Paine [21]. Then, the mathematical model was subjected to another test known as validation, where the results of the model were compared with data from temperature measurements from a roof under outdoor ambient conditions. To validate the numerical model that solved the heat conduction equation, experimental data from a previous publication of the author were used [22]. The model was validated for two cases: a conventional gray roof and a white reflective roof. In the experiment performed by Hernández-Pérez et al. [22] with two test cells, the exterior surface temperature of the roofs was measured every 10 min for five days. Figure 3 shows the temperatures obtained by solving the model and the experimental temperatures. This figure shows that the model satisfactorily reproduced the behavior of the gray roof and the white reflective roof. The maximum deviations of the temperatures obtained for the gray and white roofs were 5.5% and 4.6%, respectively. Therefore, this model can be used to study the thermal performance of concrete roof slabs in different weather conditions. One advantage that the current model brought is that most of the validated tools available were validated for a single day [17,23]. In contrast, the model presented by the author was validated by using data from an experimental test performed for five consecutive days. In other words, this tool was validated by comparing its results with 720 temperature values for each type of roof.

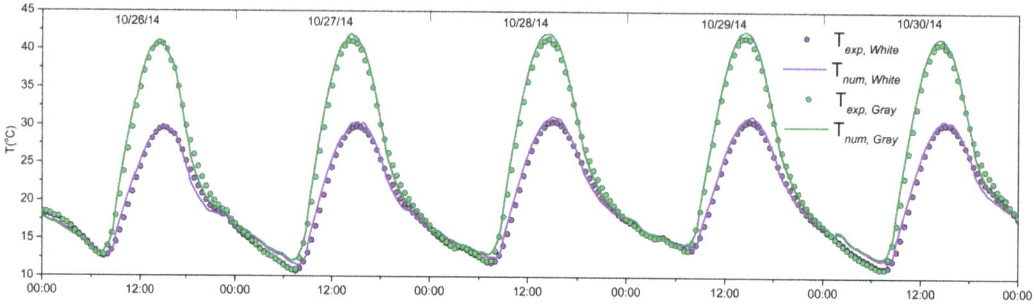

Figure 3. Comparison of the experimental data from Hernández-Pérez et al. [22] and the data obtained with the numerical model.

6. Weather Data

Once the roof model was verified and validated, its thermal behavior was studied under the outdoor environment of different cities. Data from four representative towns in Mexico with warm weather were used. Table 1 shows the four major cities with their corresponding type of weather. The weather data used for the simulations were filtered from files provided by the National Meteorological Service-National Commission of Water (Servicio Meteorológico Nacional-Comisión Nacional del Agua (SMN-CONAGUA)), the public body responsible for providing information on the state of the weather that prevails or affects the territory of this country. The SMN-CONAGUA obtains the weather data files from weather stations situated in each city. The data files contained recorded several variables recorded every ten minutes for a whole year. Still, only solar radiation, wind speed, and air temperature for the simulations were used, as shown in the Results Section. The thermal behavior of the roofs was analyzed for the week with the highest temperatures of the year 2018.

Table 1. Representative cities of the zones with warm weather in Mexico.

City	Weather Type
Hermosillo, Sonora	Warm dry
Monterrey, N.L.	Warm semi-dry
Villahermosa, Tabasco	Warm humid
Mérida, Yucatán	Warm semi-humid

7. Properties of Traditional and Reflective Coatings

Table 2 shows the values of solar absorptance and thermal emittance of the four coatings analyzed in this research. There were two traditional coatings, gray and terracotta, and two solar reflective coatings, White # 1 and White # 2. These optical properties of the coatings were measured in two previous works of the author [12,22]. The solar reflectance was measured using a spectrophotometer with an integrating sphere following the procedure indicated in the standard ASTM E 903-12. After obtaining the solar reflectance, the solar absorptance was calculated because this property was required in the boundary conditions of the model as presented above. A portable ambient temperature emissometer was used, according to the standard ASTM C1371 (2015): laboratory or field measurement of hemispherical thermal emittance with a portable emissometer (indirect technique using calibrated references) the thermal emittance obtained for all coating samples obtained at ambient temperature. The emittance of the coatings can be considered a constant value independent of temperature because the coatings were opaque materials that could be considered as paint, and according to Duffie and Beckman [19], the emittance of these materials only changed slightly at very high temperatures.

Table 2. Optical properties of the coatings.

Coating	Solar Absorptance (α)	Thermal Emittance (ε)
Gray	0.67	0.87
Terracotta	0.70	0.88
White # 1	0.20	0.90
White # 2	0.16	0.89

8. Results

This section presents the concrete roof's behavior with four coatings in terms of the temperature of the exterior surface, the temperature of the interior surface, and the heat flux traveling through the roofs. Further, the cumulative heat gain was obtained by determining the area under the heat flux curve for each day. This section shows first the results for the single roof configuration and then the corresponding results for the insulated roof configuration.

8.1. Single Roof

As mentioned above, the roofs' thermal behavior was analyzed using the weather data of the week with the warmest outdoor air temperatures of 2018 using weather data from four representative towns in Mexico with a warm climate. The detailed results of Hermosillo are shown here because it was the town with the highest outdoor temperature. At the end of this subsection, a summary table presents the results of the thermal evaluation of the single roof in all cities.

For Hermosillo Sonora, the days selected were from 30 May to 5 June 2018 because this was the week with the highest temperatures of the year. Figure 4a shows the solar irradiation and the wind speed during the seven days analyzed. The maximum solar radiation was around 1000 W/m^2. According to the behavior of solar irradiance, all days selected were clear days without clouds. The maximum wind speed reached around 5 m/s. Figure 4b shows the air temperature; due to the type of weather of this city, the ambient air reached very high temperatures, with an average maximum temperature of 45 °C.

Figure 4b presents the temperature of the exterior surface of the roofs and the ambient air temperature during the selected week. Conventional roofs had a similar behavior, and on the other hand, reflective roofs maintained a similar behavior concerning the temperature of the exterior surface. This effect occurred because the solar reflectance of conventional coatings was very similar. The exterior surface of the roofs reached its maximum temperature between 14:30 and 15:00 h. The exterior surface of the single terracotta roof (STR), the single gray roof (SGR), Single White Roof #1 (SWR1), and Single White Roof #2 (SWR2) reached on average 61, 59, 45, and 43 °C, respectively. These values indicated that applying the terracotta coating, on average, increased the temperature of the exterior surface by 2 °C compared to the SGR. In contrast, SWR1 and SWR2 reduced the temperature of the exterior surface by on average 14 and 16 °C, respectively. Furthermore, if the average maximum temperatures of reflective roofs were compared with the average maximum temperature of ambient air (45 °C), SWR1 had the same maximum temperature as the ambient air. In contrast, SWR2 reached a temperature of 2 °C lower than the maximum air temperature. On the other hand, the SGR and STR reached a temperature of 14 and 16 °C higher than the ambient air temperature.

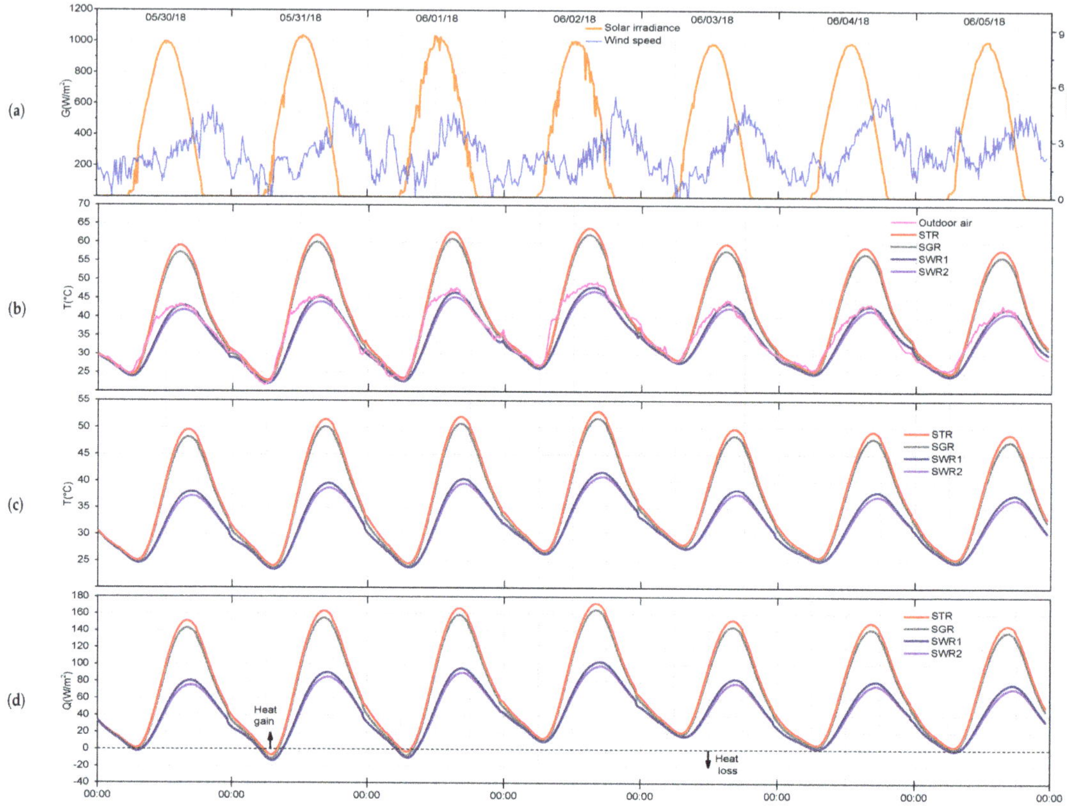

Figure 4. Thermal behavior of the single roof with traditional and solar reflective coatings in Hermosillo: (**a**) solar irradiance and wind speed; (**b**) temperature of exterior surface and outdoor air; (**c**) temperature of interior surface; (**d**) heat flux of the roofs.

Figure 4c presents the temperature of the single roofs' interior surface. The interior surface of the roofs reached their maximum temperature between 16:00 and 16:40 h. The temperature of the interior surface of the STR, SGR, SWR1, and SWR2 reached a maximum temperature of 51, 49, 38, and 37 °C, respectively. Therefore, the influence of a reflective coating on the interior surface temperature was obtained by comparing the previous temperature values. SWR1 decreased the temperature of the interior surface by around 10 °C compared to the SGR, while SWR2 reduced the temperature of this surface by around 11 °C.

Figure 4d shows the behavior of the heat flux of the roofs in Hermosillo during the seven days analyzed. The maximum heat flux traveling through the roofs occurred between 16:00 and 16:50 h. The STR, SGR, SWR1, and SWR2 had an average peak heat flux of 157, 150, 87, and 82 W/m^2, respectively. These values indicated that the peak heat flux crossing SWR1 and SWR2 was 41% and 45% smaller than that corresponding to the SGR, while the heat flux of the STR was 5% greater than that of the SGR. The cumulative heat gain for one day or daily heat gain was obtained by calculating the area under the heat flux curve for each day. During the seven days analyzed, the STR had an average heat gain of 1793 W·h/(m^2-day), the SGR a gain of 1675 W·h/(m^2-day), the SWR1 of 1045 W·h/(m^2-day), and the SWR2 roof of 993 W·h/(m^2-day). Thus, by calculating the percentage difference between the average daily heat gain, it was found that the TSR located in Hermosillo had a 7% greater heat gain than the SGR, while the SWR1 and SWR2 roofs had 38 and 41% less heat gain than the gray roof.

A similar procedure for the other three cities was developed to perform the simulations; the week with the warmest outdoor air temperatures of 2018 was selected. The weather data from this week were introduced to the simulation tool. Table 3 presents a summary of the results obtained from the evaluation of the single roof with traditional and solar reflective coatings in the four cities of Mexico. The table shows the average peak surface temperatures (T_{es} and T_{is}), the average peak heat flux (Q_R), and the average daily heat gain of the roofs (HG). Taking the SGR as a reference, the percentage differences between the peak heat flux of this roof and the other cases are also given within parenthesis (%) in the table, and the same is done for the heat gain. Table 3 demonstrates that SWR1 and SWR2 were able to reduce T_{es} between 11 and 16 °C compared to the gray roof. Thus, the peak heat flux crossing the SGR can be shaved between 42 and 57% due to a white reflective coating application. Further, SWR1 and SWR2 reduced HG between 41 and 85%. On the other hand, the STR reached a maximum temperature 2 °C above the SGR temperature in all cities. These higher temperatures caused an increase in the daily heat gain between 7 and 11%.

Table 3. Summary of the thermal evaluation of a single roof in four cities of Mexico (Average peak values for the different variables).

City	Roof	T_{es} (°C)	T_{is} (°C)	Q_R ($\frac{W}{m^2}$)	HG ($\frac{W \cdot h}{m^2-day}$)
Monterrey	SGR	51	43	113	1109
	SWR1	38	34	56 (−50%)	563 (−49%)
	SWR2	37	33	52 (−54%)	520 (−53%)
	STR	53	45	121 (+7%)	1213 (+9%)
Hermosillo	SGR	59	49	150	1675
	SWR1	45	39	87 (−42%)	1045 (−38%)
	SWR2	43	38	82 (−45%)	993 (−41%)
	STR	61	51	157 (+5%)	1793 (+7%)
Villahermosa	SGR	50	43	111	1097
	SWR1	37	34	53 (−52%)	550 (−50%)
	SWR2	36	33	48 (−57%)	506 (−54%)
	STR	52	44	120 (+8%)	1219 (+11%)
Mérida	SGR	53	43	123	1148
	SWR1	39	34	61 (−50%)	569 (−50%)
	SWR2	38	33	56 (−54%)	511 (−54%)
	STR	55	44	131 (+7%)	1255 (+9%)

8.2. Insulated Roof

The insulated roof was studied also in Hermosillo as it was the warmest city among the selected locations. As in the previous section, first, the detailed analysis of this city is presented, and then, the results for other cities are summarized at the end of this subsection.

Figure 5b shows the behavior of the temperature of the external surface of the insulated roofs and the temperature of the ambient air during the seven days considered. Insulated roofs with a conventional color had a similar behavior, and on the other hand, the insulated roofs with reflective coating maintained a similar behavior regarding the temperature of the exterior surface. The exterior surface of the roofs reached its maximum temperature between 13:30 and 14:00 h. The exterior surface of the insulated terracotta roof (ITR) reached on average 72 °C, the insulated gray roof (IGR) 69°, Insulated White Roof #1 (IWR1) 50 °C, and Insulated White Roof #2 (IWR2) 49 °C. These temperatures indicated that ITR had on average a temperature of the exterior surface about 3 °C higher than the IGR, while IWR1 and IWR2 reduced the temperature of the exterior surface by 19 and 20 °C on average, respectively. If the average maximum temperatures of white reflective roofs were compared with the average maximum temperature of ambient air (45 °C), IWR1 had a maximum temperature of 5 °C above the maximum air temperature, while the IWR2 roof reached a temperature 4 °C higher than the maximum air temperature. On the other hand, the IGR and ITR reached a temperature of 23 and 26 °C higher than the maximum ambient air temperature, respectively.

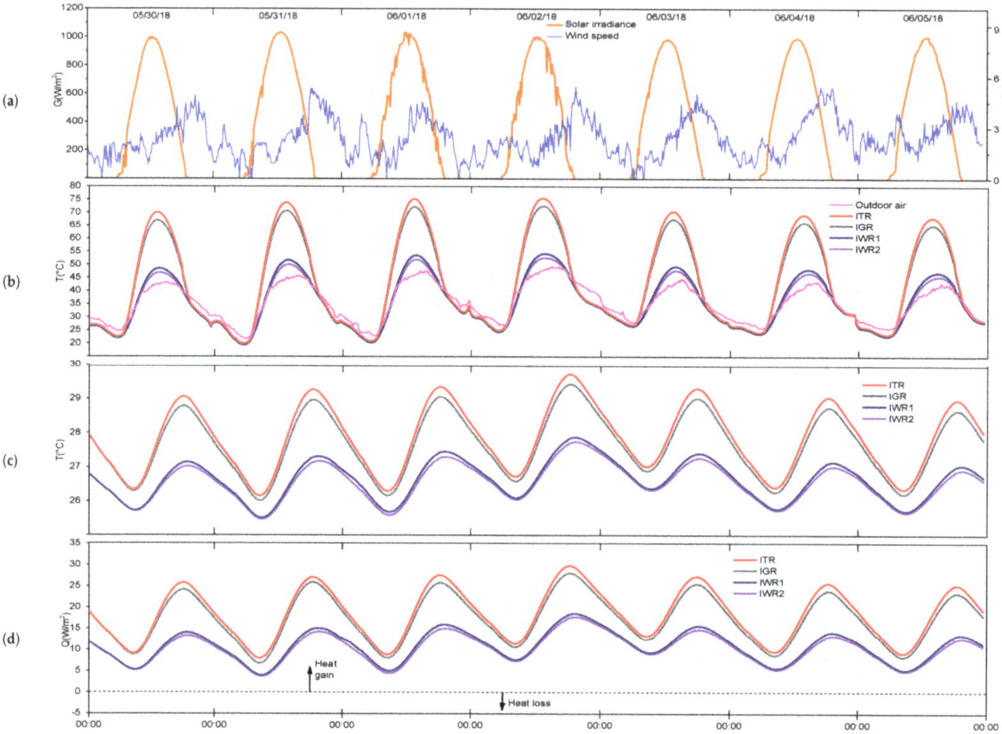

Figure 5. Thermal behavior of the insulated roof with traditional and reflective coatings in Hermosillo: (**a**) solar irradiance and wind speed; (**b**) temperature of exterior surface and outdoor air; (**c**) temperature of interior surface; (**d**) heat flux of the roofs.

Figure 5c shows the temperature of the interior surface of the four roofs. Due to thermal insulation, the temperatures of the roofs' interior surface had a small oscillation between day and night compared to the cases without insulation. These surfaces reached their maximum temperature between 17:30 and 18:10 h. The temperature of the interior surface of the ITR and that of the IGR reached a maximum temperature of 29.3 °C and 29 °C, while the surface temperature of IWR1 and IWR2 reached 27.4 °C and 27.2 °C. This figure demonstrates that the insulation caused the interior temperature of the roofs to remain relatively constant.

Figure 5d shows the behavior of the heat flux of the insulated roofs in Hermosillo. The maximum heat flux traveling through the roofs occurred between 17:30 and 18:10 h. Insulated traditional roofs (ITR and IGR) had an average peak heat flux of 27 and 25.3 W/m^2, while insulated white reflective roofs, IWR1 and IWR2, had a peak heat flux of 15.4 and 14.6 W/m^2. As mentioned above, the total heat gain of the roof over a day was determined by calculating the area under the heat flux curve of each day. The ITR had an average heat gain of 304 $W{\cdot}h/(m^2\text{-day})$, the IGR of 273 $W{\cdot}h/(m^2\text{-day})$, IWR2 of 140 $W{\cdot}h/(m^2\text{-day})$, and IWR2 of 128 $W{\cdot}h/(m^2\text{-day})$. The ITR located in Hermosillo had a 6% higher heat gain than the IGR, while IWR1 and IWR2 had a 37 and 40% lower heat gain than the IGR.

Finally, the insulated roofs' thermal behavior was simulated for the remaining cities following the same procedure used for Hermosillo. Table 4 presents a summary of the results obtained from the evaluation of the insulated roofs in the four cities of Mexico. The table shows the values for the average peak temperature of the exterior (T_{es}) and the interior surface (T_{is}), the average peak heat flux of the roofs (Q_R), and the average daily heat gain (HG). Taking the IGR as the control case, the differences between the peak temperatures (T_{es} and T_{is}) were calculated, along with the peak heat flux (Q_R) and the heat gain (HG). The white reflective roofs were able to reduce T_{es} between 17 and 21°C compared to the IGR. Therefore, they could reduce the Q_R that crossed the roofs by a factor ranging between 39 and 54%. Further, these roofs had an HG between 37 and 56% smaller than the ISG. On the other hand, the ITR reached a maximum temperature of 3 °C above the IGR. This temperature increment caused an increase of the Q_R between 7% and 15%. Moreover, the ITR increased the HG by about 11 and 33%.

Table 4. Summary of the thermal evaluation of insulated roof in four cities of Mexico (Average peak values for the different variables).

City	Roof	T_{es} (°C)	T_{is} (°C)	Q_R ($\frac{W}{m^2}$)	HG ($\frac{W \cdot h}{m^2-day}$)
Monterrey	IGR	60	27.7	17.1	260
	IWR1	43	26.3	9 (−47%)	128 (−51%)
	IWR2	42	26.2	8 (−53%)	117 (−55%)
	ITR	63	28	19 (+12%)	291 (+12%)
Hermosillo	IGR	69	29	25	439
	IWR1	50	27.4	15.4 (−39%)	276(−37%)
	IWR2	49	27.2	14.6 (−42%)	262 (−40%)
	ITR	72	29.3	27 (+7%)	459 (+6%)
Villahermosa	IGR	58	27.7	13	273
	IWR1	41	26.3	7 (−46%)	140 (−49%)
	IWR2	40	26.1	6 (−54%)	128 (−53%)
	ITR	61	28	15 (+15%)	304 (+11%)
Mérida	IGR	63	27.9	19	284
	IWR1	44	26.4	10 (−47%)	139 (−51%)
	IWR2	42	26.2	9 (−53%)	126 (−56%)
	ITR	66	28.2	21 (+11%)	318 (+12%)

9. Discussion

This section discusses the comparison of some of the results obtained in this work with other research available in the literature. Alqalaf and Alawadhi [23] evaluated the thermal effectiveness of a white reflective coating on the exterior surface of a concrete roof in Kuwait. The authors of the previous research built a test cell to perform a series of experiments and then developed a numerical model validated with experimental data. They simulated white and gray roof thermal performance in the season with the highest solar radiation and outdoor air temperature. The results showed that the temperature of the interior surface of the white roof was 6 °C lower than the temperature of the gray roof. Further, the white roof caused a reduction of the heat flux of 50%. Because Kuwait's climate is warm and dry, such as the climate of Hermosillo, it is worth comparing the results of the current research with those obtained in [23]. Besides the type of climate, the indoor air temperature and the roofs' thermophysical properties studied in the previous research were very similar to the values used in this work. Table 3 indicates that the temperature of the interior surface of SWR1 was 10 °C lower than the temperature of the SGR. At the same time, the heat flux of SWR1 was 42% smaller than the corresponding to the SGR. Here, we mention only SWR1 because this roof had the same absorptance (α = 0.2) as the white roof studied in [23]. However, the solar absorptance of the gray roof analyzed by Alqalaf and Alawadhi (α = 0.8) was greater than the absorptance of the SGR considered here (α = 0.67). Therefore, this main factor for the reduction of the heat flux presented in [23] was more significant than the reduction of the heat flux presented in the current research. On the other hand, as can be noticed above, the temperature reduction presented by [23] was smaller, but this difference occurred because the thickness of the roof analyzed in Alqalaf and Alawadhi was 0.15 m, which was greater than the thickness of the single roof; therefore, the thermal inertia of the roofs was the other factor that caused the slightly different results. Thus, the information presented above demonstrated that the findings of this research work were consistent with what other researchers have reported.

Another contribution of this research that is important to discuss is the influence of thermal insulation on the roofs' thermal performance. This effect can be obtained by comparing the results presented in Tables 3 and 4. Because the thermal insulation caused the roofs to have an indoor surface temperature with small oscillations, the heat flux crossing the insulated roof was very small compared to the flux of single roofs. Thus, the *HG* of the insulated roofs was around four-times smaller than that corresponding to single roofs regardless of the coating and the city. For instance, using the results for Hermosillo, by comparing the *HG* of the SGR with the *HG* of the IGR, it can be noticed that the first value was 3.8-times greater than the second value. Therefore, thermal insulation could have an essential contribution in reducing heat gains. On the other hand, another action that the results of this research can evaluate is the comparison between the SGR and IWR2; this is the comparison of the traditional roof configuration (SGR) with the roof configuration with two retrofitting techniques (thermal insulation and reflective coating). Using again Hermosillo as an example, the *HG* of SWR was equal to 1675 $W{\cdot}h/(m^2\text{-day})$ (Table 3), and the *HG* of IWR2 was equal to 262 $W{\cdot}h/(m^2\text{-day})$ (Table 4). Comparing the two previous values indicated that a roof with thermal insulation and a solar reflective coating could have a daily heat gain up to 6.4-times smaller than a gray roof without insulation. This result is important; however, thermal insulation installation could be more complex and more expensive than applying a reflective coating. Thus, a life cycle cost analysis is needed to find the more cost-effective configuration of roofs.

10. Conclusions

A computational tool was used to simulate the thermal behavior of insulated and non-insulated concrete slab roofs with traditional and solar reflective coatings in four cities with warm climates in Mexico. This simulation tool is a computer model based on the finite volume method that numerically solves the heat conduction equation in an unsteady state. The simulations were done using the weather data for the week with the highest outdoor air temperature. Two traditional and two solar reflective coatings installed on the exterior surface of the roofs were considered, and the following was concluded:

Regarding the simulation of the single roofs, SWR2 was the best configuration to minimize the heat transfer. Due to the small solar absorptance of the coating, SWR2 presented a peak exterior surface temperature up to 16 °C lower than the temperature of the SGR. Further, the peak interior surface temperature of SWR2 was up to 11 °C lower than the SGR. Thus, SWR2 diminished the heat flux and the daily heat gains up to 57% and 54%, respectively.

The insulated roofs simulations indicated that the surface temperature reduction of the exterior surface due to the reflective coatings was more significant than the single roofs. IWR2 was the configuration with the best thermal performance. The maximum temperature reduction provided by IWR2 was 19 °C lower than the temperature of the IGR. IWR2 provided a maximum interior surface temperature reduction 1.6 °C, which was very small. This effect occurred because the thermal insulation maintained the interior surface with small oscillations. IWR2 reduced the peak heat flux and the daily heat gain up to 54% and 15%, respectively.

This research highlights the importance of selecting the type of coating to be used in building roofs well. In the terracotta coating, this color had a solar absorptance equal to 0.7, which was just a little higher than the absorptance of the gray color of bare concrete ($\alpha = 0.67$). The difference in the solar absorptance for these two roofs may seem insignificant, but as shown in the Results Section, the STR had a daily heat gain between 9 and 11% higher than the SGR. Similarly, the ITR had a daily heat gain between 5 and 12%

higher than the IGR. These results demonstrated that even a small increment in the solar absorptance could cause a significant increment in the daily heat gain of the roofs.

Because in buildings situated in warm climates, the roof is a source of unwanted heat, applying a coating with a lower absorptance causes a lower amount of energy to be absorbed by the roof's exterior surface. Therefore, the heat traveling through the roof structure is reduced, and then, the roof exhibits better thermal performance. Since Solar Reflective Coating # 2 was the material with the smallest solar absorptance, this coating improved both the single and insulated roof thermal behavior. Finally, because most buildings in Mexico have bare gray or terracotta roofs, there is a great potential for using reflective coatings as a retrofitting technique in this country. This research demonstrated that white reflective coatings are an excellent alternative to improve the thermal performance of roofs, which could lead to energy savings and mitigating greenhouse gas emissions from buildings.

Funding: This research received no external funding.

Acknowledgments: The author acknowledges CONACYT-Mexico for the support given through the System of National Researchers program (Sistema Nacional de Investigadores, SNI). The author also acknowledges the Servicio Meteorológico Nacional-Comisión Nacional del Agua (SMN-CONAGUA) for providing the weather data used for the simulations. Finally, the author is grateful to Jesús Xamán from CENIDET for the Finite Volume Method course.

Conflicts of Interest: The author declares no conflict of interest.

Abbreviations

FVM	Finite volume method
IGR	Insulated gray roof
ITR	Insulated terracotta roof
IWR1	Insulated White Roof #1
IWR2	Insulated White Roof #2
LGS-ADI	Line-by-line Gauss–Seidel method with alternating directions
SGR	Single gray roof
STR	Single terracotta roof
SWR1	Single White Roof #1
SWR2	Single White Roof #2
TDMA	Tridiagonal matrix algorithm

Nomenclature

$a_p, a_e, a_w, a_n, a_s, b$	coefficients of the discretized equation
c_p	specific heat, J/(kg·K)
G	solar radiation, W/m^2
h	heat transfer coefficient, $W/(m^2 \cdot K)$
HG	daily heat gain, $W \cdot h/(m^2\text{-day})$
N_x	number of nodes in direction x
N_y	number of nodes in direction y
Q_R	heat flux through the roofs, W/m^2
t	time, s
T	temperature, °C
v	wind speed, m/s
W	width of the roof, m
y_1	thickness of the roof, m
x, y	coordinates, m
$\Delta x, \Delta y$	grid spacing, m

Greek	
α	solar absorptance
ϕ	dependent variable in the general differential equation
λ	thermal conductivity, W/(m·K)
ρ	density, kg/m^3
σ	Stefan–Boltzmann constant, kg/(m^3)
ε	thermal emittance
Γ	diffusion coefficient of the generalized differential equation
Subscripts	
es	exterior surface
in	indoor
is	interior surface
out	outdoor
sky	sky

References

1. International Energy Agency (IEA). *2019 Global Status Report for Buildings and Construction*; Global Alliance for Building and Construction; International Energy Agency: Paris, France, 2019.
2. Rajcic, V.; Perkovic, N.; Bedon, C.; Barbalic, J.; Zarnic, R. Thermal and energy-efficiency assessment of hybrid CLT–glass facade elements. *Appl. Sci.* **2020**, *10*, 3071. [CrossRef]
3. Wang, Z.; Tian, Q.; Jia, J. Numerical study on performance optimization of an energy-saving insulated window. *Sustainability* **2021**, *13*, 935. [CrossRef]
4. Gullbrekken, L.; Grynning, S.; Gaarder, J.E. Thermal performance of insulated constructions—Experimental studies. *Buildings* **2019**, *9*, 49. [CrossRef]
5. Saber, H.H.; Hajiah, A.E.; Alshehri, S.A.; Hussain, H.J. Investigating the effect of dust accumulation on the solar reflectivity of coatings materials for cool roofs applications. *Energies* **2021**, *14*, 445. [CrossRef]
6. Triano-Juárez, J.; Macias-Melo, E.V.; Hernández-Pérez, I.; Aguilar-Castro, K.M.; Xamán, J. Thermal behavior of a phase change material in a building roof with and without reflective coating in a warm humid zone. *J. Build. Eng.* **2020**, *32*, 101648. [CrossRef]
7. Miszczuk, A.; Heim, D. Parametric study of air infiltration in residential buildings—The effect of local conditions on energy demand. *Energies* **2021**, *14*, 127. [CrossRef]
8. Hernández-Pérez, I.; Álvarez, G.; Xamán, J.; Zavala-Guillén, I.; Arce, J.; Simá, E. Thermal performance of reflective materials applied to exterior building components. *Energy Build.* **2014**, *80*, 81–105. [CrossRef]
9. Todeschi, V.; Mutani, G.; Baima, L.; Nigra, M.; Robiglio, M. Smart solutions for sustainable cities—The re-coding experience for harnessing the potential of urban rooftops. *Appl. Sci.* **2020**, *10*, 7112. [CrossRef]
10. Pisello, A.L.; Cotana, F. The thermal effect of an innovative cool roof on residential buildings in Italy: Results from two years of continuous monitoring. *Energy Build.* **2014**, *69*, 154–164. [CrossRef]
11. Qin, Y.; He, Y.O.; Wu, B.; Ma, Z.; Zhang, X. Regulating top albedo and bottom emissivity of concrete roof tiles for reducing building heat gains. *Energy Build.* **2017**, *156*, 218–224. [CrossRef]
12. Hernández-Pérez, I.; Zavala-Guillén, I.; Xamán, J.; Belman-Flores, J.M.; Macias-Melo, E.V.; Aguilar-Castro, K.M. Test box experiment to assess the impact of waterproofing materials on the energy gain of building roofs in Mexico. *Energy* **2019**, *158*, 115847. [CrossRef]
13. Algarni, S. Potential for cooling load reduction in residential buildings using cool roofs in the harsh climate of Saudi Arabia. *Energy Environ.* **2008**, *30*, 235–253 [CrossRef]
14. Piselli, C.; Pisello, A.L.; Saffari, M.; de Gracia, A.; Cotana, F.; Cabeza, L.F. Cool roof impact on building energy need: The roleof thermal insulation with varying climate conditions. *Energies* **2019**, *12*, 3354. [CrossRef]
15. Dominguez-Delgado, A.; Dominguez-Torres, H.; Dominguez-Torres, C.A. Energy and economic life cycle assessment of cool roofs applied to the refurbishment of social housing in southern Spain. *Sustainability* **2020**, *12*, 5602. [CrossRef]
16. Tong, S.; Li, H.; Zingre, K.T.; Wan, M.P.; Chang, V.W.C.; Wong, S.K.; Toh, W.B.T.; Lee, I.Y.L. Thermal performance of concrete-based roofs in tropical climate. *Energy Build.* **2014**, *78*, 392–401. [CrossRef]
17. Zingre, K.T.; Wan, M.P.; Tong, S.; Li, H.; Chang, V.W.C.; Wong, S.K.; Toh, W.B.T.; Lee, I.Y.L. Modeling of cool roof heat transfer in tropical climate. *Renew. Energy* **2016**, *75*, 210–223. [CrossRef]
18. Hahn, D.W.; Ozisik, M.N. *Heat Conduction*; John Wiley and Sons: Hoboken, NJ, USA, 2012.
19. Duffie, J.A.; Beckman, W.A. *Solar Engineering of Thermal Processes*; John Wiley and Sons: Hoboken, NJ, USA, 1980.
20. Han, J.; Lin, L.; Yang, H. Investigation on the thermal performance of different lightweight roofing structures and its effect on space cooling load. *Appl. Therm. Eng.* **2009**, *29*, 2491–2499. [CrossRef]
21. Chen, K.C.; Payne, U.J. Analytical solution for heat conduction in a two-material-layer slab with linearly temperature Dependent Conductivity. *J. Heat Transfer* **1993**, *113*, 237–239.

22. Hernández-Pérez, I.; Xamán, J.; Macias-Melo, E.V.; Aguilar-Castro, K.M.; Zavala-Guillén, I.; Hernández-López, I.; Simá, E. Experimental thermal evaluation of building roofs with conventional and reflective coatings. *Energy Build.* **2018**, *158*, 569–579. [CrossRef]
23. Alqalaf, H.J.; Alawadhi, E.M. Thermal analysis of a concrete roof with an outdoor reflective surface. *J. Energy Eng.* **2015**, *142*, 05015001. [CrossRef]

MDPI
St. Alban-Anlage 66
4052 Basel
Switzerland
Tel. +41 61 683 77 34
Fax +41 61 302 89 18
www.mdpi.com

Applied Sciences Editorial Office
E-mail: applsci@mdpi.com
www.mdpi.com/journal/applsci

www.ingramcontent.com/pod-product-compliance
Lightning Source LLC
Chambersburg PA
CBHW061823210926
51850CB00006B/792
9783036571362